# Advanced Complex Analysis

## A Comprehensive Course in Analysis, Part 2B

# Advanced Complex Analysis

## A Comprehensive Course in Analysis, Part 2B

Barry Simon

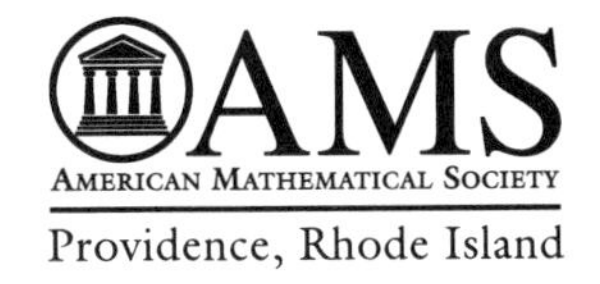


Providence, Rhode Island

2010 *Mathematics Subject Classification.* Primary 30-01, 33-01, 34-01, 11-01; Secondary 30C55, 30D35, 33C05, 60J67.

---

For additional information and updates on this book, visit
**www.ams.org/bookpages/simon**

---

**Library of Congress Cataloging-in-Publication Data**

Simon, Barry, 1946–
Advanced complex analysis / Barry Simon.
pages cm. — (A comprehensive course in analysis ; part 2B)
Includes bibliographical references and indexes.
ISBN 978-1-4704-1101-5 (alk. paper)
1. Mathematical analysis—Textbooks. I. Title.

QA300.S526 2015
515—dc23

2015015258

---



10 9 8 7 6 5 4 3 2 1 20 19 18 17 16 15

*To the memory of Cherie Galvez*

*extraordinary secretary, talented helper, caring person*

*and to the memory of my mentors,*
*Ed Nelson (1932-2014) and Arthur Wightman (1922-2013)*

*who not only taught me Mathematics*
*but taught me how to be a mathematician*

# Contents

# Preface to the Series

Young men should prove theorems, old men should write books.

—*Freeman Dyson*, quoting G. H. Hardy[1]

Reed–Simon[2] starts with "Mathematics has its roots in numerology, geometry, and physics." This puts into context the division of mathematics into algebra, geometry/topology, and analysis. There are, of course, other areas of mathematics, and a division between parts of mathematics can be artificial. But almost universally, we require our graduate students to take courses in these three areas.

This five-volume series began and, to some extent, remains a set of texts for a basic graduate analysis course. In part it reflects Caltech's three-terms-per-year schedule and the actual courses I've taught in the past. Much of the contents of Parts 1 and 2 (Part 2 is in two volumes, Part 2A and Part 2B) are common to virtually all such courses: point set topology, measure spaces, Hilbert and Banach spaces, distribution theory, and the Fourier transform, complex analysis including the Riemann mapping and Hadamard product theorems. Parts 3 and 4 are made up of material that you'll find in some, but not all, courses—on the one hand, Part 3 on maximal functions and $H^p$-spaces; on the other hand, Part 4 on the spectral theorem for bounded self-adjoint operators on a Hilbert space and det and trace, again for Hilbert space operators. Parts 3 and 4 reflect the two halves of the third term of Caltech's course.

[1] Interview with D. J. Albers, The College Mathematics Journal, **25**, no. 1, January 1994.

[2] M. Reed and B. Simon, *Methods of Modern Mathematical Physics, I: Functional Analysis*, Academic Press, New York, 1972.

While there is, of course, overlap between these books and other texts, there are some places where we differ, at least from many:

(a) By having a unified approach to both real and complex analysis, we are able to use notions like contour integrals as Stietljes integrals that cross the barrier.

(b) We include some topics that are not standard, although I am surprised they are not. For example, while discussing maximal functions, I present Garcia's proof of the maximal (and so, Birkhoff) ergodic theorem.

(c) These books are written to be keepers—the idea is that, for many students, this may be the last analysis course they take, so I've tried to write in a way that these books will be useful as a reference. For this reason, I've included "bonus" chapters and sections—material that I do not expect to be included in the course. This has several advantages. First, in a slightly longer course, the instructor has an option of extra topics to include. Second, there is some flexibility—for an instructor who can't imagine a complex analysis course without a proof of the prime number theorem, it is possible to replace all or part of the (non-bonus) chapter on elliptic functions with the last four sections of the bonus chapter on analytic number theory. Third, it is certainly possible to take all the material in, say, Part 2, to turn it into a two-term course. Most importantly, the bonus material is there for the reader to peruse long after the formal course is over.

(d) I have long collected "best" proofs and over the years learned a number of ones that are not the standard textbook proofs. In this regard, modern technology has been a boon. Thanks to Google books and the Caltech library, I've been able to discover some proofs that I hadn't learned before. Examples of things that I'm especially fond of are Bernstein polynomials to get the classical Weierstrass approximation theorem, von Neumann's proof of the Lebesgue decomposition and Radon–Nikodym theorems, the Hermite expansion treatment of Fourier transform, Landau's proof of the Hadamard factorization theorem, Wielandt's theorem on the functional equation for $\Gamma(z)$, and Newman's proof of the prime number theorem. Each of these appears in at least some monographs, but they are not nearly as widespread as they deserve to be.

(e) I've tried to distinguish between central results and interesting asides and to indicate when an interesting aside is going to come up again later. In particular, all chapters, except those on preliminaries, have a listing of "Big Notions and Theorems" at their start. I wish that this attempt to differentiate between the essential and the less essential didn't make this book different, but alas, too many texts are monotone listings of theorems and proofs.

(f) I've included copious "Notes and Historical Remarks" at the end of each section. These notes illuminate and extend, and they (and the Problems) allow us to cover more material than would otherwise be possible. The history is there to enliven the discussion and to emphasize to students that mathematicians are real people and that "may you live in interesting times" is truly a curse. Any discussion of the history of real analysis is depressing because of the number of lives ended by the Nazis. Any discussion of nineteenth-century mathematics makes one appreciate medical progress, contemplating Abel, Riemann, and Stieltjes. I feel knowing that Picard was Hermite's son-in-law spices up the study of his theorem.

On the subject of history, there are three cautions. First, I am not a professional historian and almost none of the history discussed here is based on original sources. I have relied at times—horrors!—on information on the Internet. I have tried for accuracy but I'm sure there are errors, some that would make a real historian wince.

A second caution concerns looking at the history assuming the mathematics we now know. Especially when concepts are new, they may be poorly understood or viewed from a perspective quite different from the one here. Looking at the wonderful history of nineteenth-century complex analysis by Bottazzini–Grey[3] will illustrate this more clearly than these brief notes can.

The third caution concerns naming theorems. Here, the reader needs to bear in mind Arnol'd's principle:[4] *If a notion bears a personal name, then that name is not the name of the discoverer* (and the related Berry principle: *The Arnol'd principle is applicable to itself*). To see the applicability of Berry's principle, I note that in the wider world, Arnol'd's principle is called "Stigler's law of eponymy." Stigler[5] named this in 1980, pointing out it was really discovered by Merton. In 1972, Kennedy[6] named Boyer's law *Mathematical formulas and theorems are usually not named after their original discoverers* after Boyer's book.[7] Already in 1956, Newman[8] quoted the early twentieth-century philosopher and logician A. N. Whitehead as saying: "Everything of importance has been said before by somebody who

[3] U. Bottazzini and J. Gray, *Hidden Harmony—Geometric Fantasies. The Rise of Complex Function Theory*, Springer, New York, 2013.

[4] V. I. Arnol'd, *On teaching mathematics*, available online at `http://pauli.uni-muenster.de/~munsteg/arnold.html`.

[5] S. M. Stigler, *Stigler's law of eponymy*, Trans. New York Acad. Sci. **39** (1980), 147–158.

[6] H. C. Kennedy, *Classroom notes: Who discovered Boyer's law?*, Amer. Math. Monthly **79** (1972), 66–67.

[7] C. B. Boyer, *A History of Mathematics*, Wiley, New York, 1968.

[8] J. R. Newman, *The World of Mathematics*, Simon & Schuster, New York, 1956.

did not discover it." The main reason to give a name to a theorem is to have a convenient way to refer to that theorem. I usually try to follow common usage (even when I know Arnol'd's principle applies).

I have resisted the temptation of some text writers to rename things to set the record straight. For example, there is a small group who have attempted to replace "WKB approximation" by "Liouville–Green approximation", with valid historical justification (see the Notes to Section 15.5 of Part 2B). But if I gave a talk and said I was about to use the Liouville–Green approximation, I'd get blank stares from many who would instantly know what I meant by the WKB approximation. And, of course, those who try to change the name also know what WKB is! Names are mainly for shorthand, not history.

These books have a wide variety of problems, in line with a multiplicity of uses. The serious reader should at least skim them since there is often interesting supplementary material covered there.

Similarly, these books have a much larger bibliography than is standard, partly because of the historical references (many of which are available online and a pleasure to read) and partly because the Notes introduce lots of peripheral topics and places for further reading. But the reader shouldn't consider for a moment that these are intended to be comprehensive—that would be impossible in a subject as broad as that considered in these volumes.

These books differ from many modern texts by focusing a little more on special functions than is standard. In much of the nineteenth century, the theory of special functions was considered a central pillar of analysis. They are now out of favor—too much so—although one can see some signs of the pendulum swinging back. They are still mainly peripheral but appear often in Part 2 and a few times in Parts 1, 3, and 4.

These books are intended for a second course in analysis, but in most places, it is really previous exposure being helpful rather than required. Beyond the basic calculus, the one topic that the reader is expected to have seen is metric space theory and the construction of the reals as completion of the rationals (or by some other means, such as Dedekind cuts).

Initially, I picked "A Course in Analysis" as the title for this series as an homage to Goursat's *Cours d'Analyse*,[9] a classic text (also translated into English) of the early twentieth century (a literal translation would be

[9]E. Goursat, *A Course in Mathematical Analysis: Vol. 1: Derivatives and Differentials, Definite Integrals, Expansion in Series, Applications to Geometry. Vol. 2, Part 1: Functions of a Complex Variable. Vol. 2, Part 2: Differential Equations. Vol. 3, Part 1: Variation of Solutions. Partial Differential Equations of the Second Order. Vol. 3, Part 2: Integral Equations. Calculus of Variations*, Dover Publications, New York, 1959 and 1964; French original, 1905.

"of Analysis" but "in" sounds better). As I studied the history, I learned that this was a standard French title, especially associated with École Polytechnique. There are nineteenth-century versions by Cauchy and Jordan and twentieth-century versions by de la Vallée Poussin and Choquet. So this is a well-used title. The publisher suggested adding "Comprehensive", which seems appropriate.

It is a pleasure to thank many people who helped improve these texts. About 80% was TeXed by my superb secretary of almost 25 years, Cherie Galvez. Cherie was an extraordinary person—the secret weapon to my productivity. Not only was she technically strong and able to keep my tasks organized but also her people skills made coping with bureaucracy of all kinds easier. She managed to wind up a confidant and counselor for many of Caltech's mathematics students. Unfortunately, in May 2012, she was diagnosed with lung cancer, which she and chemotherapy valiantly fought. In July 2013, she passed away. I am dedicating these books to her memory.

During the second half of the preparation of this series of books, we also lost Arthur Wightman and Ed Nelson. Arthur was my advisor and was responsible for the topic of my first major paper—perturbation theory for the anharmonic oscillator. Ed had an enormous influence on me, both via the techniques I use and in how I approach being a mathematician. In particular, he taught me all about closed quadratic forms, motivating the methodology of my thesis. I am also dedicating these works to their memory.

After Cherie entered hospice, Sergei Gel'fand, the AMS publisher, helped me find Alice Peters to complete the TeXing of the manuscript. Her experience in mathematical publishing (she is the "A" of A K Peters Publishing) meant she did much more, for which I am grateful.

This set of books has about 150 figures which I think considerably add to their usefulness. About half were produced by Mamikon Mnatsakanian, a talented astrophysicist and wizard with Adobe Illustrator. The other half, mainly function plots, were produced by my former Ph.D. student and teacher extraordinaire Mihai Stoiciu (used with permission) using Mathematica. There are a few additional figures from Wikipedia (mainly under WikiCommons license) and a hyperbolic tiling of Douglas Dunham, used with permission. I appreciate the help I got with these figures.

Over the five-year period that I wrote this book and, in particular, during its beta-testing as a text in over a half-dozen institutions, I received feedback and corrections from many people. In particular, I should like to thank (with apologies to those who were inadvertently left off): Tom Alberts, Michael Barany, Jacob Christiansen, Percy Deift, Tal Einav, German Enciso, Alexander Eremenko, Rupert Frank, Fritz Gesztesy, Jeremy Gray,

Leonard Gross, Chris Heil, Mourad Ismail, Svetlana Jitomirskaya, Bill Johnson, Rowan Killip, John Klauder, Seung Yeop Lee, Milivoje Lukic, Andre Martinez-Finkelshtein, Chris Marx, Alex Poltoratski, Eric Rains, Lorenzo Sadun, Ed Saff, Misha Sodin, Dan Stroock, Benji Weiss, Valentin Zagrebnov, and Maxim Zinchenko.

Much of these books was written at the tables of the Hebrew University Mathematics Library. I'd like to thank Yoram Last for his invitation and Naavah Levin for the hospitality of the library and for her invaluable help.

This series has a Facebook page. I welcome feedback, questions, and comments. The page is at `www.facebook.com/simon.analysis`.

Even if these books have later editions, I will try to keep theorem and equation numbers constant in case readers use them in their papers.

Finally, analysis is a wonderful and beautiful subject. I hope the reader has as much fun using these books as I had writing them.

# Preface to Part 2

Part 2 of this five-volume series is devoted to complex analysis. We've split Part 2 into two pieces (Part 2A and Part 2B), partly because of the total length of the current material, but also because of the fact that we've left out several topics and so Part 2B has some room for expansion. To indicate the view that these two volumes are two halves of one part, chapter numbers are cumulative. Chapters 1–11 are in Part 2A, and Part 2B starts with Chapter 12.

The flavor of Part 2 is quite different from Part 1—abstract spaces are less central (although hardly absent)—the content is more classical and more geometrical. The classical flavor is understandable. Most of the material in this part dates from 1820–1895, while Parts 1, 3, and 4 largely date from 1885–1940.

While real analysis has important figures, especially F. Riesz, it is hard to single out a small number of "fathers." On the other hand, it is clear that the founding fathers of complex analysis are Cauchy, Weierstrass, and Riemann. It is useful to associate each of these three with separate threads which weave together to the amazing tapestry of this volume. While useful, it is a bit of an exaggeration in that one can identify some of the other threads in the work of each of them. That said, they clearly did have distinct focuses, and it is useful to separate the three points of view.

To Cauchy, the central aspect is the differential and integral calculus of complex-valued functions of a complex variable. Here the fundamentals are the Cauchy integral theorem and Cauchy integral formula. These are the basics behind Chapters 2–5.

For Weierstrass, sums and products and especially power series are the central object. These appear first peeking through in the Cauchy chapters (especially Section 2.3) and dominate in Chapters 6, 9, 10, and parts of Chapter 11, Chapter 13, and Chapter 14.

For Riemann, it is the view as conformal maps and associated geometry. The central chapters for this are Chapters 7, 8, and 12, but also parts of Chapters 10 and 11.

In fact, these three strands recur all over and are interrelated, but it is useful to bear in mind the three points of view.

I've made the decision to restrict some results to $C^1$ or piecewise $C^1$ curves—for example, we only prove the Jordan curve theorem for that case.

We don't discuss, in this part, boundary values of analytic functions in the unit disk, especially the theory of the Hardy spaces, $H^p(\mathbb{D})$. This is a topic in Part 3. Potential theory has important links to complex analysis, but we've also put it in Part 3 because of the close connection to harmonic functions.

Unlike real analysis, where some basic courses might leave out point set topology or distribution theory, there has been for over 100 years an acknowledged common core of any complex analysis text: the Cauchy integral theorem and its consequences (Chapters 2 and 3), some discussion of harmonic functions on $\mathbb{R}^2$ and of the calculation of indefinite integrals (Chapter 5), some discussion of fractional linear transformations and of conformal maps (Chapters 7 and 8). It is also common to discuss at least Weierstrass product formulas (Chapter 9) and Montel's and/or Vitali's theorems (Chapter 6).

I also feel strongly that global analytic functions belong in a basic course. There are several topics that will be in one or another course, notably the Hadamard product formula (Chapter 9), elliptic functions (Chapter 10), analytic number theory (Chapter 13), and some combination of hypergeometric functions (Chapter 14) and asymptotics (Chapter 15). Nevanlinna theory (Chapter 17) and univalents functions (Chapter 16) are almost always in advanced courses. The break between Parts 2A and 2B is based mainly on what material is covered in Caltech's course, but the material is an integrated whole. I think it unfortunate that asymptotics doesn't seem to have made the cut in courses for pure mathematicians (although the material in Chapters 14 and 15 will be in complex variable courses for applied mathematicians).

*Chapter 12*

# Riemannian Metrics and Complex Analysis

At that moment I left Caen where I then lived, to take part in a geological expedition organized by the École des Mines. The circumstances of the journey made me forget my mathematical work; arrived at Coutances we boarded an omnibus for I don't know what journey. At the moment when I put my foot on the step the idea came to me, without anything in my previous thoughts having prepared me for it; that the transformations I had made use of to define the Fuchsian functions were identical with those of non-Euclidean geometry. I did not verify this, I did not have the time for it, since scarcely had I sat down in the bus than I resumed the conversation already begun, but I was entirely certain at once. On returning to Caen I verified the result at leisure to salve my conscience.

—*H. Poincaré, Science et méthode*, Flammarion, Paris, 1909[1]

**Big Notions and Theorems:** Conformal Metric, Conformal Distance, Pullback, Curvature, Invariance of Curvature, Euclidean Metric, Spherical Metric, Poincaré Metric, Hyperbolic Geometric, Ahlfors–Schwarz Lemma, Ahlfors–Liouville Theorem, Robinson's Metric, Picard's theorems, Bergman Space, Bergman Kernel, Bergman Projection, Bell's Formula, Bergman Metric, Jordan Domain, Hopf Lemma, Bell's Lemma, Painlevé Theorem

In this chapter, we'll explore the use of some ideas of Riemannian geometry in complex analysis (see Section 1.6 of Part 2A for background on Riemannian geometry). The key notion will be a conformal metric. Given a region, $\Omega$, a *conformal metric function* (conformal metric for short) is a

[1] as translated in [**159**].

$C^2$ function, $\lambda(z)$, from $\Omega$ to $(0,\infty)$. It will define a *conformal Riemann metric* as $\lambda^2(z)\,(dz)^2$, that is, given vectors, $u, w \in \mathbb{R}^2$ and $z_0 \in \Omega$, we think of $u, w \in T_{z_0}(\Omega)$, the tangent space at $z_0$, viewed as a point in $\mathbb{C}$ as $\mathbb{R}^2$, and let $\langle u, w\rangle = \lambda^2(z_0)\sum_{j=1}^2 u_j w_j$ so $\|u\|_\lambda = \lambda(z_0)\|u\|_{\text{Euc}}$ where $\|\cdot\|_{\text{Euc}}$ is the Euclidean norm. It is called conformal because angles are the same as in the underlying Euclidean metric. While elsewhere in these volumes we've used metric for a distance function, because metric is already used for $\lambda$, we call the underlying Riemannian geodesic distance

$$\rho_\lambda(z_0, z_1) = \inf\left(\int_0^1 \lambda(\gamma(s))|\gamma'(s)|\,ds \;\middle|\; \gamma(0) = z_0,\ \gamma(1) = z_1\right) \tag{12.0.1}$$

a *conformal distance* function.

In Section 12.1, we discuss transformation properties of $\lambda$ under analytic maps and introduce a quantity, the curvature

$$\kappa_\lambda(z) = \frac{[-\Delta(\log(\lambda(z)))]}{\lambda(z)^2} \tag{12.0.2}$$

which, remarkably, is invariant under coordinate transformation. We demanded $\lambda$ be $C^2$ so that $\kappa$ is defined.

We have already seen two natural conformal metrics. $\lambda(z) \equiv 1$ for $\Omega = \mathbb{C}$ yields the *Euclidean metric* and $\lambda(z) = (1+|z|^2)^{-1}$ yields the *spherical metric* on $\widehat{\mathbb{C}}$ studied in Section 6.5 of Part 2A. Section 12.2 will discuss a third natural metric, $\lambda(z) = (1-|z|^2)^{-1}$ on $\mathbb{D}$, the *hyperbolic metric* or *Poincaré metric*. It will have the remarkable property that the isometries (distance-preserving bijections) in this metric are precisely $\mathbb{A}\text{ut}(\mathbb{D})$ and complex conjugates of maps in $\mathbb{A}\text{ut}(\mathbb{D})$. We'll see that analytic maps of $\mathbb{D}$ to itself are contractions in the Poincaré metric. These three metrics are essentially the unique constant curvature, geodesically complete metrics.

That $f\colon \mathbb{D} \to \mathbb{D}$ is a contraction in the hyperbolic metric is a form of the Schwarz inequality. Section 12.3 will present a vast extension of the Schwarz inequality due to Ahlfors that involves $f\colon \Omega \to \Omega'$ and relates two conformal metrics, one on $\Omega$ and one on $\Omega'$. The key object will be the curvature. We'll see that Liouville's theorem can be viewed as a consequence of the Ahlfors–Schwarz principle.

Section 12.4 will provide yet another proof of Picard's theorem. The key will be the existence of a conformal metric on $\mathbb{C}\setminus\{0,1\}$ whose curvature is negative and bounded away from 0. When combined with the Ahlfors–Schwarz principle and Marty's theorem, we'll get Montel's three-point theorem, and so Picard's theorem.

By the Riemann mapping theorem, the hyperbolic metric can be transferred to any simply connected region. In fact, one can define, in a natural way, a metric, the *Bergman metric*, on any bounded region $\Omega$, and we will do this in Section 12.5.

One first defines the *Bergman kernel* as follows: One looks at $\mathfrak{A}_2(\Omega)$, the functions, $f$, holomorphic on $\Omega$ with

$$\int_\Omega |f(z)|^2 \, d^2z < \infty \tag{12.0.3}$$

This will be closed in $L^2(\Omega, d^2z)$, by the CIF, and $f \mapsto f(z_0)$ will be a bounded linear function for each $z_0$. Thus, there is a function, $k_{z_0} \in \mathfrak{A}_2(\Omega)$ with

$$f(z_0) = \langle k_{z_0}, f \rangle \tag{12.0.4}$$

One sets $k_{z_0}(z) \equiv K(z, z_0)$, a function analytic in $z$ and anti-analytic in $z_0$. This is the Bergman kernel.

The Bergman metric is given by

$$\lambda_B(z) = \frac{\partial^2}{\partial z \partial \bar{z}} \log(K(z,z)) \tag{12.0.5}$$

In Section 12.6, we explore regularity of Riemann mappings at the boundary, using the Bergman kernel. The connection of this to Riemannian geometry is weak since we'll use the Bergman kernel, not the Bergman metric.

**Notes and Historical Remarks.** We'll discuss history and references to individual topics in the separate sections. Here, I note that there are two books by Krantz [**233, 234**] that explore the full gamut of the subject sampled in this chapter. In particular, [**233**] discusses several other metrics (Carathéodory, Kobayashi). The geometric theory discussed here is especially useful when moved to several complex variables, and some ideas were initially developed there.

## 12.1. Conformal Metrics and Curvature

In this section, we study the pullback of conformal metrics, define the curvature, and show it is invariant under pullback. That it agrees with the curvature as geometrically defined will play no great role but is conceptually important, so we discuss it in the Notes.

**Definition.** A *conformal metric*, $\lambda(z)$, on a region $\Omega$ is a strictly positive $C^2$ function on $\Omega$ which defines a *conformal distance*, $\rho_\lambda$, by (12.0.1).

**Definition.** Let $f\colon \Omega \to \Omega'$ be a conformal map, that is, $f$ is analytic and $f'$ is nonvanishing. Let $\lambda$ be a conformal metric on $\Omega'$. The *pullback* of $\lambda$ under $f$ is the conformal metric $f^*\lambda$, defined by

$$(f^*\lambda)(z) = \lambda(f(z))|f'(z)| \tag{12.1.1}$$

Notice that if $\gamma$ is a curve from $z_0$ to $z_1$ in $\Omega$, then

$$\int_0^1 \lambda(f(\gamma(s)))\left|\frac{d}{ds}\, f(\gamma(s))\right| ds = \int_0^1 (f^*\lambda)(\gamma(s))|\gamma'(s)|\, ds \tag{12.1.2}$$

so, at least for $z_0$ near $z_1$, we have

$$\rho_{f^*\lambda}(z_0, z_1) = \rho_\lambda(f(z_0), f(z_1)) \tag{12.1.3}$$

If $f$ is a bijection, this holds globally (which is why it holds locally, since $f$ is always a local bijection), but, for example, if $z_0 \neq z_1$ and $f(z_0) = f(z_1)$, it fails. Essentially, not every curve $\tilde{\gamma}$ from $f(z_0)$ to $f(z_1)$ is an $f \circ \gamma$. (12.1.3) shows why (12.1.1) is a reasonable definition.

**Definition.** The *curvature* of a conformal metric, $\lambda$, is defined by

$$\kappa_\lambda(z) = \frac{[-\Delta(\log(\lambda(z)))]}{\lambda(z)^2} \tag{12.1.4}$$

Since $\lambda$ is $C^2$, $\kappa_\lambda$ is continuous. For us, its critical property is the invariance expressed in

**Theorem 12.1.1** (Invariance of the Curvature). *If $f\colon \Omega \to \Omega'$ is conformal and $\lambda$ is a conformal metric on $\Omega'$, then*

$$\kappa_{f^*\lambda}(z) = \kappa_\lambda(f(z)) \tag{12.1.5}$$

**Proof.** The chain rule says that if $f$ is analytic and $h$ is a $C^2$ function of $\text{Ran}(f)$ to $\mathbb{R}$,

$$\frac{\partial}{\partial z}\, h(f(z)) = \frac{\partial h}{\partial f}\,\frac{\partial f}{\partial z} + \frac{\partial h}{\partial \bar{f}}\,\frac{\partial \bar{f}}{\partial z} = \frac{\partial h}{\partial f}\,\frac{\partial f}{\partial z} \tag{12.1.6}$$

and similarly,

$$\frac{\partial}{\partial \bar{z}}\, h(f(z)) = \frac{\partial h}{\partial \bar{f}}\,\frac{\partial \bar{f}}{\partial \bar{z}} = \frac{\partial h}{\partial \bar{f}}\,\overline{\left(\frac{\partial f}{\partial z}\right)} \tag{12.1.7}$$

By (2.1.38) of Part 2A, $\Delta = 4\bar{\partial}\partial$, so since $\bar{\partial}(\frac{\partial f}{\partial z}) = 0$, we have

$$\Delta(h \circ f) = [(\Delta h) \circ f]\, |f'(z)|^2 \tag{12.1.8}$$

Since $\log|f'(z)|^2 = \log(f'(z)) + \overline{\log(f'(z))}$ is harmonic, we have, by (12.1.1), that

$$\begin{aligned} \Delta \log(f^*\lambda)(z) &= \Delta(\log(\lambda \circ f)) + \tfrac{1}{2}\,\Delta(\log|f'|^2) \\ &= \Delta(\log(\lambda \circ f)) \\ &= [(\Delta \log(\lambda)) \circ f]\, |f'(z)|^2 \end{aligned} \tag{12.1.9}$$

by (12.1.8). Thus,

$$\kappa_{f^*\lambda}(z) = -\frac{\Delta \log(f^*\lambda)(z)}{(f^*\lambda)(z)^2}$$

$$= -\frac{[(\Delta \log(\lambda)) \circ f]|f'(z)|^2}{(\lambda \circ f)(z)^2\,|f'(z)|^2}$$
$$= \kappa_\lambda(f(z)) \tag{12.1.10}$$

$\square$

**Example 12.1.2** (Euclidean Metric). Let $\Omega = \mathbb{C}$, $\lambda(z) \equiv 1$. Geodesics in this standard Euclidean metric are straight lines and $\mathbb{C}$ is geodesically complete (see Section 1.6 of Part 2A) in this metric. $\mathbb{C}$ is the natural domain in that no smaller region is geodesically complete. Clearly, $\kappa_\lambda \equiv 0$ since $\log(\lambda) = 0$. Isometries are precisely $z \mapsto e^{i\theta}z + a$ and $z \mapsto e^{i\theta}\bar{z} + a$. $\square$

**Example 12.1.3** (Spherical Metric). Let $\Omega = \mathbb{C}$ and let the conformal metric be given by

$$\sigma(z) = \frac{1}{1+|z|^2} = (1+z\bar{z})^{-1} \tag{12.1.11}$$

Thus,

$$\partial \log(\sigma) = -\partial \log(1+z\bar{z}) = -\frac{\bar{z}}{1+\bar{z}z} \tag{12.1.12}$$

and

$$-\Delta \log(\sigma) = -4\bar{\partial}(\partial \log(\sigma))$$
$$= 4\left[\frac{(1+\bar{z}z) - \bar{z}z}{(1+\bar{z}z)^2}\right]$$
$$= 4\sigma(z)^2 \tag{12.1.13}$$

Thus,

$$\kappa_\sigma = 4 \tag{12.1.14}$$

This metric has constant positive curvature.

As noted in (6.5.14) of Part 2A, the underlying induced distance is

$$\rho_\sigma(z,w) = \arctan\left(\frac{|z-w|}{|1+\bar{w}z|}\right) \tag{12.1.15}$$

What we called the spherical metric in Section 6.5 of Part 2A is equivalent, but not identical, to this.

Stereographic projection is an isometry of $\mathbb{C}$ to $\mathbb{S}^2 \setminus \{n\}$ with the usual metric. Geodesics through 0 are straight lines and, in general, are images of great circles under stereographic projections. $\mathbb{C}$ is not geodesically complete, but $\widehat{\mathbb{C}}$ is, with the unique $C^\infty$ extension of the Riemann metric $\sigma(z)^2\,(dz)^2$ to $\widehat{\mathbb{C}}$. Isometries are precisely the FLTs induced by elements of $\mathbb{SU}(2)$ and their complex conjugates. $\square$

**Notes and Historical Remarks.** Two high points in the use of conformal metrics are the work of Poincaré and Ahlfors discussed in the next two sections. While the use of curvature in this context is associated with Ahlfors, he stated in his paper that the formula (12.0.4) for the Gaussian curvature was well known, but no reference was given!

Notice that if $c$ is a constant, then $\kappa_{c\lambda} = c^{-2}\kappa_\lambda$. The more common choice of spherical metric $\sigma(z) = 2/(1+|z|^2)$ has constant curvature $+1$.

Curvature is discussed briefly in Section 1.6 of Part 2A. In the conformal case ($g = \det(g_{ij})$),

$$g_{11} = g_{12} = \sqrt{g} = \lambda^2 \tag{12.1.16}$$

so (1.6.14) of Part 2A and

$$\frac{1}{\lambda^2}\frac{\partial}{\partial x}\lambda^2 = 2\frac{\partial}{\partial x}\log\lambda \tag{12.1.17}$$

lead immediately to

$$K = \frac{\Delta(\log\lambda)}{\lambda^2} \tag{12.1.18}$$

Since $K$ is invariant under coordinate change, we see why Theorem 12.1.1 holds.

## 12.2. The Poincaré Metric

> One geometry cannot be more true than another; it can only be more convenient.
>
> —*H. Poincaré* [**319**]

$\mathbb{C}$, $\widehat{\mathbb{C}}$, and $\mathbb{D}$ are the three simply connected Riemann surfaces. In the last section, we saw that $\mathbb{C}$ and $\widehat{\mathbb{C}}$ have natural conformal metrics, in which they are geodesically complete and in which the set of isometries is transitive (which implies the curvature is constant). In this section, we'll see that $\mathbb{D}$ also possesses a natural metric in which it is geodesically complete. Moreover, unlike for $\mathbb{C}$ or $\widehat{\mathbb{C}}$, *every* biholomorphic bijection is an isometry in this metric. This is the Poincaré metric and $\mathbb{D}$ becomes a model for the Lobachevsky (hyperbolic) plane. We'll also see that every analytic map of $\mathbb{D}$ to $\mathbb{D}$ is a contraction in this metric. We'll also transfer the metric and geometry to $\mathbb{C}_+$.

Recall (see Section 7.4 of Part 2A) that all elements of $\mathbb{A}\mathrm{ut}(\mathbb{D})$ have the form

$$f(z) = \frac{\alpha z + \beta}{\bar\beta z + \bar\alpha} \tag{12.2.1}$$

where $|\alpha|^2 - |\beta|^2 = 1$. Thus,

$$f'(z) = \frac{1}{(\bar\beta z + \bar\alpha)^2} \tag{12.2.2}$$

while $|\bar{\beta}z + \bar{\alpha}|^2 - |\alpha z + \beta|^2 = 1 - |z|^2$, so

**Proposition 12.2.1.** *If $f \in \mathbb{A}\mathrm{ut}(\mathbb{D})$, then*

$$|f'(z)| = \frac{1 - |f(z)|^2}{1 - |z|^2} \tag{12.2.3}$$

**Remark.** We'll see below (see the remark after Theorem 12.2.4) that this also follows from the Schwarz lemma.

This has a neat expression in terms of an invariant conformal metric. Define the Poincaré metric to be the conformal metric on $\mathbb{D}$ with metric function

$$\pi(z) = (1 - |z|^2)^{-1} \tag{12.2.4}$$

Thus, (12.2.3) becomes

$$\pi(f(z))|f'(z)| = \pi(z) \tag{12.2.5}$$

which, by (12.2.1), implies

**Theorem 12.2.2.** *Let $\pi$ be the Poincaré metric on $\mathbb{D}$ and $f \in \mathbb{A}\mathrm{ut}(\mathbb{D})$. Then*

$$f^*\pi = \pi \tag{12.2.6}$$

Invariance of a metric in this sense implies that the map, if a bijection, is an isometry. We want to prove that an inequality implies a contraction. In this, $f$ need not even be a local bijection. We use the definition of (12.1.1) even if $f'$ has some zeros. $f^*\lambda$ will then not be strictly positive and may not even be $C^1$ at zeros of $f'$, but it is still a well-defined continuous function.

**Theorem 12.2.3.** *Let $f\colon \Omega \to \Omega'$ be analytic and let $\lambda$ and $\tilde{\lambda}$ be conformal metrics on $\Omega$ and $\Omega'$, respectively.*

(a) *If for some $c > 0$,*

$$f^*\tilde{\lambda} \le c\lambda \tag{12.2.7}$$

*then for all $z, w \in \Omega$,*

$$\rho_{\tilde{\lambda}}(f(z), f(w)) \le c\rho_\lambda(z, w) \tag{12.2.8}$$

(b) *If $f$ is a bijection and, for some $c > 0$,*

$$f^*\tilde{\lambda} = c\lambda \tag{12.2.9}$$

*then*

$$\rho_{\tilde{\lambda}}(f(z), f(w)) = c\rho_\lambda(z, w) \tag{12.2.10}$$

**Proof.** If $\gamma$ is a curve from $z$ to $w$, (12.1.2) holds. Minimizing over all curves from $z$ to $w$, we get

$$\inf\biggl(\int_0^1 \tilde\lambda(\tilde\gamma(s))\biggl|\frac{d\tilde\gamma}{ds}\biggr|\,ds \biggm| \tilde\gamma = f\circ\gamma;\ \gamma(0)=z,\ \gamma(1)=w\biggr) \le c\rho_\lambda(z,w) \tag{12.2.11}$$

Since the set of curves minimized over is a subset of all curves from $f(z)$ to $f(w)$, we get (12.2.8).

If $f$ is a bijection and $g$ is its inverse, then (12.2.9) implies

$$g^*\lambda = c^{-1}\tilde\lambda \tag{12.2.12}$$

Thus, (12.2.8) holds, and so does

$$\rho_\lambda(g(u),g(s)) \le c^{-1}\rho_{\tilde\lambda}(u,v) \tag{12.2.13}$$

which implies (12.2.10). □

Returning to the Poincaré metric, we want to note

**Theorem 12.2.4.** *Let $f\colon \mathbb{D}\to\mathbb{D}$ be analytic. Let $\pi$ be the Poincaré conformal metric on $\mathbb{D}$. Then*

$$f^*\pi \le \pi \tag{12.2.14}$$

**Proof.** We start with the Schwarz–Pick lemma (3.6.19) of Part 2A:

$$\left|\frac{f(z)-f(w)}{1-\overline{f(z)}f(w)}\right| \le \frac{|z-w|}{|1-\bar z w|} \tag{12.2.15}$$

which, for $z\ne w$, implies

$$\left|\frac{f(z)-f(w)}{z-w}\right| \le \left|\frac{1-\overline{f(z)}f(w)}{1-\bar z w}\right| \tag{12.2.16}$$

Taking $w$ to $z$ yields

$$|f'(z)| \le \frac{1-|f(z)|^2}{1-|z|^2} \tag{12.2.17}$$

which is equivalent to (12.2.14). □

**Remark.** If $f$ is a bijection and $g$ is its inverse, (12.2.17) and $g'(f(z))f'(z) = 1$ yields the opposite inequality for $f$ to (12.2.17), and so the promised alternate proof of (12.2.3).

Theorem 12.2.3 together with Theorems 12.2.2 and 12.2.4 yield

**Theorem 12.2.5.** *Let $\rho_\pi$ be the conformal distance in the Poincaré metric. If $f\in\mathbb{A}\mathrm{ut}(\mathbb{D})$, then for all $z,w\in\mathbb{D}$,*

$$\rho_\pi(f(z),f(w)) = \rho_\pi(z,w) \tag{12.2.18}$$

*If $f$ is an analytic function from $\mathbb{D}$ to $\mathbb{D}$, then*

$$\rho_\pi(f(z), f(w)) \le \rho_\pi(z, w) \tag{12.2.19}$$

We'll prove shortly an almost converse to (12.2.18). We are about to describe the geometry of the Poincaré unit disk both qualitatively and quantitatively. We need the following preliminaries. Since $\frac{d\tanh z}{dz} = \cosh^{-2} z = 1 - \tanh^2 z$, we see that

$$\operatorname{arctanh} u = \int_0^u \frac{dx}{1-x^2} \tag{12.2.20}$$

Since $\tanh z$ is strictly monotone going from $0$ at $z = 0$ to $1$ at $z = \infty$, $\operatorname{arctanh} u$ is strictly monotone from $[0, 1)$ to $[0, \infty)$.

We also define an orthocircle in $\mathbb{D}$ to be the intersection of $\overline{\mathbb{D}}$ with a circle so that the circle intersects $\partial\mathbb{D}$ at a $90^\circ$ angle. Since any $f \in \mathbb{A}\mathrm{ut}(\mathbb{D})$ extends to a conformal map of a neighborhood of $\mathbb{D}$, such $f$'s map orthocircles to orthocircles. It is not hard to see (Problem 1) that orthocircles containing $0$ are exactly straight lines through $0$. Thus, since for any $w$ there is an $f \in \mathbb{A}\mathrm{ut}(\mathbb{D})$ with $f(w) = 0$, we see for any distinct $z, w \in \mathbb{D}$, there is precisely one orthocircle through $z$ and $w$.

**Theorem 12.2.6.** *Place the Poincaré metric on $\mathbb{D}$.*

(a) *Geodesics through $0$ are precisely straight lines. In particular, there is exactly one geodesic between $0$ and any $z \in \mathbb{D}$.*

(b) $$\rho_\pi(z, 0) = \operatorname{arctanh}(|z|) \tag{12.2.21}$$

(c) *Geodesics are precisely the orthocircles. There is a unique geodesic through any pair of points in $\mathbb{D}$.*

(d) *$\mathbb{D}$ is geodesically complete in the Poincaré metric.*

(e) $$\rho_\pi(z, w) = \operatorname{arctanh}\left[\frac{|z-w|}{|1-\bar{z}w|}\right] \tag{12.2.22}$$

(f) $$\{w \mid \rho_\pi(w, 0) = r\} = \{w \mid |w| = \tanh r\} \tag{12.2.23}$$

(g) *$\{w \mid \rho_\pi(w, z) = r\}$ is a circle.*

(h) *For any $w_1, w_2, z$ in $\mathbb{D}$, there is at most one other $\zeta \in \mathbb{D}$ with $\rho(\zeta, w_j) = \rho(z, w_j)$ for $j = 1, 2$.*

(i) *If $PB(w_1, w_2) = \{z \mid \rho(z, w_1) = \rho(z, w_2)\}$ (for perpendicular bisection), then for $0 < a < 1$,*

$$PB(a, -a) = \{iy \mid |y| < 1\} \tag{12.2.24}$$

*If $\operatorname{Re} w > 0$, then $\rho(w, a) < \rho(w, -a)$.*

(j) *For any two distinct $w_1, w_2$ in $\mathbb{D}$, there is an $f \in \mathbb{A}\mathrm{ut}(\mathbb{D})$ so $f(w_1) \in (0, 1)$ and $f(w_2) = -f(w_1)$.*

(k) *For any distinct $w_1, w_2$ in $\mathbb{D}$, $PB(w_1, w_2)$ is an orthocircle with $w_1$ and $w_2$ separated by the orthocircle. $\{z \mid \rho(w_1, z) < \rho(w_2, z)\}$ is the side of the orthocircle containing $w_1$.*

**Remarks.** 1. (12.2.21) is equivalent to

$$e^{-2\rho_\pi(0,z)} = \frac{1-|z|}{1+|z|} \tag{12.2.25}$$

2. The disk with "straight lines" as orthocircles obeys all the axioms of Euclidean geometry, except for the parallel postulate: If $z \in \mathbb{D}$ and $L$ is an orthocircle through $z$ and $w \notin L$, there are multiple orthocircles through $w$ disjoint from $L$ (see Problem 1).

**Proof.** (a), (b) Let $\gamma$ be a $C^1$ path from 0 to $z \neq 0$. If $\gamma(s_0) = 0$ for some $s_0 > 0$, we can clearly get a shorter length $\gamma$ by restricting to $[s_0, 1]$. Thus, we can suppose $\gamma(s) \neq 0$ for $s \in (0,1]$, so

$$\gamma(s) = r(s)e^{i\theta(s)} \tag{12.2.26}$$

for $r$ and $\theta$ $C^1$ on $(0,1]$. Then,

$$L(\gamma) = \int_0^1 \sqrt{\dot{r}(s)^2 + r(s)^2\dot{\theta}(s)}\,(1-|r(s)|^2)^{-1}\,ds$$

This can clearly be made smaller by keeping $r(s)$ the same and replacing $\theta(s)$ by the constant value $\theta(1)$, that is, any minimum distance path has $\theta$ constant, and thus has length

$$\rho_\pi(z,0) = \int_0^{|z|} \frac{ds}{1-s^2} = \operatorname{arctanh}(|z|) \tag{12.2.27}$$

by (12.2.1). Up to reparametrization, this is unique.

(c) This is implied by (a) and the use of elements of $\mathbb{A}\mathrm{ut}(\mathbb{D})$ mapping any $w$ to 0.

(d) By the isometries, we need only show that every geodesic starting at 0 can be extended indefinitely. This follows from the fact that $\lim_{|z|\uparrow 1} \operatorname{arctanh}(|z|) = \infty$.

(e) If $f_w(z) = (z-w)/(1-\bar{w}z)$, then $f$ is an isometry with $f_w(w) = 0$. So $\rho_\pi(z,w) = \rho(f_w(z), 0)$, proving (12.2.22).

(f) is immediate from (b).

(g) follows from (f) and the fact that $\mathbb{A}\mathrm{ut}(\mathbb{D})$ maps circles to circles.

(h) Two circles that intersect do so in one or two points, so this follows from (g).

(i) If $\rho(z,w_1)=\rho(z,w_2)$, by the triangle inequality, $\rho(z,w_1)\geq\frac{1}{2}(\rho(w_1,w_2))$. Thus, if $z\in PB(a,-a)$, there is $y\in(-1,1)$ so $\rho(z,w_j)=\rho(iy,w_j)$. If $y\neq 0$, $-iy$ also has those distances, so by (h), $z$ must be $y$ or $-y$. For $y=0$, uniqueness of geodesic implies it is the only point with $\rho(a,0)=\rho(0,-a)$ (Problem 2).

(j) Let $w_3$ be the unique point on the geodesic from $w_1$ to $w_2$ with $\rho(w_3,w_1)=\rho(w_3,w_2)$. Let $g\in\mathbb{A}\mathrm{ut}(\mathbb{D})$ map $w_3$ to 0. Then the geodesic maps to a geodesic through 0, so a straight line. Thus, $g$ maps $w_1$ to $z$ and $w_2$ to $-z$. By following $g$ by a rotation about 0, we can map $w_1$ to $a$, and so $w_2$ to $-a$.

(k) follows from (i) and (j). □

We've already seen that any $f\in\mathbb{A}\mathrm{ut}(\mathbb{D})$ is an isometry in the Poincaré distance, and since $z\to\bar{z}$ does the same, so does $z\mapsto\overline{f(z)}$ for $z\in\mathbb{A}\mathrm{ut}(\mathbb{D})$. Remarkably, these are all the isometries.

**Theorem 12.2.7.** *Let $f\colon\mathbb{D}\to\mathbb{D}$ be an isometry in the Poincaré metric. Then either $f$ or $\bar{f}$ is an element of $\mathbb{A}\mathrm{ut}(\mathbb{D})$.*

**Proof.** First find $g\in\mathbb{A}\mathrm{ut}(\mathbb{D})$, so $g\circ f(0)=0$. Thus, $g\circ f$ maps $\{z\mid |z|=\frac{1}{2}\}$ setwise to itself. So by adding a rotation to $g$, we can also suppose $(g\circ f)(\frac{1}{2})=\frac{1}{2}$. It follows then (Problem 3) that $g\circ f$ maps $(-1,1)$ pointwise to itself, so it suffices (Problem 4) to see that if $f$ is an isometry and $f$ leaves $(-1,1)$ pointwise fixed, then either $f(z)=z$ or $f(z)=\bar{z}$. Since only $z=\frac{i}{2}$ and $z=-\frac{i}{2}$ have $\pi(z,\frac{1}{2})=\pi(\frac{i}{2},\frac{1}{2})$ and $\pi(z,-\frac{1}{2})=\pi(\frac{i}{2},-\frac{1}{2})$, we can suppose (by replacing $f$ by $\bar{f}$) that $f(\frac{i}{2})=\frac{i}{2}$, and we must then show $f$ is the identity.

Let $w\in\mathbb{C}_+\cap\mathbb{D}$. Then $f(w)=w$ or $f(w)=\bar{w}$ since these are the only two points $\tilde{w}$ with $\pi(\tilde{w},\frac{1}{2})=\pi(w,\frac{1}{2})$ and $\pi(\tilde{w},-\frac{1}{2})=\pi(w,-\frac{1}{2})$. Let $L$ be the line from $\frac{i}{2}$ to $w\in\mathbb{C}_+\cap\mathbb{D}$. Then $f[L]\cap(-1,1)=\emptyset$, so $f[L]$ must lie in $\mathbb{C}_+\cap\mathbb{D}$, so $f(w)\neq\bar{w}$, that is, $f$ is the identity. □

**Theorem 12.2.8.** *The curvature of the Poincaré conformal metric is $-4$ for all $z$.*

**Proof.** A straightforward calculation (Problem 5), essentially identical to the calculation of Example 12.1.3 with some sign changes. □

Finally, we want to say something about the Poincaré metric on $\mathbb{C}_+$. Temporarily, we use $\pi_{\mathbb{D}}$ for the Poincaré conformal metric on $\mathbb{D}$ and $\pi_{\mathbb{C}_+}$ for its image under

$$\varphi(z)=\frac{z-i}{z+i} \tag{12.2.28}$$

which is a bijection of $\mathbb{C}_+$ to $\mathbb{D}$. Thus,

$$\pi_{\mathbb{C}_+}(z) = (\varphi^* \pi_{\mathbb{D}})(z) \tag{12.2.29}$$

so (Problem 6)

$$\pi_{\mathbb{C}_+}(z) = \frac{1}{\operatorname{Im} z} \tag{12.2.30}$$

the Poincaré conformal metric on $\mathbb{C}_+$. Since (Problem 6)

$$\frac{|\varphi(z) - \varphi(w)|}{|1 - \overline{\varphi(z)}\,\varphi(w)|} = \left| \frac{z - w}{\bar{z} - w} \right| \tag{12.2.31}$$

we see that

$$\tanh(\rho_{\pi_{\mathbb{C}_+}}(z, w)) = \left| \frac{z - w}{\bar{z} - w} \right| \tag{12.2.32}$$

Geodesics are precisely "orthocircles" which now mean circles or lines that meet $\mathbb{R}$ orthogonally. This is obvious since $\varphi$ is conformal and maps geodesics to geodesics. This justifies our use of "geodesic triangles" in Section 8.3 of Part 2A.

The analog of Theorem 12.2.5 is that if $F$ is a Herglotz function, that is, $F\colon \mathbb{C}_+ \to \mathbb{C}_+$, then

$$\rho_{\pi_{\mathbb{C}_+}}(F(z), F(w)) \le \rho_{\pi_{\mathbb{C}_+}}(z, w) \tag{12.2.33}$$

**Notes and Historical Remarks.** Poincaré introduced his metric in [**316**]. He realized this provided an elegant, explicit model for hyperbolic non-Euclidean geometry. The key figures in the earlier developments of hyperbolic geometry are Gauss, Bolyai, Lobachevsky, and Beltrami; see Stillwell [**369**, Ch. 18] or Milnor [**278**] for the involved history.

Henri Poincaré (1854–1912) was a French mathematician and physicist, usually regarded as the greatest mathematician of the late nineteenth century. He got a first degree at École Polytechnique but then a degree in mining engineering. He made his first great discoveries on automorphic functions, Fuchsian groups, and hyperbolic geometry while working as an engineer. Then, under the influence of his teacher, Hermite, he moved to a math appointment in Paris in 1881, remaining there for the rest of his life. He died of a pulmonary embolism at the age of only 58. Gray [**161**] is a biography.

Poincaré was a key figure in the development of topology—both homotopy and homology theory—and in the development of the qualitative theory of differential equations. He is usually regarded as a father of chaos theory. He revolutionized celestial mechanics and was also a seminal figure in the theory of special relativity, developing many of the ideas independently of Einstein. He was also a popularizer of science, easily the most famous French scientist of the first decade of the twentieth century.

**Problems**

1. (a) Prove that the only orthocircles through 0 are straight lines. (*Hint*: Prove two distinct orthocircles that intersect on $\partial\mathbb{D}$ have no other intersections.)

   (b) If $C$ is an orthocircle with $0 \notin C$, prove there are uncountably many diameters disjoint from $C$. (*Hint*: Consider the diameters tangent to $C$.)

   (c) Prove that if $C$ is any orthocircle and $z \notin C$, there are uncountably many orthocircles through $z$ disjoint from $C$.

2. If $L$ is a geodesic from $z$ to $w$ and $u \in L$, prove there is no other $v \neq u$ with $\rho_\pi(z,v) = \rho_\pi(z,u)$ and $\rho_\pi(w,v) = \rho_\pi(w,u)$. (*Hint*: Use uniqueness of geodesics.)

3. If $f$ is an isometry of $\mathbb{D}$ in the Poincaré metric and $f(u) = u$, $f(v) = v$, then $f$ leaves the geodesic through $u$ and $v$ fixed. (*Hint*: Use uniqueness of geodesics.)

4. If $g, f \in \mathbb{A}\mathrm{ut}(\mathbb{D})$, prove $\tilde{g}(z) = \overline{g(\bar{z})}$ is also in $\mathbb{A}\mathrm{ut}(\mathbb{D})$ and that $\overline{g \circ \tilde{f}} = \tilde{g} \circ f$.

5. Compute the curvature of $\pi$.

6. (a) With $\varphi$ given by (12.2.28), prove that $(\varphi_* \pi_D)(z) = 1/\operatorname{Im} z$.

   (b) Verify (12.2.31).

7. The pseudohyperbolic "metric" on $\mathbb{C}_+$ is defined by

$$\gamma(z,w) = \frac{|z-w|}{\sqrt{\operatorname{Im} z}\,\sqrt{\operatorname{Im} w}} \tag{12.2.34}$$

   (a) Prove that there is a strictly monotone function $\Psi$ from $(0,\infty)$ to $(0,\infty)$ so that $\gamma(z,w) = \Psi(\rho_{\pi_{\mathbb{C}_+}}(z,w))$. (*Hint*: $4\gamma(z,w)^{-1} = (|z-w|/|\bar{z}-w|)^{-2} - 1$.)

   (b) Prove for any Herglotz function, $F$, that $\gamma(F(z),F(w)) \leq \gamma(z,w)$.

   **Remark.** We put "metric" in quotes because the triangle inequality may not hold. However, the formula for $\gamma$ is somewhat simpler than for $\rho$.

8. Let $f$ be an isometry on a metric space $X$. Suppose $f(x) \neq x$. Prove that $f^{[n]}(x)$ cannot cannot have a convergent limit. What does this say about fixed points of hyperbolic and parabolic elements of $\mathbb{A}\mathrm{ut}(\mathbb{D})$?

9. This problem will prove the following: If $f$ is an analytic map of $\mathbb{D}$ into $\mathbb{D}_r(0)$ with $r < 1$, then $f$ has a unique fixed point.

   (a) Prove for some $\varepsilon > 0$ that $\pi(f(x), f(y)) \leq (1-\varepsilon)\pi(x,y)$.

   (b) Prove that $f$ has a unique fixed point.

**Remarks.** 1. This theorem is due to Farkas [**130**] and Ritt [**329**]; the proof in this problem is due to Earle–Hamilton [**112**].

2. See Problem 5 of Section 3.3 of Part 2A for a Rouché theorem alternate proof of this result.

## 12.3. The Ahlfors–Schwarz Lemma

In this section, as preparation for the next, we prove a generalization of the Schwarz lemma due to Ahlfors which is expressed in terms of curvature. We'll show its power by proving the Liouville theorem—this is not surprising since we've seen (see Problem 9 in Section 3.6 of Part 2A) that the regular Schwarz lemma implies Liouville's theorem, but it is relevant since we'll interpret the little Picard theorem as a Liouville-type theorem.

**Theorem 12.3.1** (Ahlfors–Schwarz Lemma). *Let $f\colon \mathbb{D} \to \Omega$ be analytic. Let $\lambda$ be a conformal metric on $\Omega$ whose curvature $\kappa_\lambda$ obeys*

$$\kappa_\lambda(z) \le -4 \tag{12.3.1}$$

*for all $z \in \Omega$. Let $\pi$ be a Poincaré metric on $\mathbb{D}$. Then*

$$(f^*\lambda)(z) \le \pi(z) \tag{12.3.2}$$

*for all $z \in \mathbb{D}$.*

**Proof.** Fix $0 < r < 1$. Define $\eta_r$ on $\mathbb{D}_r(0)$ by

$$\eta_r(z) = \frac{(f^*\lambda)(z)}{\pi_r(z)} \tag{12.3.3}$$

where $\pi_r$ is the conformal metric on $\mathbb{D}_r(0)$,

$$\pi_r(z) = \frac{1}{r^2 - |z|^2} \tag{12.3.4}$$

If we prove that for all $r$, when $z \in \mathbb{D}_r(0)$,

$$\eta_r(z) \le 1 \tag{12.3.5}$$

then taking $r \uparrow 1$ yields (12.3.2).

Since $f^*\lambda$ is continuous on $\overline{\mathbb{D}_r(0)}$, it is bounded. Since $\pi_r(z)^{-1} \to 0$ as $|z| \to r$, we see the nonnegative function $\eta_r(z)$ must take its maximum value at some point $z_0 \in \mathbb{D}_r(0)$. If $\eta_r(z_0) = 0$, then $\eta_r(z) \equiv 0$, so (12.3.5) holds. Thus, we can suppose $\eta_r(z_0) > 0$, so $f'(z_0) \neq 0$ and $f^*\lambda$ is a conformal metric near $z_0$. Thus, $\eta_r$ is $C^2$ near $z_0$ and $\log(\eta)$ has a maximum at $z_0$. Thus,

$$\Delta(\log(\eta))(z) \le 0 \tag{12.3.6}$$

since all second-directional derivatives are nonpositive.

By (12.1.4), for $\rho = \lambda$ or $\pi$, $\Delta(\log(\rho)) = -\rho^2\kappa_\rho$, so (12.3.6) becomes

$$\begin{aligned} 0 &\geq \kappa_{\pi_r}(z_0)\pi_r(z_0)^2 - \kappa_{f^*\lambda}(z_0)(f^*\lambda)(z_0)^2 \\ &\geq \kappa_{\pi_r}(z_0)\pi_r(z_0)^2 + 4f^*\lambda(z_0)^2 \end{aligned} \tag{12.3.7}$$

by (12.3.1) and $\kappa_{f^*\lambda}(z_0) = \kappa_\lambda(f(z_0))$ (by Theorem 12.1.1).

Let $g\colon \mathbb{D}_r \to \mathbb{D}$ by $g(z) = z/r$. Then

$$(g^*\pi)(z) = r^{-2}\left(\frac{1}{1-(\frac{z}{r})^2}\right) = \pi_r(z) \tag{12.3.8}$$

so, by Theorem 12.2.8,

$$\kappa_{\pi_r}(z_0) = -4 \tag{12.3.9}$$

Thus, (12.3.7) becomes

$$0 \geq 4f^*\lambda(z_0)^2 - 4\pi_r(z_0)^2 \tag{12.3.10}$$

so $\eta_r(z_0) \leq 1$, which proves (12.3.5). □

For any $\alpha > 0$, let $\pi_\alpha(z) = 1/(\alpha^2 - |z|^2)$ on $\mathbb{D}_\alpha(0)$. As in the last proof, $\kappa_{\pi_\alpha} \equiv -4$, so the above result extends to $f\colon \mathbb{D}_\alpha(0) \to \Omega$—one need only replace $\pi$ by $\pi_\alpha$. If $\sup\kappa_\rho \leq -B$ instead of $-4$, then we let $\lambda = \sqrt{B/4}\,\rho$, so $\kappa_\lambda = \frac{4}{B}\kappa_\rho \leq -4$ and $f^*\rho = \sqrt{4/B}\,f^*\lambda$. We conclude:

**Corollary 12.3.2** (Scaled Ahlfors–Schwarz Lemma). *Let $\rho$ be a conformal metric on $\Omega$ with*

$$\kappa_\rho(z) \leq -B \tag{12.3.11}$$

*for some $B > 0$ and all $z \in \Omega$. Let $f\colon \mathbb{D}_\alpha(0) \to \Omega$ be analytic. Then for all $z \in \mathbb{D}_\alpha(0)$,*

$$(f^*\rho)(z) \leq \frac{2}{\sqrt{B}}\,\pi_\alpha(z) \tag{12.3.12}$$

As an immediate corollary, we have

**Theorem 12.3.3** (Ahlfors–Liouville Theorem). *Let $\Omega$ have a conformal metric, $\rho$, where (12.3.11) holds for some $B > 0$ and all $z \in \Omega$. Let $f\colon \mathbb{C} \to \Omega$ be entire. Then $f$ is constant.*

**Proof.** Fix $z_0 \in \mathbb{C}$. For each $\alpha > |z|$, we have (12.3.12). Since for $z_0$ fixed, $\lim_{\alpha\to\infty} \pi_\alpha(z_0) = 0$, we conclude

$$|f'(z_0)|^2\rho(f(z_0)) = 0 \tag{12.3.13}$$

Since $\rho$ is strictly positive, $f'(z_0) = 0$. Since $z_0$ is arbitrary, $f$ is constant. □

Since $\mathbb{D}_R(0)$ has $\pi_R$ as metric with $\kappa_{\pi_R} = -4$, this implies Liouville's theorem.

**Notes and Historical Remarks.** The Ahlfors–Schwarz lemma is from Ahlfors [**8**] published in 1938. In his 1982 collected works, Ahlfors remarked that this paper "has more substance than I was aware of." In the paper, he used his lemma to prove Schottky's and Bloch's theorem, obtaining a lower bound on the Bloch constant within .01% of the current best lower bound (although it is about 10% smaller than the best upper bound, conjectured to be the actual value). Given the relation of these theorems to Picard's theorems, it is not surprising that the next year, Robinson used Ahlfors' work to prove Picard theorems, as we'll describe in the next section.

Lars Ahlfors (1907–96) was born in what is now Helsinki, Finland and was trained by Lindelöf and Nevanlinna, two of the greatest Finnish complex analysts. Ahlfors won one of the first Fields Medals in 1936 for his work on Denjoy's conjecture and Nevanlinna theory. Under Carathéodory's influence, he went to Harvard in 1935, but returned to Helsinki in 1938 when a position opened at the university there. Given the war, this was, in retrospect, a poor choice. The Finnish universities closed during the war but, for medical reasons, Ahlfors couldn't serve in the army, so he was able to work. After a brief time in Zurich near the end of the war, he returned to Harvard in 1946, staying until his retirement in 1977.

Without a doubt, Ahlfors was the greatest complex analyst born in the twentieth century. His work includes not only the results of this section, but critical contributions to Fuchsian groups, quasiconformal and conformal mapping, value distribution theory, and Riemann surfaces.

## 12.4. Robinson's Proof of Picard's Theorems

Using the Ahlfors–Schwarz lemma, Robinson found simple proofs of Picard's theorems. Their key is:

**Theorem 12.4.1.** *There exists a conformal metric* $\lambda$ *on* $\mathbb{C}\setminus\{0,1\}$ *that obeys*

(i) $\lambda(z) \to \infty \quad \text{as } z \to 0, 1.$ (12.4.1)

(ii) $\lambda(z)^{-1} = o(|z|^2) \quad \text{as } z \to \infty.$ (12.4.2)

(iii) *For some* $B > 0$ *and all* $z \in \mathbb{C}\setminus\{0,1\}$,

$$\kappa_\lambda(z) \le -B \tag{12.4.3}$$

We'll prove this at the end of the section by giving an explicit formula for $\lambda$. While the great Picard theorem implies the little one, the proof of the little one is so immediate, we give it first.

**Proof of Theorem 11.3.1 of Part 2A (Picard's Little Theorem).** By the standard argument, it suffices to show an entire function with values in $\mathbb{C} \setminus \{0,1\}$ is constant. This is immediate from Theorems 12.3.3 and 12.4.1. $\square$

**Proof of Theorem 11.3.2 of Part 2A (Picard's Great Theorem).** As in the discussion in Section 11.3 of Part 2A (see following the proof of Theorem 11.3.6 of Part 2A), it suffices to prove Montel's three-value theorem (Theorem 11.3.6 of Part 2A), so we let $f_n \in \mathfrak{A}(\Omega)$ be a sequence with values in $\mathbb{C} \setminus \{0,1\}$ which we have to show is normal. By Marty's theorem (Theorem 6.5.6 of Part 2A), it suffices to prove for any disk $\mathbb{D}_r(z_0) \subset \Omega$, that for any $r' < r$,

$$\sup_{n, z \in \mathbb{D}_{r'}(z_0)} |f_n^\sharp(z)| < \infty \tag{12.4.4}$$

where $f^\sharp$ is the spherical derivative.

Notice that, by (12.1.1),

$$f^\sharp(z) = (f^*\sigma)(z) \tag{12.4.5}$$

where $\sigma$ is the spherical metric. Thus, (12.4.4) is equivalent to

$$\sup_{n, z \in \mathbb{D}_{r'}(z_0)} f_n^*\sigma(z) < \infty \tag{12.4.6}$$

On $\mathbb{C} \setminus \{0,1\}$, consider

$$\eta(z) = \frac{\sigma(z)}{\lambda(z)} \tag{12.4.7}$$

where $\lambda$ is the conformal metric of Theorem 12.4.1. By (12.4.1), $\eta(z) \to 0$ as $z \to 0, 1$, and by (12.4.2), $\eta(z) \to 0$ as $z \to \infty$. Since $\eta$ is continuous, we conclude it is bounded on $\widehat{\mathbb{C}}$ if set to 0 at $0, 1, \infty$. Thus, for some $M$ finite,

$$\sigma(z) \le M\lambda(z) \tag{12.4.8}$$

which implies

$$f_n^*\sigma \le M f_n^*\lambda \tag{12.4.9}$$

By Corollary 12.3.2, with $B$ the constant in Theorem 12.4.1 and $\pi_{r,z_0}(z) = 1/[r^2 - |z - z_0|^2]$, we have for all $n$ that

$$f_n^*\sigma \le \frac{2M}{\sqrt{B}}\, \pi_{r,z_0} \tag{12.4.10}$$

which implies (12.4.6). $\square$

**Proof of Theorem 12.4.1.** Let

$$\lambda(z) = \frac{(1 + |z|^{1/3})^{1/2}}{|z|^{5/6}} \, \frac{(1 + |z-1|^{1/3})^{1/2}}{|z-1|^{5/6}} \tag{12.4.11}$$

Then $\lambda(z) = O(|z|^{-5/6})$ at $z = 0$ and $O(|z-1|^{-5/6})$ at $z = 1$, so (12.4.1) holds. At $\infty$, $\lambda(z)^{-1} = O(|z|^{1/3})$, so (12.4.2) holds.

A straightforward calculation (Problem 1) shows that

$$\kappa_\lambda(z) = -\frac{1}{18}\left[\frac{|z-1|^{5/3}}{(1+|z|^{1/3})^2} + \frac{|z|^{5/3}}{(1-|z-1|^{1/3})^2}\right]\frac{1}{(1+|z|^{1/3})(1+|z-1|^{1/3})} \tag{12.4.12}$$

Thus, $\kappa_\lambda(z) < 0$ for all $z$ in $\mathbb{C}\setminus\{0,1\}$ is continuous, and

$$\lim_{z\to 0}\kappa_\lambda(z) = \lim_{z\to 1}\kappa_\lambda(z) = -\frac{1}{36} \tag{12.4.13}$$

$$\lim_{z\to\infty}\kappa_\lambda(z) = -\infty \tag{12.4.14}$$

So, by continuity and compactness of $\widehat{\mathbb{C}}$, we have (12.4.3). □

**Notes and Historical Remarks.** Robinson [**332**], a year after Ahlfors, first used the choice of a suitable negative curvature conformal metric on $\mathbb{C}\setminus\{0,1\}$ to prove Picard's theorems. Narasimhan–Nievergelt [**286**] discusses how to go from the Robinson metric to Landau's theorem and Schottky's theorem (Theorem 11.4.2 of Part 2A).

**Problems**

1. (a) Prove $\Delta(\log|z|^k) = 0$ for any $k$.

   (b) Prove that

$$\Delta\log(1+|z|^{1/3})^{1/2} = \frac{1}{18}\,\frac{1}{|z|^{5/3}(1+|z|^{1/3})^2}$$

   (c) Prove (12.4.12).

   (*Hint for* (a) and (b): $\Delta = 4\partial\bar\partial$ and $|z|^2 = z\bar z$.)

## 12.5. The Bergman Kernel and Metric

We've seen that there is a natural conformal metric on $\mathbb{D}$ in which analytic bijections are isometries. By using a Riemann mapping $f\colon \Omega \to \mathbb{D}$ and pulling back the Poincaré metric, we can define a conformal metric on any simply connected $\Omega$ in which automorphisms are isometries. In this section, for any bounded $\Omega$, we'll construct in an intrinsic way a natural metric, called the Bergman metric, that agrees with the Poincaré metric (up to a constant) when $\Omega$ is simply connected but which is always defined and in which maps in $\mathbb{A}\mathrm{ut}(\Omega)$ are isometries. This metric will depend on the Bergman kernel, an object we'll use in the next section. We begin by describing it.

On $\Omega$, a bounded region in $\mathbb{C}$, we let $d^2z = dxdy$ be a two-dimensional Lebesgue measure and we define the *Bergman space*, $\mathfrak{A}_2(\Omega)$, by

$$\mathfrak{A}_2(\Omega) = \mathfrak{A}(\Omega) \cap L^2(\Omega, d^2z) \tag{12.5.1}$$

that is, functions, $f$, which are analytic and

$$\|f\|_2 \equiv \left( \int_\Omega |f(z)|^2 d^2z \right)^{1/2} < \infty \tag{12.5.2}$$

Of course, such $f$ are defined for all $z$, not just for a.e. $z$. We'll see shortly that $\mathfrak{A}_2(\Omega)$ is a closed subspace. As a preliminary, we define on $\Omega$,

$$r_\Omega(z) = \operatorname{dist}(z, \mathbb{C} \setminus \Omega) \tag{12.5.3}$$

and note that

**Proposition 12.5.1.** *For all $f \in \mathfrak{A}_2(\Omega)$ and $z \in \Omega$,*

$$|f(z)| \le \frac{1}{\sqrt{\pi}\, r_\Omega(z)} \|f\|_2 \tag{12.5.4}$$

**Proof.** By the Cauchy integral formula for $r < r_\Omega(z)$,

$$f(z) = \int_0^{2\pi} f(z + re^{i\theta}) \frac{d\theta}{2\pi} \tag{12.5.5}$$

so multiplying by $r$ and integrating,

$$|f(z)| \le \frac{1}{\pi r_\Omega(z)^2} \int_{|w-z|<r_\Omega(z)} |f(w)|\, d^2w \tag{12.5.6}$$

$$\le \left( \frac{1}{\pi r_\Omega(z)^2} \int_{|w-z|<r_\Omega(z)} |f(w)|^2\, d^2w \right)^{1/2} \tag{12.5.7}$$

by the Cauchy–Schwarz inequality with $f = 1 \cdot f$. Since $\{w \mid |w - z| < r_\Omega(z)\} \subset \Omega$, (12.5.7) proves (12.5.4). □

(12.5.4) says that $L^2$ convergence implies uniform convergence on compacts, so $L^2$ limits of functions in $\mathfrak{A}_2(\Omega)$ lie in $\mathfrak{A}_2(\Omega)$ by the Weierstrass theorem (Theorem 3.1.5 of Part 2A), that is,

**Corollary 12.5.2.** *$\mathfrak{A}_2(\Omega)$ is a closed subspace of $L^2(\Omega, d^2z)$.*

The *Bergman projection*, $P_\Omega$, is the orthogonal projection of $L^2(\Omega)$ to $\mathfrak{A}_2(\Omega)$. (12.5.4) also says that $z \to f(z)$ is a bounded linear functional on $\mathfrak{A}_2$, so by the Riesz representation theorem (see Theorem 3.3.3 in Part 1):

**Corollary 12.5.3.** *For each $z \in \Omega$, there exists a function, $k_z \in \mathfrak{A}_2(\Omega)$, with*

$$\|k_z\|_{L^2} \le \pi^{-1/2} r_\Omega(z)^{-1} \tag{12.5.8}$$

*so that*

$$f(z) = \langle k_z, f \rangle_{L^2(\Omega, d^2z)} \tag{12.5.9}$$

We define the *Bergman kernel*, $K_\Omega(z,w)$ on $\Omega\times\Omega$ (sometimes we write $K(z,w)$) by

$$K_\Omega(z,w) \equiv \langle k_z, k_w\rangle_{L^2} \tag{12.5.10}$$

Here are many of its properties:

**Theorem 12.5.4.** (a) *We have*

$$K_\Omega(z,w) = k_w(z) = \overline{k_z(w)} = \overline{K_\Omega(w,z)} \tag{12.5.11}$$

(b) $$|K_\Omega(z,w)| \le \frac{1}{\pi r_\Omega(z) r_\Omega(w)} \tag{12.5.12}$$

(c) *For any orthonormal basis* $\{\varphi_n\}_{n=1}^\infty$ *of* $\mathfrak{A}_2(\Omega)$, *we have*

$$\sum_{n=1}^\infty |\varphi_n(z)|^2 = K_\Omega(z,z) \le \frac{1}{\pi r_\Omega(z)^2} \tag{12.5.13}$$

(d) $K_\Omega(z,z)$ *is continuous.*

(e) $z\to k_z$ *is (norm) continuous in* $\mathfrak{A}_2(\Omega, d^2z)$.

(f) $$K_\Omega(z,w) = \sum_{n=1}^\infty \varphi_n(z)\,\overline{\varphi_n(w)} \tag{12.5.14}$$

*where the sum is uniformly convergent on compacts of* $\Omega\times\Omega$.

(g) $K_\Omega(z,w)$ *is jointly analytic in* $z$ *and* $\bar w$.

(h) *The Bergman projection is given by*

$$(P_\Omega h)(z) = \int K_\Omega(z,w) h(w)\, d^2w \tag{12.5.15}$$

*In particular, if* $f\in\mathfrak{A}_2(\Omega)$,

$$f(z) = \int K_\Omega(z,w) f(w)\, d^2w \tag{12.5.16}$$

(i) $$\int K_\Omega(z,w) K_\Omega(w,u)\, d^2w = K_\Omega(z,u) \tag{12.5.17}$$

**Remarks.** 1. Because of (12.5.16), $K_\Omega$ is sometimes called the *reproducing kernel.*

2. This construction and some of its properties are special cases of the general construction of $L^2$ reproducing Hilbert spaces explored in Problem 6 of Section 3.3 and Problem 8 of Section 3.4 of Part 1.

3. Our proof of (d) below is related to the theorem of Hartogs that a separately analytic function of two variables is jointly analytic.

4. As an alternate to (d), (e) that uses a Cauchy estimate in place of the Cauchy formula, one can follow the argument in Problem 4 of Section 3.3 of Part 1. By (12.5.8), if $z_n\to z$, we have a uniform bound on $k'_{z_n}(z)$ on a disk containing $z$ and $\{z_n\}$ for $n$ large. That shows $k_{z_n}(z) - k_{z_n}(z_n)\to 0$ as

$n \to \infty$, which leads to a proof that $\|k_z - k_{z_n}\| \to 0$ which implies continuity of $\langle k_z, k_w \rangle$.

**Proof.** (a) Taking $f = k_w$ in (12.5.9) gives the first and last equalities. Since $\overline{\langle f, g \rangle} = \langle g, f \rangle$, we get the middle equality.

(b) follows from the Cauchy–Schwarz inequality and (12.5.8).

(c) Since $|\varphi_n(z)| = |\langle k_z, \varphi_n \rangle| = |\langle \varphi_n, k_z \rangle|$, the equality is the Parseval relation (see Theorem 3.4.1 of Part 1) and the inequality is (12.5.12).

(d) Fix $z_0$ and note, since $k_w(z)$ is analytic in $z$, that for $z \in \mathbb{D}_{r(z_0)/2}$,

$$k_w(z) = \frac{1}{2\pi i} \oint_{|\zeta - z_0| = r(z_0)/2} \frac{k_w(\zeta)}{\zeta - z}\, d\zeta \tag{12.5.18}$$

Since $k_w(z) = \overline{k_z(w)}$, we see that for $z, w \in \mathbb{D}_{r(z_0)/2}$,

$$k_w(z) = \frac{1}{(2\pi i)^2} \int_{|\zeta - z_0| = |\omega - z_0| = r(z_0)/2} \frac{k_\omega(\zeta)}{(\zeta - z)(\bar\omega - \bar w)}\, d\zeta\, d\bar\omega \tag{12.5.19}$$

Since (12.5.12) implies that $|k_\omega(\zeta)| \le 4/\pi r(z_0)^2$ on the region of integration, we conclude for $z, w \in \mathbb{D}_{r(z_0)/2}$,

$$\begin{aligned} &|K(z,z) - K(w,w)| \\ &\le \left[\frac{r(z_0)}{2}\right]^2 \frac{4}{\pi r(z_0)^2} \sup \left[ \left| \frac{1}{\zeta - z} \frac{1}{\bar\omega - \bar z} - \frac{1}{\zeta - w} \frac{1}{\bar\omega - \bar w} \right| \right] \end{aligned} \tag{12.5.20}$$

where the sup is over the region of integration in (12.5.19). This implies continuity in $\mathbb{D}_{r(z_0)/4}(z_0)$.

(e) By (12.5.9), $z \to k_z$ is weakly continuous. Since $\|k_z\|^2 = K_\Omega(z,z)$, $z \to \|k_z\|$ is continuous by (d). Thus, $z \to k_z$ is norm continuous by Theorem 3.6.2 of Part 1.

(f) is a Fourier expansion of $k_z$ and $k_w$ in the $\varphi_n$ basis. The uniform convergence follows from the norm continuity of $k_w$.

(g) $\sum_{n=1}^N \varphi_n(z) \overline{\varphi_n(w)}$ is jointly analytic in $z$ and $w$, so the uniform convergence implies joint analyticity of $K_\Omega$.

(h) Pick a basis $\{\varphi_n\}_{n=1}^\infty$ for $\mathfrak{A}_2(\Omega)$. Let $h$ in $L^2$ have compact support in $\Omega$. By the uniform convergence of (12.5.14), we can interchange sum and integral to see

$$\text{RHS of (12.5.15)} = \sum_{n=1}^\infty \langle \varphi_n, h \rangle \varphi_n(z) = (P_\Omega h)(z) \tag{12.5.21}$$

Since the functions of compact support in $\Omega$ are dense in $L^2(\Omega, d^2z)$ and both sides of (12.5.15) are continuous in $L^2$ (the right side uses continuity of $k_w(z)$ in $L^2$ norm), we get (12.5.15) in general.

(i) (12.5.17) is (12.5.16) for $f = K_\Omega(\,\cdot\,, u)$. $\square$

**Example 12.5.5** (Bergman Kernel for $\mathbb{D}$). In $L^2(\mathbb{D})$,

$$\langle z^n, z^m \rangle_{L^2} = \int r^{n+m} e^{i(m-n)\theta} r\,dr\,d\theta = \frac{2\pi}{n+m+2}\,\delta_{mn} \tag{12.5.22}$$

Since polynomials in $z$ are dense in $\mathfrak{A}_2(\mathbb{D})$ (Problem 1),

$$\varphi_n(z) = \left(\frac{n+1}{\pi}\right)^{1/2} z^n, \qquad n = 0, 1, 2, \dots \tag{12.5.23}$$

is an orthonormal basis for $\mathfrak{A}_2(\Omega)$. Thus, by (12.5.14),

$$K_{\mathbb{D}}(z,w) = \frac{1}{\pi} \sum_{n=0}^\infty (n+1)(\bar{w}z)^n \tag{12.5.24}$$

$$= \frac{1}{\pi}\,\frac{1}{(1-\bar{w}z)^2} \tag{12.5.25}$$

(by Problem 2). See Problems 3 and 5 for other proofs of (12.5.25). $\square$

Next, we turn to showing conformal invariance of $K_\Omega$.

**Theorem 12.5.6.** *Let $\Omega_1, \Omega_2$ be two bounded regions in $\mathbb{C}$ and $F\colon \Omega_1 \to \Omega_2$ an analytic bijection. For $\varphi \in \mathfrak{A}_2(\Omega_2)$ (respectively, $L_2(\Omega_2)$), define $U\varphi$ on $\Omega_1$ by*

$$(U\varphi)(z) = F'(z)\varphi(F(z)) \tag{12.5.26}$$

*Then $U\varphi \in \mathfrak{A}_2(\Omega_1)$ (respectively, $L^2(\Omega_1)$) and $U$ is a unitary map of $L^2(\Omega_2)$ onto $L^2(\Omega_1)$ and of $\mathfrak{A}_2(\Omega_2)$ onto $\mathfrak{A}_2(\Omega_1)$.*

**Proof.** That $U$ is an isometry just says that

$$\int_{\Omega_1} |\varphi(F(z))|^2 |F'(z)|^2\, d^2z = \int_{\Omega_2} |\varphi(w)|^2\, d^2w \tag{12.5.27}$$

which follows from (2.2.21) of Part 2A and a Jacobian change of variables.

That $U$ is onto follows from the fact that

$$(U^{-1}\psi)(w) = (F^{-1})'(w)\psi(F^{-1}(w)) \tag{12.5.28}$$

is the inverse map to $U$ and is also an isometry. $\square$

Since $U$ is unitary and maps $\operatorname{Ran}(P_{\Omega_2})$ to $\operatorname{Ran}(P_{\Omega_1})$, we have that

$$UP_{\Omega_2} = P_{\Omega_1} U \tag{12.5.29}$$

This has an immediate consequence:

**Corollary 12.5.7.** *Let $\Omega_1, \Omega_2$ be two bounded regions in $\mathbb{C}$ and $F\colon \Omega_1 \to \Omega_2$ an analytic bijection. Then*

$$K_{\Omega_1}(z,w) = F'(z) K_{\Omega_2}(F(z),F(w))\,\overline{F'(w)} \tag{12.5.30}$$

**Proof.** By (12.5.29) and (12.5.15), for any $\varphi \in L^2(\Omega_2)$,

$$\begin{aligned}\int_{\Omega_1} & K_{\Omega_1}(z,w)F'(w)\varphi(F(w))\,d^2w \\ &= \int_{\Omega_2} F'(z)K_{\Omega_2}(F(z),u)\varphi(u)\,d^2u \\ &= \int_{\Omega_1} F'(z)K_{\Omega_2}(F(z),F(w))\varphi(F(w))|F'(w)|^2\,d^2w \end{aligned} \tag{12.5.31}$$

by a change of variables $u = F(w)$ and (2.2.21) of Part 2A. Thus, for a.e. $z$ and a.e. $w$, and then by analyticity for all $z$ and $w$,

$$K_{\Omega_1}(z,w)F'(w) = F'(z)K_{\Omega_2}(F(z),F(w))|F'(w)|^2 \tag{12.5.32}$$

which implies (12.5.28). $\square$

We'll use this result in Problem 3 to give a second computation of $K_{\mathbb{D}}$. The following formula for $F'$ with $\Omega_1 = \mathbb{D}$ will be a key to study boundary behavior of Riemann mappings in Section 12.6:

**Corollary 12.5.8** (Bell's Formula)**.** *Let $\Omega_1, \Omega_2$ be two bounded regions in $\mathbb{C}$ and $F\colon \Omega_1 \to \Omega_2$ an analytic bijection. Let $\varphi \in L^2(\Omega_2, d^2w)$ be such that $P_{\Omega_2}\varphi \equiv 1$. Then*

$$F'(z) = \bigl[P_{\Omega_1}(F'(\varphi \circ F))\bigr](z) \tag{12.5.33}$$

**Proof.** Clearly,

$$U(P_{\Omega_2}\varphi)(z) = F'(z) \tag{12.5.34}$$

so (12.5.33) is immediate from (12.5.29). $\square$

Finally, we'll define the Bergman metric.

**Proposition 12.5.9.** *For any bounded region, $\Omega$, if $g(z) = \log(K_\Omega(z,z))$, then for all $z \in \Omega$,*

$$(\partial\bar\partial g)(z) > 0 \tag{12.5.35}$$

**Proof.** By (12.5.13) and $\bar\partial\varphi_n = 0$,

$$(\bar\partial g)(z) = \frac{\sum_{n=1}^\infty \bigl[\overline{\partial\varphi_n}\,(z)\bigr]\varphi_n(z)}{\sum_{n=1}^\infty |\varphi_n(z)|^2} \tag{12.5.36}$$

Thus,

$$(\partial\bar\partial g)(z)$$
$$= \biggl\{\biggl(\sum_{n=1}^\infty |\varphi_n(z)|^2\biggr)\biggl(\sum_{n=1}^\infty |\partial\varphi_n(z)|^2\biggr) - \biggl|\sum_{n=1}^\infty [\overline{\partial\varphi_n}(z)]\varphi_n(z)\biggr|^2\biggr\}\biggl[\sum_{n=1}^\infty |\varphi_n(z)|^2\biggr]^{-2} \tag{12.5.37}$$

$$= \frac{1}{2}\biggl[\sum_{n,m=1}^\infty |\varphi_n|^2|\partial\varphi_m|^2 + |\varphi_m|^2|\partial\varphi_n|^2 - \overline{\partial\varphi_n}\,\varphi_n\overline{\varphi_m}\,\partial\varphi_m - \overline{\partial\varphi_m}\,\varphi_m\overline{\varphi_n}\,\partial\varphi_n\biggr]\Big/K_\Omega^2 \tag{12.5.38}$$

$$= \frac{1}{2}\biggl[\sum_{n,m=1}^\infty |\varphi_n\partial\varphi_m - \varphi_m\partial\varphi_n|^2\biggr]\Big/K_\Omega^2 \tag{12.5.39}$$

where (12.5.38) comes from the fact that $n$ and $m$ are dummy indices and so can be interchanged.

This is clearly nonnegative. To see it is strictly positive, we note the equality holds for any orthonormal basis. By using Gram–Schmidt on $\psi_1(z) = z - z_0$ and $\psi_2(z) = 1$ and completing to a basis, we see that we can suppose $\varphi_1(z_0) = 0$, $\partial\varphi_1(z_0) \neq 0 \neq \varphi_2(z_0)$, so $|\varphi_1(z_0)\partial\varphi_2(z_0) - \varphi_2(z_0)\partial\varphi_1(z_0)| \neq 0$, proving strict positivity. □

We define the *Bergman metric* to be the conformal metric,

$$\beta_\Omega(z) = \sqrt{\partial\bar\partial\log(K_\Omega(z,z))} \tag{12.5.40}$$

**Example 12.5.5 (continued).** By a straightforward calculation (Problem 6) from $K_{\mathbb{D}}(z,z) = \pi^{-1}(1-|z|^2)^{-2}$, we see that

$$\beta_{\mathbb{D}}(z) = \frac{\sqrt{2}}{1-|z|^2} \tag{12.5.41}$$

which is $\sqrt{2}\,\pi_{\mathbb{D}}(z)$, that is, the Bergman metric and the Poincaré metric agree up to a constant. We have met our goal of finding an intrinsic definition of metric that works for any bounded sets and is the Poincaré metric for the disk. We'll see next that it is also conformally invariant. □

**Theorem 12.5.10** (Conformal Invariance of the Bergman Metric). *Let $\Omega_1, \Omega_2$ be two bounded regions of $\mathbb{C}$ and $F\colon \Omega_1 \to \Omega_2$ an analytic bijection. Then*

$$F^*\beta_{\Omega_2} = \beta_{\Omega_1} \tag{12.5.42}$$

**Proof.** Locally, $\log|F'(z)|^2 = \log(F(z)) + \log(\overline{F'(z)})$ is killed by $\bar\partial\partial$ (i.e., is harmonic). Thus, in $\log(|F'(z)|^2 K_{\Omega_2}(F(z), F(z)))$, the $\log|F'(z)|^2$ goes away under $\bar\partial\partial$. But the $F(z)$ terms produce an $|F'(z)|^2$ anew, that is,

$$\begin{aligned}\partial_z\bar\partial_z \log(|F'(z)|^2 K_{\Omega_2}(F(z), F(z)))\\ = |F'(z)|^2 \partial_w\bar\partial_w \log(K_{\Omega_2}(w,w))|_{w=F(z)}\end{aligned} \tag{12.5.43}$$

which implies (12.5.42). □

**Notes and Historical Remarks.**

> Bergman was an extraordinarily kind and gentle man. He went out of his way to help many young people begin their careers, and he made great efforts on behalf of Polish Jews during the Nazi terror. He is remembered fondly by all who knew him.
>
> *S. Krantz* [**232**]

Bergman began the study of Hilbert spaces of analytic functions and invented his kernel in his 1921 thesis [**47**], written even before Hilbert spaces were defined (although $L^2$ was known). He developed the idea in a series of papers [**48, 49, 50**] and a book [**51**]. An early adapter and developer of the ideas was Aronszajn [**20, 21, 22**]. See Fefferman [**132**] for a survey of their use in quantum theory. In his thesis, written at the same time as Bergman's initial work, S. Bochner (1899–1982) introduced the same kernel but he didn't pursue it. Krantz [**235**] has a book on the Bergman kernel and metric.

Starting especially with the work of Fefferman (discussed in the Notes to the next section), Bergman kernel and metric ideas have become a staple in the study of several complex variables, with many papers on the method.

Stefan Bergman (1895–1977) was a Polish Jew born in the southern Polish city of Czestochowa, then part of the Russian Empire. He got engineering degrees from Breslau and Vienna, and in 1921, moved to Berlin to study applied mathematics with von Mises, who had a lifetime influence on him and his career.

His life illustrates the maxim that "may you live in interesting times" is a curse. He stayed on in Berlin until the rise of Hitler caused him to leave Germany in 1933. Initially, he went to the Soviet Union, fled there in 1937 during Stalin's purge of foreign scientists. He went to France until the German invasion when he fled once more, this time to the U.S. After periods at M.I.T., Yeshiva University, Brown, and Harvard, he was brought to Stanford in 1952 by Szegő, where he completed his career.

Bergman was born Bergmann and [**47**] appears with that name. After he came to the U.S., he dropped an "n" so [**51**] has one "n." We have systematically used the spelling he did towards the end of his life.

## Problems

1. (a) For any $f$ analytic on $\mathbb{D}$, prove that $r \to \int_0^{2\pi} |f(re^{i\theta})|^2 \frac{d\theta}{2\pi} \equiv L(r)$ is monotone in $r$. (*Hint*: First show that it suffices to prove $L(r) \le L(1)$ for $r < 1$ and $f$ analytic in a neighborhood of $\overline{\mathbb{D}}$ and then use the Poisson representation, Theorem 5.3.3 of Part 2A.)

   (b) If $f \in \mathfrak{A}_2(\mathbb{D})$ and $f_r(z) = f(rz)$ for $0 < r < 1$, prove that $\|f_r - f\|_{\mathfrak{A}_2(\mathbb{D})} \to 0$ as $r \uparrow 1$.

   (c) If $f$ is analytic in a neighborhood of $\overline{\mathbb{D}}$, prove that its Taylor series converges to $f$ in $\mathfrak{A}_2(\mathbb{D})$.

   (d) Prove the polynomials are dense in $\mathfrak{A}_2(\mathbb{D})$.

2. For $\zeta \in \mathbb{D}$, prove $\sum_{n=0}^{\infty} (n+1)\zeta^n = (1-\zeta)^2$. (*Hint*: Look at $\frac{d}{dz}(1-z)^{-1}$.)

3. (a) Show that for $\Omega = \mathbb{D}$, $k_{w=0}(z)$ is constant, and compute the constant.

   (b) Compute $K_{\mathbb{D}}(z, 0)$.

   (c) Using conformal invariance, find another proof of (12.5.25).

4. The purpose of this problem is to prove that

$$K_\Omega(z, w) = 4 \frac{\partial}{\partial z} \frac{\partial}{\partial \bar{w}} G_\Omega(z, w) \tag{12.5.44}$$

   where $\frac{\partial}{\partial z}, \frac{\partial}{\partial \bar{w}}$ are $\partial$ and $\bar{\partial}$ in the $z$ and $w$ variables, and $G_\Omega$ is a Green's function of $\Omega$. Classical Green's functions, as discussed in the Notes to Section 8.1 of Part 2A, are defined on $\Omega \times \Omega$, are real-valued, and of the form

$$G_\Omega(z, w) = \frac{1}{2\pi} \log|z - w| + F(z, w) \tag{12.5.45}$$

   where $F$ is harmonic in $w$ (respectively, $z$) for $z$ (respectively, $w$) fixed and $\lim_{z \to \partial\Omega} G_\Omega(z, w) = 0$ for all $w$ fixed and vice versa. In this problem, you may suppose $\Omega$ is a smooth Jordan domain (i.e., interior of a $C^\infty$ Jordan curve). You may also suppose the functions analytic in a neighborhood of $\overline{\Omega}$ are dense in $\mathfrak{A}_2(\Omega)$. Let

$$G_0(z, w) = \frac{1}{2\pi} \log|z - w| \tag{12.5.46}$$

   (a) Prove that

$$\frac{\partial}{\partial z} \frac{\partial}{\partial \bar{w}} G_0(z, w) = 0 \tag{12.5.47}$$

$$(\textit{Hint}: \quad G_0(z, w) = \frac{1}{4\pi} \left[\log(z - w) + \log(\bar{z} - \bar{w})\right].) \tag{12.5.48}$$

   (b) Prove that

$$\text{RHS of (12.5.44)} = 4 \frac{\partial}{\partial z} \frac{\partial}{\partial \bar{w}} F(z, w)$$

   and conclude that this right side is analytic in $z$ and anti-analytic in $w$.

(c) Prove that if $f$ is analytic in a neighborhood of $\Omega$, then

$$f(z) = -\frac{2}{i} \oint_{\partial\Omega} \left[ \frac{\partial}{\partial z} G_0(z,w) \right] f(w)\, dw \tag{12.5.49}$$

(*Hint*: Use (12.5.47) to see $\frac{\partial}{\partial z} G(z,w) = -(4\pi(w-z))^{-1}$.)

(d) Prove for $z \in \Omega$, $w \in \partial\Omega$, $\frac{\partial}{\partial z} G(z,w) = 0$ and conclude

$$f(z) = -2i \oint_{\partial\Omega} \left[ \frac{\partial}{\partial z} F(z,w) \right] f(w)\, dw \tag{12.5.50}$$

(e) Using Stokes' theorem (see (2.1.36) of Part 2A), prove that

$$f(z) = 4 \int_{\Omega} \frac{\partial}{\partial \bar{w}} \left[ \frac{\partial}{\partial z} F(z,w) f(0) \right] d^2w \tag{12.5.51}$$

(f) Prove (12.5.44).

5. This uses the previous problem.

(a) Prove the Green's function for $\mathbb{D}$ is given by

$$G_{\mathbb{D}}(z,w) = \frac{1}{2\pi} \log|z-w| - \frac{1}{2\pi} \log|1 - z\bar{w}| \tag{12.5.52}$$

(b) Using (12.5.44), find another calculation that (12.5.25) holds.

6. Prove (12.5.41).

## 12.6. The Bergman Projection and Painlevé's Conformal Mapping Theorem

Our main goal in this section is to prove the following (already stated as Theorem 8.2.8 of Part 2A):

**Theorem 12.6.1** (Painlevé's Smoothness Theorem). *Let $\gamma$ be a $C^\infty$ Jordan curve, that is, $\gamma\colon [0,1]$ is $C^\infty$ with $\gamma^{(k)}(0) = \gamma^{(k)}(1)$ for all $k$ and $\gamma'(s) \neq 0$ for all $s$. Let $\Omega^+$ (respectively, $\Omega^-$) be the inside (respectively, outside) of $\gamma$. Let $F\colon \mathbb{D} \to \Omega^+$ be a biholomorphic bijection. Then $F$ and $F^{-1}$ have $C^\infty$ extensions to $\overline{\mathbb{D}}$ and $\overline{\Omega^+}$, respectively, that is, all derivatives of $F$ (respectively, $F^{-1}$) have continuous extensions to $\overline{\mathbb{D}}$ (respectively, $\overline{\Omega^+}$).*

We call such an $\Omega^+$ a $C^\infty$ *Jordan domain.* The proof will be long because of lots of small steps (and a few big ones), so we'll leave the details of some of them to the reader. The key will be to use Bell's formula, (12.5.33), with $\Omega_1 = \mathbb{D}$, $\Omega_2 = \Omega^+$, which says that

$$F'(z) = P_{\mathbb{D}}(F'(\varphi \circ F))(z) \tag{12.6.1}$$

where $\varphi$ is any function on $\Omega^+$ with $P_{\Omega^+}\varphi = 1$.

In general, $P_{\mathbb{D}}g$ is not even bounded near $\partial\mathbb{D}$ (e.g., if $g(z) = (z-1)^{-1/3}$). This is because $K_{\mathbb{D}}(z,w)$ is not bounded; indeed, by (12.5.25),

$$\sup_z |K_{\mathbb{D}}(z,w)| = \frac{1}{\pi}(1-|w|)^{-2} \tag{12.6.2}$$

But if $g$ vanishes to order 2 (we'll be precise what that means below) at $\partial\mathbb{D}$, then $\sup_z |(P_{\mathbb{D}}g)(z)| \le \int \frac{1}{\pi}(1-|w|)^{-2}|g(w)|\, d^2w$ will be finite.

Derivatives in $z$ of $K_D(z,w)$ blow up even faster than $(1-|w|)^{-2}$ as $|w| \to 1$, but if $g$ vanishes to high enough order at $\partial\mathbb{D}$, those will also stay finite (and if $h \in \mathfrak{A}(\mathbb{D})$ has bounded derivatives of order $k+1$, we'll see all derivatives of order $k$ have continuous extensions to $\overline{\mathbb{D}}$).

Thus, the key to proving $F'$ (and its derivatives bounded on $\mathbb{D}$) will be to show that, for any $\ell$, there is $\varphi$ which is $C^\infty$ on $\overline{\Omega^+}$ so that $P_{\Omega^+}\varphi = 1$ and so that $F'(\varphi \circ F)$ vanishes to order $\ell$ on $\partial\mathbb{D}$. This, in turn, we will prove with some analysis from the fact that there are such $\varphi$ which vanish on $\partial\Omega^+$ to any prescribed order $k$, that is, the key technical fact will be (we'll precisely define vanishing to order $k$ later, but it's what you'd expect):

**Theorem 12.6.2** (Bell's Lemma). *Let $\Omega^+$ be the interior of a $C^\infty$ Jordan curve. Given any $k = 1, 2, \dots$ and $\psi \in C^\infty(\Omega^+)$ with all derivatives bounded, there exists $\varphi \in C^\infty(\Omega^+)$ so that $P_D\varphi = P_D\psi$ and so that $\varphi$ vanishes to order $k$ on $\partial\Omega^+$.*

At first sight, this might look surprising, but it is easy to see (Problem 1) for $\Omega^+ = \mathbb{D}$ that $c_\ell(1-|z|^2)^\ell = h_\ell$, which vanishes on $\partial\mathbb{D}$ to order $\ell$, has $P_{\mathbb{D}}h_\ell = 1$ if $c_\ell$ is suitably chosen—this makes it perhaps less surprising. We'll defer the proof of Bell's lemma to the end of the section.

We begin with some geometric facts. Given $\Omega^+$, a $C^\infty$ Jordan region, define $\delta_{\Omega^+}$ on $\Omega^+$ by

$$\delta_{\Omega^+}(z) = \min_{w=\operatorname{Ran}(\gamma)} |w-z| \tag{12.6.3}$$

the distance from $z$ to $\partial\Omega^+$.

**Proposition 12.6.3.** *There exists a $C^\infty$ real-valued function, $\rho$, on $\mathbb{C}$ so that*

(i) $\mp\rho(z) > 0$ *for* $z \in \Omega^\pm$ *(so* $\partial\Omega^+ = \{z \mid \rho(z) = 0\}$*).*
(ii) $\nabla\rho$ *is nonzero on* $\partial\Omega^+$ *(so* $\inf_{z\in\partial\Omega}|\nabla\rho(z)| > 0$ *by continuity).*

**Proof.** Following the construction in Proposition 4.8.2 of Part 2A, by the implicit function theorem, for any $z \in \partial\Omega^+$, there are a disk, $N_z$, and real-valued function $q_z$ on $N_z$ so $\mp q_z > 0$ on $\Omega^\pm \cap N_z$ and $|\nabla q_z| > 0$ on $\partial\Omega^+ \cap N_z$ (use a local coordinate system in which $\partial\Omega^+ \cap N_z$ is one axis).

By compactness, find $N_1, \dots, N_k$ covering $\partial\Omega^+$ and let $q_1, \dots, q_k$ be the corresponding functions. Let $q_\pm$ be defined on $\Omega^\pm$ by $q_\pm(w) = \mp 1$.

$\bigcup_{j=1}^k N_j \cup \Omega^+ \cup \Omega^- = \mathbb{C}$, so we can find (by Theorem 1.4.6 of Part 2A) $C^\infty$ functions $j_1, \dots, j_k, j_+, j_-$ so that $0 \le j_\ell \le 1$, $0 \le j_\pm \le 1$, $j_+ + j_- + \sum_{\ell=1}^k j_\ell = 1$, and $\text{supp}(j_\ell) \subset N_\ell$, $\text{supp}(j_\pm) \subset \Omega_\pm$.

Let

$$\rho = q_+ j_+ + q_- j_- + \sum_{\ell=1}^k q_\ell j_\ell \tag{12.6.4}$$

$\rho$ is $C^\infty$ and is easily seen to obey (i). On $\partial\Omega^+$, $\nabla\rho = \sum_{\ell=1}^k (\nabla q_\ell) j_\ell$. Since $\nabla q_\ell$ is in the direction normal to $\partial\Omega^+$ pointing outwards, they add up to a nonzero number. $\square$

$\rho$ is called a *defining function* for $\Omega^+$. It is easy to see (Problem 2) that there are $c_1, c_2 \in (0,\infty)$, so for all $z \in \Omega^+$,

$$-c_1\rho(z) \le \delta_{\Omega^+}(z) \le -c_2\rho(z) \tag{12.6.5}$$

The following geometric fact is left to the reader:

**Proposition 12.6.4.** *There is an $r > 0$ so that for all $z \in \partial\Omega^+$, the closed disks $D^\pm(z)$ of radius $r$ with centers $z \mp rw$, where $w$ is the outward-pointing normal to $\gamma$ at $z$ (so the disks are tangent to $\partial\Omega^+$ at $z$), obey*

$$D^\pm(z) \setminus \{z\} \subset \Omega^\pm \tag{12.6.6}$$

As a final geometric fact, we define for $z, w \in \Omega^+$

$$d_{\Omega^+}(z,w) = \inf\left\{ \int_0^1 |\gamma'(s)|\, ds \;\middle|\; \forall s\, \gamma(s) \in \Omega^+,\, \gamma(0) = z,\, \gamma(1) = w \right\}$$

the geodesic distance within $\Omega^+$. In the following, the first inequality is immediate; the other left to Problem 3.

**Proposition 12.6.5.** *For each $\Omega^+$, we have a $c > 0$ so that for all $z, w \in \Omega^+$,*

$$|z - w| \le d_{\Omega^+}(z,w) \le c|z - w| \tag{12.6.7}$$

**Definition.** For any $\Omega^+$ and $k = 0, 1, 2, \dots$, we define $C_b^k(\Omega^+)$ to be the $C^k$ functions (in real-valued sense) on $\Omega^+$ so that for any $m = 0, 1, \dots, k$ and $\ell = 0, 1, \dots, m$, $\|\frac{\partial^m}{\partial x^\ell \partial y^{m-\ell}} f\|_\infty < \infty$. We define $C^k(\overline{\Omega^+})$ to be the $C^k$ functions on $\Omega^+$ so that for any $m = 0, 1, \dots, k$ and $\ell = 0, 1, \dots, m$, $\frac{\partial^m}{\partial x^\ell \partial y^{m-\ell}} f$ has a continuous extension to $\overline{\Omega^+}$. $C_b^\infty(\Omega_+) \equiv \cap_{\ell=1}^\infty C_b^\ell(\Omega_+)$.

Clearly, by compactness, $C^k(\overline{\Omega^+}) \subset C_b^k(\Omega^+)$. Slightly more subtle is:

**Proposition 12.6.6.** *For $k = 0, 1, 2, \dots$, $C_b^{k+1}(\Omega^+) \subset C^k(\overline{\Omega^+})$.*

**Proof.** By integrating gradients along curves, we see

$$|g(z) - g(w)| \leq \big[ \sup_{y \in \Omega^+} |\nabla g(y)| \big] \, d_{\Omega^+}(z, w) \tag{12.6.8}$$

$$\leq \tilde{c}|z - w| \tag{12.6.9}$$

by (12.6.7). Therefore, if $g$ is in $C_b^1(\Omega^+)$ and $z_0$ in $\partial\Omega^+$, we can define $g(z_0)$ as a limit of a Cauchy sequence along normal directions. The extended $g$ obeys (12.6.9) and so is globally continuous on $\overline{\Omega^+}$.

Applying this argument to derivatives, we get the general result. □

Next, we want to combine Harnack's inequality, (5.4.17) of Part 2A, and tangent circles.

**Theorem 12.6.7** (Hopf's Lemma). *Let $u$ be harmonic in a bounded region, $\Omega$, and continuous on $\overline{\Omega}$. Let $\mathbb{D}_r(w_0)$ be a disk with $\mathbb{D}_r(w_0) \subset \overline{\Omega}$ containing some point, $z_0$, in $\partial\Omega^+$. Suppose $u(z_0) \geq u(w)$ for all $w \in \mathbb{D}_r(w_0)$. Then for $z$ in the segment between $w_0$ and $z_0$, we have*

$$\frac{u(z_0) - u(z)}{z_0 - z} \geq \frac{u(z_0) - u(w_0)}{2r} \tag{12.6.10}$$

**Remark.** In particular, if $u(z_0) > u(w_0)$ and $\omega$ is the outward normal to $\mathbb{D}_r(w_0)$ at $z_0$, then $\frac{\partial u}{\partial \omega}$, if it exists, is strictly positive.

**Proof.** By restricting to $\mathbb{D}_r(w_0)$, we can suppose $\Omega^+ = \mathbb{D}_r(w_0)$. By scaling, translation, and rotation, we can suppose $r = 1$, $w_0 = 0$, and $z_0 = 1$. By Harnack's inequality, (5.4.17) of Part 2A, applied to $v(z) = u(1) - u(z)$ for $x \in (0, 1)$,

$$v(x) \geq \frac{1 - x}{1 + x} v(0) \geq \frac{1 - x}{2} v(0) \tag{12.6.11}$$

since $v(0) = u(z_0) - u(w_0) \geq 0$. Note that, by hypothesis, $v$ is nonnegative, so Harnack's inequality is applicable. This is (12.6.10). □

As a corollary of this, we have:

**Proposition 12.6.8.** *If $F$ is a biholomorphic bijection of $\mathbb{D}$ to $\Omega^+$, a smooth Jordan domain, then there is a constant $c$ with*

$$\delta_{\Omega^+}(F(z)) \leq c(1 - |z|) \tag{12.6.12}$$

*for all $z$ in $\mathbb{D}$.*

**Proof.** Pick $r_0$ so small that the internal tangent disks, $D(z)$, of radius $r_0$ tangent to $\partial\Omega^+$ at $z$ lie in $\Omega^+$ for all $z \in \partial\Omega^+$, so that $\text{dist}(F(0), \Omega^+) > 2r_0$, and so that for any $w \in \widetilde{\Omega} = \{w \in \Omega^+ \mid \min_{z \in \partial\Omega^+} |w - z| < r_0\}$ is connected to the closest point in $\partial\Omega^+$ by a straight line in $\Omega^+$ (see Problem 4). By minimum distance, that means $w$ lies on a radius between the center of some $D(z)$ and $z$.

Let $G(w) = F^{-1}(w)$ on $\widetilde{\Omega}$ and $u(w) = \log|G(w)|$. $u$ is harmonic on $\widetilde{\Omega}$, continuous on its closure (if set to 1 on $\partial\Omega^+$), and takes its maximum at each point in $\partial\Omega^+$, so by (12.6.10), we have

$$-u(w) \geq C\delta_\Omega(w); \qquad C = (2r_0)^{-1} \min_{\{w|\delta_{\Omega^+}(\tilde{w})=r\}} (-u(\tilde{w})) \tag{12.6.13}$$

This implies (12.6.12) for some $c$ and all $z \in G[\widetilde{\Omega}]$. Since $G[\Omega^+ \setminus \widetilde{\Omega}]$ is compact in $\partial\mathbb{D}$, $(1 - |z|)$ is bounded away from zero there and (12.6.12) holds for a suitable $c$. □

In the right side of (12.6.1), we need an a priori bound on $F'$ as $|z| \to 1$:

**Lemma 12.6.9.** *We have that*

$$|F'(z)| \leq (1 - |z|)^{-1} \sup_{w \in \Omega^+} |w| \tag{12.6.14}$$

**Proof.** Clearly, $\|F\|_\infty = \sup_{w\in\Omega^+}|w|$. Given that $\mathbb{D}_{1-|z|}(z) \subset \mathbb{D}$. (12.6.14) is just a Cauchy estimate. □

The key to seeing that $F$ has bounded derivatives will be:

**Proposition 12.6.10.** (a) *For each* $\ell = 0, 1, 2, \dots$ *and* $\partial = \frac{\partial}{\partial z}$, *we have that*

$$\sup_{z\in\mathbb{D}} |\partial^\ell K_\mathbb{D}(z, w)| \leq \pi^{-1}(\ell + 1)!\, (1 - |w|)^{-\ell-2} \tag{12.6.15}$$

(b) *If $u$ is a Borel function on $\mathbb{D}$ obeying*

$$|u(z)| \leq C(1 - |z|)^{\ell+2} \tag{12.6.16}$$

*then* $P_\mathbb{D} u \in C_b^\ell(\mathbb{D})$.

**Proof.** (a) By (12.5.25),

$$\partial^\ell K_\mathbb{D}(z, w) = \frac{(\ell + 1)!\, \bar{w}^\ell}{\pi(1 - z\bar{w})^{\ell+2}} \tag{12.6.17}$$

Since $\min_{z\in\mathbb{D}}|1 - z\bar{w}| = 1 - |w|$ (as $z \to w/|w|$), (12.6.15) is immediate.

(b) Let $f = P_\mathbb{D} u$. By the Cauchy–Riemann equations, $f \in C_b^\ell(\mathbb{D})$ if and only if $\partial^k f$ is bounded on $\mathbb{D}$ for $k = 0, 1, \dots, \ell$. Since

$$|(\partial^k f)(z)| \leq \int |\partial^k K_\mathbb{D}(z, w)|\, |u(w)|\, d^2 w \tag{12.6.18}$$

this is immediate from (12.6.15) and (12.6.16). □

We are about to apply Bell's lemma, so we need to be precise about vanishing to order $k$.

**Definition.** Let $k \in \{1,2,\dots\}$. A $C^\infty$ function, $u$, on $\Omega$, a bounded region in $\mathbb{C}$ is said to *vanish on the boundary to order* $k$ if and only if for each $m = 0,1,\dots,k-1$, $\ell = 0,1,\dots,m$, we have ($m=0$ means $u$ with no derivatives)

$$\sup_{z\in\Omega} d_\Omega(z)^{-k+m} \left| \frac{\partial^m u}{\partial x^\ell \partial y^{m-\ell}}(z) \right| < \infty \tag{12.6.19}$$

In case $\Omega$ is a $C^\infty$ Jordan domain, it is not difficult to see (Problem 5) that a $C^\infty$ $u$ vanishes at the boundary of $\Omega^+$ to order $k$ if and only if $u \in C^k(\overline{\Omega^+})$ with all derivatives up to order $k-1$ vanishing on $\partial\Omega^+$. With this, one can understand Bell's lemma and use it to prove the first half of Theorem 12.6.1:

**Proposition 12.6.11.** *Under the hypotheses of Theorem* 12.6.1, $F \in C_b^k(\mathbb{D})$ *for all* $k$.

**Proof.** Fix $k \geq 1$. By Bell's lemma, find $\varphi \in C^\infty(\Omega^+)$ with all derivatives bounded, vanishing to order $k+2$ so that $P_\Omega\varphi = 1$. In particular,

$$|\varphi(w)| \leq C\, d_\Omega(w)^{k+2} \tag{12.6.20}$$

Thus, by (12.6.12),

$$|(\varphi\circ F)(z)| \leq C_1(1-|z|)^{k+2} \tag{12.6.21}$$

and then, by (12.6.14),

$$|F'(z)\varphi(F(z))| \leq C_2(1-|z|)^{k+1}$$

By Proposition 12.6.10, $P_{\mathbb{D}}(F'(\varphi\circ F))$ is $C_b^{k-1}(\mathbb{D})$. By Bell's formula, (12.6.1), $F'$ is $C_b^{k-1}(\varphi)$, so $F$ is in $C_b^k(\mathbb{D})$. □

By Proposition 12.6.6, $F'$ has a continuous extension to $\overline{\mathbb{D}}$. To handle the inverse, we need a lower bound on $|F'(z)|$ on all of $\mathbb{D}$, which is implied by $F'$ being nonvanishing on $\overline{\mathbb{D}}$. This will come from Hopf's lemma.

**Lemma 12.6.12.** $F'$ *extended to* $\overline{\mathbb{D}}$ *is everywhere nonvanishing.*

**Proof.** Let $z_0 \in \partial\mathbb{D}$ and $w_0 = F(z_0) \in \partial\Omega^+$. Let $D_{r_0}(a_0)$ be a disk tangent to $\partial\Omega^+$ at $w_0$ and exterior (see Proposition 12.6.4). Pick $\omega \in \partial\mathbb{D}$ so $\omega(w_0 - a_0) = |w_0 - a_0| = r_0$, and let

$$h(w) = \text{Re}\left(\frac{1}{\omega(w-a_0)}\right) \tag{12.6.22}$$

so $h(w_0) = 1/r_0$, and for $w \in \overline{\Omega^+}$, $|h(w)| < 1/r_0$, since $w_0$ is the unique point in $\mathbb{D}_{r_0}(a)\cap\overline{\Omega^+}$. Thus, $h$ is a harmonic function taking its maximum on $\overline{\Omega^+}$ at $w_0$.

Define $u$ on $\overline{\mathbb{D}}$ by

$$u(z) = h(F(z)) \tag{12.6.23}$$

It is harmonic on $\mathbb{D}$, $C^\infty$ on $\overline{\mathbb{D}}$, and takes its maximum at $z_0$. Thus, by (12.6.10),

$$\frac{\partial u}{\partial r}\bigg|_{z_0} > 0 \tag{12.6.24}$$

(defined by taking limits from inside $\mathbb{D}$). By the chain rule and the conformality of $F$ at $\partial\mathbb{D}$,

$$\frac{\partial u}{\partial r}\bigg|_{z_0} = \frac{\partial h}{\partial n}\bigg|_{w_0} \frac{\partial |F|}{\partial r}\bigg|_{z_0} \tag{12.6.25}$$

where $n$ is the normal to $\partial\Omega^+$ at $w_0$. It follows that $F'(z_0) \neq 0$. $\square$

**Proof of Theorem 12.6.1.** Let $G = F^{-1}$ and, on $\mathbb{D}$, define

$$H_k(z) = G^{(k)}(F(z)) \tag{12.6.26}$$

We'll prove inductively that $H_k \in C_b^\infty(\mathbb{D})$, which implies that each $G^{(k)}$ is bounded, so that $G \in C_b^\infty(\overline{\Omega^+})$. Since $F$ and $G$ are in $C_b^\infty$, by Proposition 12.6.6, they are in $C^\infty(\overline{\mathbb{D}})$ and $C^\infty(\overline{\Omega^+})$, proving the theorem.

For $k = 0$, $H_0(z) = z$, which is clearly in $C_b^\infty(\mathbb{D})$. By the chain rule,

$$H_k'(z) = H_{k+1}(z)F'(z) \tag{12.6.27}$$

so

$$H_{k+1}(z) = H_k'(z)(F'(z))^{-1} \tag{12.6.28}$$

Thus, derivatives of $H_{k+1}$ are polynomials in $(F')^{-1}$ and derivatives of $H_k$ and $F$. Since $(F')^{-1}$ is bounded, by the inductive hypothesis, $H_{k+1} \in C_b^\infty(\mathbb{D})$. $\square$

All that remains is the proof of Bell's lemma. We need to add to $\psi$ functions, $\eta$, with $P_\Omega \eta = 0$. The following will construct many such $\eta$'s. We use $\partial, \bar\partial$ notation defined in (2.1.29) of Part 2A.

**Proposition 12.6.13.** *If $g \in C_b^1(\Omega^+)$, then*

$$P_{\Omega^+}(\partial(\rho g)) = 0 \tag{12.6.29}$$

**Proof.** Formally, if $h \in \mathfrak{A}_2(\Omega^+)$, then

$$\begin{aligned}
\langle h, \partial(\rho g)\rangle_{L^2} &= \int \bar h(z)\partial(\rho g)(z)\, d^2z \\
&= -\int (\partial \bar h)(z)(\rho g)(z)\, d^2 z \\
&= -\int \overline{\bar\partial h}\,(z)(\rho g)(z)\, d^2 z = 0
\end{aligned} \tag{12.6.30}$$

by the Cauchy–Riemann equation, $\bar\partial h = 0$. (12.6.30) comes from an integration by parts which is formal since $h$ may not have boundary values. So we need a limiting argument.

For $\varepsilon > 0$, let

$$\Omega_\varepsilon = \{z \mid \rho(z) < -\varepsilon\} \tag{12.6.31}$$

so if $\varepsilon$ is so small that $\nabla\rho$ is nonvanishing on $\{z \mid \rho(z) \in [-\varepsilon, 0]\}$, $\partial\Omega_\varepsilon$ is the smooth set where $\rho(z) = -\varepsilon$. Thus, $g(\rho+\varepsilon)$ vanishes on $\partial\Omega_\varepsilon$, $h$ is $C^\infty$ in a neighborhood of $\overline{\Omega}_\varepsilon$, and the formal argument works (integration by parts means writing the two-dimensional integral as iterated one-dimensional and literally integrating $\frac{\partial}{\partial x}$ or $\frac{\partial}{\partial y}$ by parts). Thus, if $\chi_\varepsilon$ is the characteristic function of $\Omega_\varepsilon$, we have

$$\langle \chi_\varepsilon h, \partial[(\rho+\varepsilon)g]\rangle_{L^2} = 0 \tag{12.6.32}$$

As $\varepsilon \downarrow 0$, $\chi_\varepsilon h \to h$ in $L^2$ by dominated convergence and $\partial[(\rho+\varepsilon)g] - \partial(\rho g) = \varepsilon\,\partial g$ trivially goes to zero in $L^2$, so

$$\langle h, \partial(\rho g)\rangle = \lim_{\varepsilon\downarrow 0}\langle \chi_\varepsilon h, \partial[(\rho+\varepsilon)g]\rangle = 0 \tag{12.6.33}$$

Thus, $\partial(\rho g) \in \mathfrak{A}_2^\perp$, that is, (12.6.29) holds. □

Before writing out the details of the proof of Bell's lemma, let us informally discuss the ideas. We'll use $\varphi_k$ to denote the function we need to construct. We'll want to get $\varphi_{k+1}$ from $\varphi_k$ (and $\varphi_1$ from $\psi$) by adding something of the form $\partial(g\rho)$ with $g \in C^\infty(\overline{\Omega})$. We need $\varphi_1$ to agree with $\psi$ on $\partial\Omega^+$. Thus, if $\varphi_1 = \psi - \partial(g\rho)$, we need $\psi - g\partial\rho$ to vanish on $\partial\Omega^+$ (since $(\partial g)\rho$ automatically does). It is thus natural to pick $g = \psi/\partial\rho$. That doesn't work, since $\partial\rho$ has zeros but not near $\partial\Omega^+$, so we'll pick $g = \alpha\psi/\partial\rho$, where $\alpha$ is a $C^\infty$ function vanishing near the zeros of $\partial\rho$ and one near $\partial\Omega^+$.

We'll want to pick $\varphi_{k+1} = \varphi_k - \partial(f)$ for $f = \rho g$. We'll need $\partial f$ to vanish to order $k$ (so that $\varphi_{k+1}$ does) and to cancel the derivatives of $\varphi_k$ of order $k+1$. To insure vanishing to order $k$, it is natural to pick $f = \rho^{k+1}\theta_k$ (so $\partial f$ will have a $\rho^k$). The key realization is that because $\varphi_k$ vanishes on $\partial\Omega^+$, the only nonvanishing derivatives of order $k+1$ are $(\frac{\partial}{\partial n})^{k+1}\varphi_k$. Thus, $\theta_k$ should have (up to a $(k+1)!$) $(\frac{1}{\partial\rho/\partial n})^k(\frac{\partial}{\partial n})^k\varphi_k$. The $(\frac{\partial\rho}{\partial n})^{-k}$ is needed to cancel derivatives of $\rho$. Thus, we want to pick once and for all a first-order differential operator $\mathcal{D}$ so that any point, $z_0$, on $\partial\Omega^+$,

$$(\mathcal{D}u)(z_0) = \frac{1}{\partial\rho/\partial n(z_0)}\,\frac{\partial u}{\partial n}(z_0) \tag{12.6.34}$$

Problem 6 has such an explicit choice, but without an explicit formula, it is obvious that such an operator exists since $1/(\partial\rho/\partial n)$ is a smooth function near $\partial\Omega^+$ (because $\nabla\rho \neq 0$ on $\partial\Omega^+$ and $|\nabla\rho| = |\partial\rho/\partial n|$ since $\rho$ is constant in that tangent direction).

**Proof of Theorem 12.6.2.** Let $\varepsilon$ be picked so that $\partial\rho$, which is nonvanishing on $\partial\Omega^+$, is nonvanishing on $\{w \mid -\varepsilon < \rho(w) < 0\} = \widetilde{\Omega}_\varepsilon$. Pick $\alpha$ a $C^\infty$ function supported on $\widetilde{\Omega}_\varepsilon$ and identically 1 on $\widetilde{\Omega}_{\varepsilon/2}$. Let $\mathcal{D}$ be a first-order differential operator defined in a neighborhood of $\overline{\Omega^+}$ so (12.6.34) holds for $z_0 \in \partial\Omega^+$.

Define $\{\varphi_\ell\}_{\ell=0}^\infty$ and $\{\theta_\ell\}_{\ell=1}^\infty$ inductively by

$$\varphi_0 = \psi, \quad \theta_\ell = (\ell!\,\partial\rho)^{-1}\alpha\mathcal{D}^{\ell-1}(\varphi_{\ell-1}), \quad \varphi_\ell = \varphi_{\ell-1} - \partial(\theta_\ell\rho^\ell) \tag{12.6.35}$$

By Proposition 12.6.13, $P_{\Omega^+}(\varphi_\ell) = P_{\Omega^+}(\varphi_{\ell-1})$ for $\ell = 1, 2, \dots$, so for all $\ell$,

$$P_{\Omega^+}(\varphi_\ell) = P_{\Omega^+}(\psi) \tag{12.6.36}$$

Thus, we need only show that for $\ell \ge 1$, $\varphi_\ell$ vanishes on $\partial\Omega^+$ and all its derivatives up to order $\ell - 1$ vanish there also. Tangential derivatives vanish if $\varphi_\ell$ does, so we need only prove for $z \in \partial\Omega^+$,

$$\left[\left(\frac{\partial}{\partial n}\right)^j \varphi_\ell\right](z) = 0, \qquad j = 0, 1, \dots, \ell - 1 \tag{12.6.37}$$

By (12.6.35), we have that $\theta_1 = (\partial\rho)^{-1}\alpha\psi$, so

$$\begin{aligned}\varphi_1 &= \psi - \partial((\partial\rho)^{-1}\alpha\rho\psi) \\ &= \psi - \alpha\psi + \rho g\end{aligned}$$

with $g \in C^\infty(\overline{\Omega^+})$, so $\varphi_1$ vanishes on $\partial\Omega^+$ (where $1 - \alpha$ and $\rho = 0$).

For $\ell \ge 2$, $\rho^\ell$ and $\partial(\rho^\ell) = 0$ on $\partial\Omega^+$, so $\partial(\theta_\ell\rho^\ell) = 0$ there also, and thus by (12.6.35), $\varphi_\ell$ vanishes on $\partial\Omega^+$. Since $\partial(\theta_\ell\rho^\ell) = h_\ell\rho^{\ell-1}$ with $h_\ell \in C^\infty(\overline{\Omega^+})$, we see all derivatives of $\partial(\theta_\ell\rho^\ell)$ up to order $(\ell-2)$ vanish on $\partial\Omega^+$. By induction, this is true of $\varphi_{\ell-1}$ also, so inductively (12.6.37) holds for $j = 0, 1, \dots, \ell - 2$ and we need only prove that on $\partial\Omega^+$,

$$\left(\frac{\partial}{\partial n}\right)^{\ell-1} \varphi_{\ell-1} = \left(\frac{\partial}{\partial n}\right)^{\ell-1} \partial(\theta_\ell\rho^\ell) \tag{12.6.38}$$

At first sight, $(\frac{\partial}{\partial n})^{\ell-1}\partial(\alpha(\partial\rho)^{-1}\rho^\ell\mathcal{D}^{\ell-1}(\varphi_{\ell-1}))$ looks very complicated. There are $2\ell-1$ derivatives and some could apply to any of $\rho$, $(\partial\rho)^{-1}$, $\partial$, $\varphi_{\ell-1}$, or the coefficients of $\mathcal{D}$. It is complicated on $\Omega^+$, but on $\partial\Omega^+$, it simplifies because $\rho^\ell\varphi_{\ell-1}$ vanishes to order $(2\ell-1)$. So unless all the derivatives act on this product, one gets zeros. The derivatives in $\mathcal{D}^{\ell-1}$ must all apply to $\varphi_{\ell-1}$ and so all $\ell$ from $(\frac{\partial}{\partial n})^{\ell-1}\partial$ must apply to $\rho^\ell$. In addition, since tangential derivatives are zero, $\frac{\partial}{\partial n}$ and $\partial$ are related by a complex phase. We conclude that on $\partial\Omega^+$ (recall $\alpha = 1$ there)

$$\left(\frac{\partial}{\partial n}\right)^\ell \partial(\theta_\ell\rho^\ell) = (\ell!)^{-1}(\partial\rho)^{-1}\left(\frac{\partial\rho}{\partial n}\right)^{-\ell+1}\left(\left[\frac{\partial^{\ell-1}}{\partial n^{\ell-1}}\,\partial\right]\rho^\ell\right)\left(\frac{\partial}{\partial n}\right)^{\ell-1}\varphi_{\ell-1}$$

which, as required, is $(\frac{\partial}{\partial n})^{\ell-1}\varphi_{\ell-1}$. $\square$

**Notes and Historical Remarks.** Painlevé's theorem was proven by Painlevé [**310**] in his 1887 thesis using Green's function methods. The approach in this section has its roots in Fefferman's 1974 paper [**131**] where he proved a higher-dimensional analog for biholomorphic bijections between strictly pseudoconvex domains in $\mathbb{C}^n$ with $C^\infty$ boundaries. Fefferman relied on the analysis of geodesics in the Bergman metric.

In a series of papers, Bell [**40, 41, 42, 44**] found a related approach that instead used the Bergman projection. In particular, Bell's lemma (in a higher-dimensional form) appears in [**40**]. The translation of these ideas to one complex dimension, essentially along the lines we use here, is in Bell–Krantz [**43**].

The Hopf lemma is named after 1927 work of E. Hopf [**197**]. He studied fairly general second-order elliptic differential equations and proved a maximum principle and also strict positivity of normal derivatives at the boundary at maximum points. As we've seen, these ideas are powerful even in the context of the operator being $\Delta$. The essence of his proof of the positivity of the derivative is a Harnack inequality. Since we need explicit and uniform bounds, we leave it in the form (12.6.10) rather than taking a limit as $z \to z_0$. Eberhard Hopf (1902–83) is no relation to Heinrich Hopf (1894–1971) of the Hopf–Rinow theorem and Hopf fibration.

## Problems

1. Let $h$ be an $L^2(\mathbb{D}, d^2z)$ function, which is only a function of $|z|$. Prove that
$$P_{\mathbb{D}}h = c1, \qquad c = \frac{1}{\pi}\int h(z)\,d^2z$$
and, in particular, for any $\delta \in (0,1)$, there is a $C^\infty$ function, supported in $\overline{\mathbb{D}_\delta(0)}$ so that $P_{\mathbb{D}}(h) = 1$.

2. (a) Prove that (12.6.5) holds in the intersection of $\Omega^+$ and a neighborhood of any point in $\partial\Omega^+$.

   (b) Verify (12.6.5) on all of $\Omega^+$.

3. Prove Proposition 12.6.5.

4. Verify that an $r_0$ exists with the properties stated at the start of the proof of Proposition 12.6.8.

5. If $\Omega$ is a $C^\infty$ Jordan domain in $u \in C^\infty(\overline{\Omega})$ and for $m = 0, 1, \dots, k-1$, $\ell = 0, 1, \dots, m$, $\frac{\partial^m u}{\partial x^\ell \partial y^{m-\ell}}$ vanishes on $\partial\Omega^+$, prove that (12.6.18) holds. (*Hint*: Taylor's formula plus (12.6.7).)

6. Prove that
$$\mathcal{D}u = \frac{\alpha}{|\partial\rho|^2}\,\mathrm{Re}((\partial\rho)(\bar\partial u))$$
provides an explicit choice for the operator $\mathcal{D}$ obeying (12.6.34).

Chapter 13

# Some Topics in Analytic Number Theory

> It was said that whoever proved the Prime Number Theorem would attain immortality. Sure enough, both Hadamard and de la Vallée Poussin lived into their late nineties. It may be that there is a corollary here. It may be that the Riemann Hypothesis is false: but, should anyone manage to actually prove its falsehood—to find a zero off the critical line—he will be struck dead on the spot, and his result will never become known.
>
> —*John Derbyshire* [**100**] quoting Andrew Odlyzko.[1]

**Big Notions and Theorems:** Fermat Two-Square Theorem, Lagrange Four-Square Theorem, Jacobi Two- and Four-Square Theorems, Dirichlet Series, Prime Progression Theorem, Prime Number Theorem, Generating Function, Lambert Series, Möbius Function, Mertens' Function, Möbius Inversion, Euler Totient Function, Liouville Function, von Mangoldt Function, Landau's Theorem, Riemann Zeta Function, Euler Factorization Formula, Dirichlet $L$-Function, Bell Series, Dirichlet Character, Orthogonality Relations for Characters, Function Equation for Zeta, Chebyshev $\theta$-Function, Newman's Tauberian Theorem

Number theory is the study of the properties of the integers, $\mathbb{Z}$, often the strictly positive ones. Sometimes it is extended to include three closely

[1]The irony is that Odlyzko has done computer searches for zeros off the critical line. I note that not only did de la Vallée Poussin live to age 95 and Hadamard to age 97, but the first elementary proofs were found by Erdős and Selberg who lived to ages 83 and 90, respectively.

related objects: the rationals, $\mathbb{Q}$, the *algebraic integers* (i.e., the roots of monic polynomials with integral coefficients), and the *algebraic numbers* which can be defined either as the field of fractions of the algebraic integers or, equivalently, as roots of not necessarily monic polynomials with integral coefficients. Here we'll discuss $\mathbb{Z}$ only.

Number theory thus discusses a discrete set and, to some extent, it would thus seem unrelated to derivatives, integrals, and infinite power series. It is one of the great miracles of mathematical thought that analytic, especially complex analytic, functions are a powerful tool in understanding the integers—and this chapter will illustrate that. We'll postpone the key to understanding why this is so until we've stated the main theorems that we'll prove in this chapter. We'll assume the reader has some familiarity with the simplest elements of arithmetic, such as the notion of relatively prime numbers, modular arithmetic (i.e., $a \equiv b \bmod n$), and unique prime factorization; see Section 1.3 of Part 2A.

Given $n = 0, 1, 2, 3, \dots$ and $k = 1, 2, \dots$, we define $r_k(n)$ to be the number of points $(m_1, \dots, m_k) \in \mathbb{Z}^k$ with

$$m_1^2 + \cdots + m_k^2 = n \tag{13.0.1}$$

We count separately each sign flip (if $m_j \neq 0$) and permutations. So, for example, $r_2(1) = 4$ as $(\pm 1, 0)$ and $(0, \pm 1)$. Here are the two theorems we'll prove about $r_k(n)$:

**Theorem 13.0.1** (Jacobi Two-Square Theorem). *Let $d_j(n)$ denote the number of divisors of $n$ of the form $4k + j$ $(j = 1, 3)$. Then for $n \geq 1$,*

$$r_2(n) = 4(d_1(n) - d_3(n)) \tag{13.0.2}$$

**Theorem 13.0.2** (Jacobi Four-Square Theorem). *Let $\sigma_{\neq 4}(n)$ be the sum of all divisors of $n$ which are not themselves divisible by 4. Then*

$$r_4(n) = 8\sigma_{\neq 4}(n) \tag{13.0.3}$$

It is not immediately obvious that $d_1(n) \geq d_3(n)$ but not hard to see. Indeed (Problem 1) $d_1(n) > d_3(n)$ if and only if every prime of the form $4k + 3$ occurs in the prime factorization an even number of times (and that $d_1(n) = d_3(n)$ if not). Thus, we have

**Corollary 13.0.3** (Fermat's Two-Square Theorem). *$n \geq 1$ is a sum of two squares if and only if no single prime of the form $4k + 3$ occurs an odd number of times in the prime factorization of $n$.*

So 12 is not a sum of two squares, but $36 = 6^2 + 0^2$ is.

Since every $n$ has 1 as a divisor, $\sigma_{\neq 4}(n) \geq 1$, and thus,

**Corollary 13.0.4** (Lagrange Four-Square Theorem). *Every positive number is a sum of four squares.*

The other theorems we'll focus on involve prime numbers.

**Theorem 13.0.5** (Dirichlet's Prime Progression Theorem). *Let $m, k$ be positive relatively prime integers. Then for infinitely many $n = 0, 1, 2, \dots$, $mn + k$ is a prime.*

And the most famous theorem in number theory:

**Theorem 13.0.6** (Prime Number Theorem). *Let $\pi(x)$ be the number of primes $< x$. Then as $x \to \infty$,*

$$\frac{\pi(x)}{(x/\log x)} \to 1 \tag{13.0.4}$$

Beside these four big theorems, all proven in this chapter, Chapter 15 will have a result, Theorem 15.4.12, on the asymptotics of partitions that is often considered a part of analytic number theory.

The magic that links this to complex analysis can be described in two words: generating functions. For example, consider

$$G_k(q) = \sum_{n=0}^{\infty} r_k(n) q^n \tag{13.0.5}$$

While this need only be a formal power series, it is easy to see (Problem 2) that the radius of convergence as a function of $q$ is 1. The link of this case to what we've already studied can be seen if we note that

$$G_k(q) = (\widetilde{\theta}(q))^k \tag{13.0.6}$$

where

$$\widetilde{\theta}(q) = \sum_{n=-\infty}^{\infty} q^{n^2} \tag{13.0.7}$$

We'll also see (Propositions 13.1.1 and 13.1.2) that (13.0.2) and (13.0.3) are equivalent, respectively, to:

$$(\widetilde{\theta}(q))^2 = 1 + 4 \sum_{n=1}^{\infty} \frac{q^n}{1 + q^{2n}} \tag{13.0.8}$$

$$(\widetilde{\theta}(q))^4 = 1 + 8 \sum_{n=1}^{\infty} \frac{q^n}{(1 + (-q)^n)^2} \tag{13.0.9}$$

Section 13.1 will prove these formulae, using $q = e^{\pi i \tau}$ to move everything to the upper half-plane. $\widetilde{\theta}$ will then be essentially a Jacobi theta function and we'll use a relative of the Fuchsian group of Section 8.3 of Part 2A.

The two theorems involving primes will use a different series than a power series. These are *Dirichlet series*, that is, functions $f(s)$ of the form

$$f(s) = \sum_{n=1}^{\infty} \frac{a_n}{n^s} \tag{13.0.10}$$

where, typically, $|a_n| \leq Cn^k$, so that $f(s)$ converges absolutely and uniformly in each region $\operatorname{Re} s > k + 1 + \varepsilon$. $f(s)$ is then analytic in this region. Section 13.2 will discuss Dirichlet series. Dirichlet series are relevant to number theory because $n^{-s}m^{-s} = (nm)^{-s}$, so products of Dirichlet series have a multiplicative structure made precise by Theorem 13.2.10. For example, if $f(s)$ has the form (13.0.10) with $a_1 = 0$, $a_n > 0$ for $n \geq 2$, then $f(s)^2 = \sum_{n=1}^{\infty} c_n \, n^{-s}$ with $c_n = 0$ if and only if $n$ is a prime!

The prime number theorem will depend on the analysis of the most famous Dirichlet series—the Riemann zeta function

$$\zeta(s) = \sum_{n=1}^{\infty} \frac{1}{n^s} \tag{13.0.11}$$

To get a feel for how this is relevant, we note that, by the arguments used in Section 9.7 of Part 2A, $\zeta(s) - \int_1^{\infty} \frac{dx}{x^s}$ can be given by an expression that is absolutely convergent in $\{s \mid \operatorname{Re} s > 0\}$ and thus, defines a function continuous there. In particular,

$$\sup_{\operatorname{Re} s > 1} \left| \zeta(s) - \frac{1}{s-1} \right| < \infty \tag{13.0.12}$$

Indeed, the methods of Section 9.7 of Part 2A will allow us to extend $\zeta(s) - 1/(1-s)$ to an entire function; see Theorem 13.3.3.

The other critical preliminary input is what we'll call the *Euler factorization formula* (others call it the Euler product formula, a term we used for something else in Section 9.2 of Part 2A). It will follow fairly easily from prime factorization (see Theorem 13.3.1) that if $\mathcal{P}$ is the set of primes, then for $\operatorname{Re} s > 1$,

$$\zeta(s) = \prod_{p \in \mathcal{P}} \left(1 - \frac{1}{p^s}\right)^{-1} \tag{13.0.13}$$

Taking logs and using (Problem 3(a))

$$|-\log(1-y) - y| \leq |y|^2 \tag{13.0.14}$$

for $|y| < \frac{1}{2}$, we get from (13.0.12) that

$$\sup_{s\in(1,2)} \left| \sum_{p\in\mathcal{P}} \frac{1}{p^s} - \log(s-1) \right| < \infty \tag{13.0.15}$$

On the other hand, one can see (Problem 3(b)) that

$$\sum_{p\in\mathcal{P}} \frac{1}{p^s} = s \int_1^\infty x^{-s-1} \pi(x)\, dx \tag{13.0.16}$$

and that if

$$M(x) = \frac{\pi(x)}{(x/\log x)} \tag{13.0.17}$$

then (Problem 3(c))

$$\sum_{\substack{p\in\mathcal{P} \\ p\geq k}} \frac{1}{p^s} = B_k(s) + s \int_{k(s-1)}^\infty \frac{e^{-u} M(e^{u/(s-1)})}{u}\, du \tag{13.0.18}$$

where $B_k(s)$ is bounded as $s \to 1$ for any fixed $k$.

(13.0.15) and (13.0.18) easily show that if $M(x) \to c$ as $x \to \infty$, then $c = 1$. In the Abelian/Tauberian viewpoint we discuss in the Notes, the result above that existence of the limit implies the known asymptotics is called an Abelian theorem. Theorems that go in the opposite direction but require extra conditions are known as Tauberian theorems. There is an approach to the prime number theorem that relies on subtle Tauberian theorems, depending on the Wiener Tauberian theorem for $L^1$. It will be discussed in Sections 6.11 and 6.12 of Part 4.

Instead, we'll present a more complex–variable-focused proof in Section 13.5 based on our analysis of $\zeta$ in Section 13.3 and an elegant Tauberian theorem of Newman whose proof depends only on the Cauchy integral formula. A key element of the proof will be that $\zeta$ has an analytic continuation in a neighborhood of $Q \equiv \{z \mid \operatorname{Re} z \geq 1,\ z \neq 1\}$ (indeed, in all of $\mathbb{C} \setminus \{1\}$) with no zeros in $Q$. This is relevant because it is $\log(\zeta(s))$ that matters because of (13.0.13), and the lack of zeros means it too is analytic in a neighborhood of $Q$.

Lack of zeros of suitable analytic functions will also be a key to the proof of the prime progression theorem in Section 13.4. To illustrate, let us discuss the case $m = 4$. If $\mathcal{P}_1^{(4)}$ and $\mathcal{P}_3^{(4)}$ are the primes of the form $4n+1$ or $4n+3$, let us show that

$$\sum_{p\in\mathcal{P}_1^{(4)}} \frac{1}{p} = \infty, \qquad \sum_{p\in\mathcal{P}_3^{(4)}} \frac{1}{p} = \infty \tag{13.0.19}$$

Define for $n \in \mathbb{Z}$,

$$\chi(n) = \begin{cases} 1, & n \equiv 1 \bmod 4 \\ -1, & n \equiv 3 \bmod 4 \\ 0, & n \equiv 0 \bmod 2 \end{cases} \tag{13.0.20}$$

Since $3 \equiv -1 \bmod 4$,

$$\chi(nk) = \chi(n)\chi(k) \tag{13.0.21}$$

Also define the Dirichlet $L$-function associated to $\chi$,

$$L_\chi(s) = \sum_{n=1}^{\infty} \frac{\chi(n)}{n^s} \tag{13.0.22}$$

Because of (13.0.21), we have

$$L_\chi(s) = \prod_{p\in\mathcal{P}} \left(1 - \frac{\chi(p)}{p^s}\right)^{-1} \tag{13.0.23}$$

that is, the proof of (13.0.13) extends to $L_\chi$. We'll give a complete proof in Section 13.3. Notice that formally,

$$L_\chi(1) = 1 - \tfrac{1}{3} + \tfrac{1}{5} - \tfrac{1}{7} + \dots \tag{13.0.24}$$

and this sum is conditionally convergent. By the same argument that was used in Problem 8 of Section 2.3 of Part 2A,

$$\lim_{s\downarrow 1} L_\chi(s) = 1 - \tfrac{1}{3} + \tfrac{1}{5} - \tfrac{1}{7} + \dots \in (0,\infty) \tag{13.0.25}$$

since $(1-\frac{1}{3}), (\frac{1}{5}-\frac{1}{7}), \dots$ are all positive. Indeed, by expanding a geometric series, the sum is $\int_0^1 \frac{dx}{1+x^2} = \arctan(1) = \frac{\pi}{4}$. Thus,

$$\lim_{s\downarrow 1} \log(L_\chi(s)) \in (-\infty,\infty)$$

that is, by (13.0.23) and $\sum_{p\in\mathcal{P}} |1/p^s|^2 < \infty$,

$$\lim_{s\downarrow 1} \left| \sum_{p\in\mathcal{P}} \frac{\chi(p)}{p^s} \right| < \infty \tag{13.0.26}$$

Since

$$\sum_{p\in\mathcal{P}_{2\pm 1}^{(4)}} \frac{1}{p^s} = \frac{1}{2}\left[\sum_{\substack{p\in\mathcal{P}\\ p\neq 2}} \frac{1}{p^s} \mp \sum_{p\in\mathcal{P}} \frac{\chi(p)}{p^s}\right] \tag{13.0.27}$$

we see that (13.0.18) holds by (13.0.15) and (13.0.26). This proves Dirichlet's prime progression theorem for $m = 4$. The general proof in Section 13.4 will use similar ideas. We'll need to discuss analogs of $\chi$, functions on modular classes relatively prime to $m$ obeying (13.0.21) ("characters"), and the key will be that $\lim_{s\downarrow 1} L_\chi(s) \neq 0$.

**Notes and Historical Remarks.** Euler, Gauss, and Jacobi are the grandfathers of the application of methods of function theory applied to generating functions in number theory, but Dirichlet and especially Riemann should be viewed as its fathers (see the later sections for references). Dirichlet used real-variable methods and, in particular, only considered Dirichlet $L$-functions for real values of $s$. It was Riemann who emphasized the use of contour integrals and other elements of complex analysis. For a particularly striking, but simple, application of generating functions as formal power series, see Problem 6 of Section 15.4.

One can ask if there is a prime number counting-type result for $\mathcal{P}_k^{(m)}$ and the answer is yes. For all $k$ relatively prime to $m$, $\lim_{x\to\infty}[\#\{p \le x \mid p \in \mathcal{P}_k^{(m)}\}/(x/\log x)]$ exists and the limits are equal, which means it is $1/\varphi(m)$, where $\varphi$ is the Euler totient function, that is, the number of integers in $[0, m-1]$ relatively prime to $m$. This follows from the fact that $\mathcal{P} \setminus \bigcup_{(k,m)=1} \mathcal{P}_k^{(m)}$ is the finite set of prime factors of $m$ (see Problem 2 of Section 13.2).

Abelian and Tauberian theorems are pairs that involve interchange of sum or integral with limits or, even more precisely, with existence of a limit of sums or integrals with convergence of the sum or integrals. The initial pair concerned $S_N = \sum_{n=0}^N a_n$ and $f(r) = \sum_{n=0}^\infty a_n r^n$. In 1826, Abel [**1**] proved (see Problem 8 of Section 2.3 of Part 2A) that

$$\lim S_n = \alpha \Rightarrow \lim_{r\uparrow 1} f(r) = \alpha \tag{13.0.28}$$

The converse is not true in general; for example, if $a_n = (-1)^n$, $S_N$ has no limit but $\lim_{r\uparrow 1} f(r) = \frac{1}{2}$. However, in 1897, Tauber [**380**] proved the original Tauberian theorem (see Problem 4)

$$\lim_{r\uparrow 1} f(r) = \alpha \;\&\; \lim_{n\to\infty} na_n = 0 \Rightarrow \lim S_N = \alpha \tag{13.0.29}$$

Since then, following Hardy–Littlewood, one uses the term Tauberian theorem for results that are partial converses of easier theorems relying on some reasonable additional hypotheses where an absolute converse is not valid. Littlewood later showed $\lim na_n = 0$ could be replaced by $na_n$ bounded (see Section 6.11, especially Problems 4–7 of Part 4).

We note that Tauber's paper was just after Hadamard and de la Vallée Poussin initially proved the prime number theorem, so their proofs did not use Tauberian theorems. The majority of modern approaches, but certainly not all, rely on Tauberian theorems. One disadvantage of Tauberian proofs is that they do not normally provide explicit error estimates. Alfred Tauber (1866–1942) failed to find a mathematical academic job and spent most of his life in Vienna working for an insurance company. He was sent to the

Theresienstadt concentration camp on June 28, 1942 and is presumed killed shortly thereafter.

Most Tauberian theorems provide conditions under which convergence of some average imply convergence of the universal object, and to some, this kind of result is called a Tauberian theorem even if there is no Abelian partner.

The complicated-looking integral we found in (13.0.18) is usually replaced by something simpler by the introduction of subsidiary functions to $\pi(x)$ and an easy proof that asymptotics of $\pi(x)$ is equivalent to asymptotics of the subsidiary functions. For example, in the proof we'll give in Section 13.5, we use a function of Chebyshev,

$$\theta(x) = \sum_{p \le x} \log p \tag{13.0.30}$$

We'll first prove that (13.0.4) is equivalent to $\lim_{x\to\infty} \theta(x)/x = 1$.

This chapter will only cherry-pick the subject of analytic number theory. For additional textbook material, see [**18, 19, 30, 78, 135, 214, 298, 321, 374, 382**].

### Problems

1. (a) Let $m$ and $k$ be relatively prime and $n = mk$. Prove that $d_1(n) = d_1(m)d_1(k) + d_3(m)d_3(k)$ and $d_3(n) = d_1(m)d_3(k) + d_3(m)d_1(k)$. Conclude that if $\delta(n) = d_1(n) - d_3(n)$, then $\delta(n) = \delta(m)\delta(k)$.

   (b) If $n = p_1^{k_1} \dots p_m^{k_m}$ is the prime factorization of $n$, prove that $\delta(n) = \delta(p_1^{k_1}) \dots \delta(p_m^{k_m})$.

   (c) Prove $\delta(2^\ell) = 1$ (recall 1 counts as a divisor). If $p$ is prime and $p \equiv 1 \pmod 4$, prove that $\delta(p^\ell) = \ell + 1$. If $p$ is prime and $p \equiv 3 \pmod 4$, prove that $\delta(p^\ell) = 1$ if $\ell$ is even and 0 if $\ell$ is odd.

   (d) Prove that $\delta(n) > 0$ (respectively, $= 0$) if all primes $p \equiv 3 \pmod 4$ occurs any even number of times in the prime factorization of $n$ (respectively, some such prime occurs an odd number of times).

2. Prove for any $k \ge 1$, $r_k(n^2) \ge 1$ and conclude that the radius of convergence of $G_k(q)$, given by (13.0.5), is at most 1. Prove that $r_k(n) \le (2\sqrt{n})^k$ and conclude that the radius of convergence is at least 1. Conclude that the radius of convergence is exactly 1.

3. (a) For $|y| < \frac{1}{2}$, prove (13.0.14).

   (b) Prove (13.0.16).

   (c) Prove (13.0.18).

4. (a) Let $f(x) = \sum_{n=0}^{\infty} a_n x^n$ where $\sup_n |a_n| < \infty$. Let $S_N = \sum_{n=0}^{N} a_n$. Prove that for $0 < x < 1$,

$$|S_N - f(x)| \leq (1-x) \sum_{n\equiv 0}^{N} |na_n| + \frac{1}{N} \sum_{n \equiv N+1}^{\infty} |na_n| x^n$$

(*Hint*: $1 - x^n \leq n(1-x)$.)

(b) Prove that

$$\left| S_N - f\left(1 - \frac{1}{N}\right)\right| \leq \frac{1}{N} \sum_{n=0}^{N} |na_n| + \sup_{n \geq N+1} |na_n|$$

(c) If $na_n \to 0$ and $\lim_{x\uparrow 1} f(x) = \alpha$, prove that $\lim_{N\to\infty} S_N = \alpha$. (This is Tauber's theorem [**380**].)

5. The proof of an infinitude of primes in *Euclid's Elements* supposes there are finitely many primes $p_1, p_2, \dots, p_\ell$ and looks at prime factors of $p_1 p_2 \dots p_\ell + 1$ to get a contradiction. This problem will find similarly elementary proofs for primes of the form $p \equiv 1 \pmod 4$ or $p \equiv 3 \pmod 4$. Thus, the proof in this section can be replaced, but there is no elementary analog known that works for all $m$ and $k$. Also, Euclid-type proofs give very weak lower bounds on $N(x)$.

   (a) If $p_1, \dots, p_\ell$ were all the primes of the form $4k+3$, prove that $4p_1 \dots p_\ell - 1$ has to have a distinct prime factor of the form $4n+3$, and so conclude $\mathcal{P}_3^{(4)}$ is infinite.

   (b) For this, you'll need the fact proven in Problem 6 that if $p = 4k+3$ is prime, $x^2 + 1 \equiv 0 \pmod p$ has no solutions ($-1$ is not a quadratic residue mod $p$). Prove that if $N$ is a positive odd integer and $x^2 + 1 \equiv 0 \pmod N$ has a solution and $p$ is a prime divisor of $N$, then $p \equiv 1 \pmod 4$.

   (c) Suppose there are only finitely many primes $p_1, \dots, p_\ell$ of the form $4k+1$. Let $N = (2p_1 \dots p_\ell)^2 + 1$. Prove that $x^2 + 1 \equiv 0 \pmod N$ has a solution and conclude that there must be a prime of the form $4k+1$ not among $p_1, \dots, p_\ell$.

6. (A problem in non-analytic number theory needed in Problem 5.) You'll prove $x^2 + 1 \equiv 0 \pmod p$ has a solution for an odd prime $p$ if and only if $p \equiv 1 \pmod 4$, a result of Euler. You'll consider $\mathbb{Z}_p^\times$, the set of invertible elements of $\mathbb{Z}_p = \mathbb{Z}/p\mathbb{Z}$, that is, the conjugacy classes $[1], \dots, [p-1]$. We say $a$ is a *quadratic residue* mod $p$ if and only if $x^2 - a \equiv 0$ has a solution.

   (a) Prove exactly $(p-1)/2$ elements are quadratic residues. (*Hint*: How many solutions does $x^2 - a \equiv 0$ have in $\mathbb{Z}_p^\times$ if it has any?)

(b) Prove Fermat's theorem that $b \in \mathbb{Z}_p^\times$ implies $b^{p-1} \equiv 1$. Conclude that if $a$ is a quadratic residue, then $a^{\frac{1}{2}(p-1)} \equiv 1$.

(c) Prove $a^{\frac{1}{2}(p-1)} \equiv 1$ has at most $\frac{1}{2}(p-1)$ solutions and conclude (Euler's criterion) that $a$ is a quadratic residue if and only if $a^{\frac{1}{2}(p-1)} \equiv 1$. (*Hint*: You need the fact that polynomials of degree $d$ in $\mathbb{Z}_p[X]$ have at most $d$ roots by factorization.)

(d) Show $(-1)^{\frac{1}{2}(p-1)} \equiv 1 \pmod{p}$ if and only if $p \equiv 1 \pmod 4$.

## 13.1. Jacobi's Two- and Four-Square Theorems

Here we'll prove Theorems 13.0.1 and 13.0.2. There will be three parts to the proof. First, we'll reduce the results to equality of generating functions—essentially proving that powers of $\widetilde{\theta}(q)$ (given by (13.0.7)) are given by certain Lambert series. Then we'll need to prove this equality by moving the functions to $\mathbb{C}_+$ by a change of variables, $q = e^{\pi i \tau}$. Since $\widetilde{\theta}$ is nonvanishing on $\mathbb{D}$, the ratio will be analytic on $\mathbb{C}_+$, and we'll want to prove it is 1.

The final step in the proof will be to show the ratios are invariant under a large subgroup of $\mathbb{SL}(2,\mathbb{Z})$ with fundamental domain a hyperbolic triangle close to the one studied in Section 8.3 of Part 2A, and this part will also show that the ratio goes to 1 at the three corners of this triangle.

The middle part of the proof will be a general theorem that uses the maximum principle to show that any function with this set of invariances and which goes to 1 at all three corners is identically one. We begin with

**Proposition 13.1.1.** *We have that*

$$\sum_{n=1}^{\infty} (d_1(n) - d_3(n)) q^n = \sum_{n=1}^{\infty} \frac{q^n}{1+q^{2n}} \tag{13.1.1}$$

**Proof.** $d_j(n)$ is a sum of 1's over all numbers of the form $4k+j$ dividing $n$. Since we also sum over $n = \ell(4k+j)$, we are summing over $\ell = 1, 2, \dots$ and $k = 0, 1, 2 \dots$. Thus,

$$\sum_{n=1}^{\infty} d_j(n) q^n = \sum_{\ell=1}^{\infty} \sum_{k=0}^{\infty} q^{(4k+j)\ell} = \sum_{n=1}^{\infty} \frac{q^{j\ell}}{1-q^{4\ell}} \tag{13.1.2}$$

so

$$\text{LHS of (13.1.1)} = \sum_{\ell=1}^{\infty} \frac{q^\ell - q^{3\ell}}{1-q^{4\ell}} = \sum_{\ell=1}^{\infty} q^\ell \, \frac{1-q^{2\ell}}{1-q^{4\ell}} = \sum_{\ell=1}^{\infty} \frac{q^\ell}{1+q^{2\ell}} \tag{13.1.3}$$

proving (13.1.1). $\square$

**Proposition 13.1.2.** *We have that*

$$\sum_{n=1}^{\infty} \sigma_{\neq 4}(n) q^n = \sum_{n=1}^{\infty} \frac{q^n}{(1+(-q)^n)^2} \tag{13.1.4}$$

**Proof.** We have that

$$\frac{x}{(1-x)^2} = x \frac{d}{dx}\left(\frac{1}{1-x}\right) = x \frac{d}{dx} \sum_{n=0}^{\infty} x^n = \sum_{n=1}^{\infty} n x^n \tag{13.1.5}$$

and thus

$$\sum_{k=0}^{\infty} \frac{q^{2k+1}}{(1-q^{2k+1})^2} = \sum_{k=0}^{\infty} \sum_{m=1}^{\infty} m q^{(2k+1)m} \tag{13.1.6}$$

and

$$\sum_{k=1}^{\infty} \frac{q^{2k}}{(1+q^{2k})^2} = \sum_{k=1}^{\infty} \sum_{m=1}^{\infty} m q^{(2k)m} (-1)^{m+1} \tag{13.1.7}$$

This implies that

$$\text{RHS of (13.1.4)} = \sum_{n=1}^{\infty} \tilde{\sigma}(n) q^n \tag{13.1.8}$$

where

$$\widetilde{\sigma}(n) = \sum_{d|n} s(n,d) d \tag{13.1.9}$$

where

$$s(n,d) = \begin{cases} 1 & \text{if } n/d \text{ is odd} \\ 1 & \text{if } n/d \text{ is even and } d \text{ is odd} \\ -1 & \text{if } n/d \text{ is even and } d \text{ is even} \end{cases} \tag{13.1.10}$$

If $n$ is odd, all $n/d$ are odd and $\widetilde{\sigma}(n) = \sigma_1(n) = \sigma_{\neq 4}(n)$ since $n$ has no divisors divisible by 4. If $n = 2^\ell r$ with $r$ odd and $\ell \geq 1$ and if $r_1 \mid r$, then $d = 2^\ell r_1$ and $d = r_1$ have $s(n,d) = 1$ and $d = 2^j r_1$, $j = 1, \dots, \ell - 1$ have $s(n,d) = -1$. Thus,

$$\widetilde{\sigma}(n) = \sum_{r_1|r} (1 + 2^\ell - 2^{\ell-1} - \cdots - 2) r_1 = \sum_{r_1|r} 3 r_1 = \sigma_{\neq 4}(n) \tag{13.1.11}$$

□

With $\theta(x)$ given by (10.5.44) of Part 2A, we let

$$\theta^\sharp(\tau) \equiv \tilde{\theta}(e^{\pi i \tau}) = \sum_{n=-\infty}^{\infty} e^{i\pi\tau n^2} \tag{13.1.12}$$

Since $q^n/(1+q^{2n}) = 1/(q^n+q^{-n})$ with $q = e^{\pi i\tau}$, we see that Theorem 13.0.1 is equivalent to proving

$$\theta^\sharp(\tau)^2 = G_2(\tau) \equiv 1 + 2\sum_{n=1}^\infty \frac{1}{\cos(\pi n\tau)} \tag{13.1.13}$$

and using $q^n/(1+(-q^n))^2 = 1/(q^{n/2}+(-1)^n q^{-n/2})^2$, then Theorem 13.0.2 is equivalent to proving

$$\theta^\sharp(\tau)^4 = G_4(\tau) \equiv 1 + 2\sum_{k=1}^\infty \frac{1}{\cos^2(\pi k\tau)} - 2\sum_{k=1}^\infty \frac{1}{\sin^2(\pi(k-\frac{1}{2})\tau)} \tag{13.1.14}$$

Next, we'll prove the theorem on when certain functions on $\mathbb{C}_+$ are 1, which we'll apply to the ratios $\widetilde{\theta}(e^{\pi i\tau})^{-2}[1+4\sum_{n=1}^\infty e^{n\pi i\tau}(1+e^{2n\pi i\tau})^{-1}]$ and $\widetilde{\theta}(e^{\pi i\tau})^{-4}[1+8\sum_{n=1} e^{n\pi i\tau}/(1+(-e^{\pi i\tau})^n)^2]$ to complete the proofs of Jacobi's theorems.

**Theorem 13.1.3.** *Suppose $f(\tau)$ is analytic in $\mathbb{C}_+$ and obeys*

(i) $f(\tau+2) = f(\tau)$ (13.1.15)

(ii) $f\left(-\dfrac{1}{\tau}\right) = f(\tau)$ (13.1.16)

(iii) $\lim_{\operatorname{Im}\tau\to\infty} f(\tau) = 1$ (13.1.17)

(iv) $\lim_{\operatorname{Im}\tau\to\infty} f\left(1-\dfrac{1}{\tau}\right) = 1$ (13.1.18)

*Then $f(\tau)\equiv 1$.*

**Remark.** The reader should review the discussion in Section 8.3 of Part 2A since the geometry is very close. The region $\mathcal{G}$ below is the image under $\tau \to 2(\tau-\frac{1}{2})$ of the region $\mathcal{F}^\sharp$.

**Proof.** Let $\mathcal{G}$ be the hyperbolic triangle

$$\mathcal{G} = \{\tau \mid -1 < \operatorname{Re}\tau < 1,\ |\tau| > 1\} \tag{13.1.19}$$

Let $g(\tau) = g(\tau+2)$ and $h(\tau) = -1/\tau$, each of which maps $\mathbb{C}_+$ to $\mathbb{C}_+$. We claim the images of $\overline{\mathcal{G}}$ (closure in $\mathbb{C}_+$) under the group generated by $g$ and $h$ cover $\mathbb{C}_+$. For define the reflections $\mathcal{S}_0, \mathcal{S}_1, \mathcal{S}_2, \mathcal{S}_3$ by

$$\mathcal{S}_0 = -\bar\tau, \quad \mathcal{S}_1(\tau) = 2-\bar\tau, \quad \mathcal{S}_2(\tau) = -2-\bar\tau, \quad \mathcal{S}_3(\tau) = \frac{1}{\bar\tau} \tag{13.1.20}$$

$\mathcal{S}_1, \mathcal{S}_2, \mathcal{S}_3$ are reflections in the three sides of $\mathcal{G}$, so as in Section 8.3 of Part 2A, the images of $\overline{\mathcal{G}}$ under the group generated by $\mathcal{S}_1, \mathcal{S}_2, \mathcal{S}_3$ cover $\mathbb{C}_+$.

Note next that

$$(\mathcal{S}_1\mathcal{S}_0)(\tau) = (\mathcal{S}_0\mathcal{S}_{-1})(\tau) = g(\tau), \quad (\mathcal{S}_0\mathcal{S}_1)(\tau) = (\mathcal{S}_{-1}\mathcal{S}_0)(\tau) = g^{-1}(\tau) \tag{13.1.21}$$

$$(\mathcal{S}_0\mathcal{S}_3)(\tau) = (\mathcal{S}_3\mathcal{S}_0)(\tau) = h(\tau) \tag{13.1.22}$$

Thus, every $\pi$ in the group generated by $\mathcal{S}_1, \mathcal{S}_2, \mathcal{S}_3$ has the form $\tilde{\pi}$ or $\tilde{\pi}\mathcal{S}_0$ where $\tilde{\pi}$ is in the group generated by $g$ and $h$. Since $\mathcal{S}_0(\overline{\mathcal{G}}) = \overline{\mathcal{G}}$, we conclude that the images of $\overline{\mathcal{G}}$ under the group generated by $g$ and $h$ cover $\mathbb{C}_+$.

By (13.1.15) and (13.1.16), $f$ is invariant under this group, and so the values of $f$ on $\overline{\mathcal{G}}$ are the same as on all of $\mathbb{C}_+$. Let $\mathcal{G}^\sharp = \overline{\mathcal{G}} \cup \{1, -1, \infty\}$ and extend $f$ to $\mathcal{G}^\sharp$ by setting it to 1 at $\pm 1$ and $\infty$. By (13.1.11) and $f(-1-1/\tau) = f(1-1/\tau)$, we see $f$ extended is continuous on $\mathcal{G}^\sharp$. Suppose $f(z_0) \neq 1$ for some $z_0$ in $\overline{\mathcal{G}}$. Then we can find a rotation about $w = 1$ (i.e., $R(w) = 1 + e^{i\theta}(w-1)$) so that if $g = Rf$, then $|g(z_0)| > 1$. $g$ is also continuous on $\overline{\mathcal{G}} \cup \{1, -1, \infty\}$, so it takes its maximum at a point in $\overline{\mathcal{G}}$ (since at $\{1, -1, \infty\}$, it is $1 < |g(z_0)|$). Since $g$ is analytic in $\mathbb{C}_+$ with values the same as in $\overline{\mathcal{G}}$, this violates the maximum principle. □

**Proposition 13.1.4.** *Let $\theta^\sharp$ be given by (13.1.12). Then*

(a) $$\theta^\sharp(\tau + 2) = \theta^\sharp(\tau) \tag{13.1.23}$$

(b) $$\theta^\sharp\left(-\frac{1}{\tau}\right) = \sqrt{\frac{\tau}{i}}\, \theta^\sharp(\tau) \tag{13.1.24}$$

(c) $$\lim_{\operatorname{Im}\tau\to\infty} \theta^\sharp(\tau) = 1 \tag{13.1.25}$$

(d) $$\lim_{\operatorname{Im}\tau\to\infty} \theta^\sharp\left(1 - \frac{1}{\tau}\right)\left[2\sqrt{\frac{\tau}{i}}\, e^{i\pi\tau/4}\right]^{-1} = 1 \tag{13.1.26}$$

**Proof.** (a) is immediate from (13.1.12) and (c) from the fact that as $\operatorname{Im}\tau \to \infty$, all terms in (13.1.12) go to zero exponentially fast, except for $n = 0$.

In terms of $\theta$ given by (10.5.44) of Part 2A,

$$\theta^\sharp(\tau) = \theta(w = 0 \mid \tau) \tag{13.1.27}$$

so (13.1.24) is just (10.5.46) of Part 2A for $w = 0$.

Since $(-1)^{n^2} = (-1)^n$,

$$\theta^\sharp(1+\tau) = \sum_{n=-\infty}^{\infty} (-1)^n e^{i\pi\tau n^2} = \theta(\tfrac{1}{2} \mid \tau) \tag{13.1.28}$$

Thus, by (10.5.46) of Part 2A,

$$\begin{aligned} \theta^\sharp\left(1 - \frac{1}{\tau}\right) &= \theta\left(\frac{1}{2} \,\middle|\, -\frac{1}{\tau}\right) \\ &= \sqrt{\frac{\tau}{i}}\, e^{\pi i\tau/4}\theta(\tfrac{1}{2}\tau \mid \tau) \\ &= \sqrt{\frac{\tau}{i}}\, e^{\pi i\tau/4} \sum_{n=-\infty}^{\infty} e^{i\pi\tau(n^2+n)} \end{aligned} \tag{13.1.29}$$

Since $n^2+n = (n+\frac{1}{2})^2 - \frac{1}{4}$, $n^2+n > 0$ if $n \neq 0, 1$, so $\sum_{n=-\infty}^{\infty} e^{i\pi\tau(n^2+n)} \to 2$ as $\operatorname{Im}\tau \to \infty$, so (13.1.29) implies (13.1.26). $\square$

**Proposition 13.1.5.** *Let $G_2(\tau)$ be given by* (13.1.13). *Then*

(a) $G_2(\tau+2) = G_2(\tau)$ (13.1.30)

(b) $G_2\left(-\dfrac{1}{\tau}\right) = \dfrac{\tau}{i}\, G_2(\tau)$ (13.1.31)

(c) $\displaystyle\lim_{\operatorname{Im}\tau\to\infty} G_2(\tau) = 1$ (13.1.32)

(d) $\displaystyle\lim_{\operatorname{Im}\tau\to\infty} G_2(\tau)\big[4\frac{\tau}{i}\, e^{i\pi\tau/2}\big]^{-1} = 1$ (13.1.33)

**Proof.** (a) and (c) are immediate (since $|\cos z| \geq \frac{1}{2}(e^{\operatorname{Im} z} - 1)$). For (b) and (d), we need the fact that

$$\sum_{n=-\infty}^{\infty} \frac{e^{2\pi i\alpha n}}{\cosh(\pi n/t)} = t \sum_{n=-\infty}^{\infty} \frac{1}{\cosh(\pi(n+\alpha)t)} \tag{13.1.34}$$

We proved this via a Poisson summation formula in Example 6.6.14 in Section 6.6 of Part 1. An argument via contour integration and Poisson summation formula can be found in Problem 1.

Taking $\alpha = 0$ and analytically continuing $t \to i\tau$ yields (13.1.31). Taking $\alpha = \frac{1}{2}$ and using

$$G_2(1+\tau) = \sum_{n=-\infty}^{\infty} (-1)^n \frac{1}{\cos(\pi n\tau)} \tag{13.1.35}$$

and (13.1.34) for $\alpha = \frac{1}{2}$ yields

$$G_2\left(1 - \frac{1}{\tau}\right) = \frac{\tau}{i} \sum_{n=-\infty}^{\infty} \frac{1}{\cos(\pi(n+\frac{1}{2})\tau)} \tag{13.1.36}$$

As $\tau \to \infty$, the dominant terms in the sum come from $n = 0, -1$ to give

$$G_2\left(1 - \frac{1}{\tau}\right) = \frac{2\tau}{i}\, \frac{2}{e^{-i\pi\tau/2}} + O(|\tau| e^{-3\pi \operatorname{Im}\tau/2}) \tag{13.1.37}$$

proving (13.1.33). $\square$

**Proof of Theorem 13.0.1.** Let

$$f(\tau) = \frac{G_2(\tau)}{\theta^\sharp(\tau)^2} \tag{13.1.38}$$

Since $\theta^\sharp$ is nonvanishing in $\mathbb{C}_+$ (by (10.5.6) of Part 2A, $\Theta(z \mid q)$ only vanishes if $z = q^{2n-1}$ ($n \in \mathbb{Z}$) and $\theta^\sharp(\tau) = \Theta(z = 1 \mid q = e^{\pi i\tau})$), $f(\tau)$ is analytic in $\mathbb{C}_+$. By Propositions 13.1.4 and 13.1.5, $f$ obeys (i)–(iv) of Theorem 13.1.3. Thus, $f(\tau) \equiv 1$, that is, (13.1.13) holds. $\square$

**Proposition 13.1.6.** *Let $G_4(\tau)$ be given by (13.1.14). Then*

(a) $G_4(\tau+2) = G_4(\tau)$ (13.1.39)

(b) $G_4\left(-\frac{1}{\tau}\right) = -\tau^2 G_4(\tau)$ (13.1.40)

(c) $\lim_{\operatorname{Im}\tau\to\infty} G_4(\tau) = 1$ (13.1.41)

(d) $\lim_{\operatorname{Im}\tau\to\infty} G_4\left(1-\frac{1}{\tau}\right)[-16\tau^2 e^{i\pi\tau}]^{-1} = 1$ (13.1.42)

**Proof.** We'll link $G_4$ to the Weierstrass $\wp$-function. By (10.4.87) of Part 2A,

$$\wp(z \mid \tau) - \wp(w \mid \tau) = \pi^2 \sum_k \left[\frac{1}{\sin^2(\pi z - k\pi\tau)} - \frac{1}{\sin^2(\pi w - k\pi\tau)}\right] \tag{13.1.43}$$

Comparing this to the definition of $G_4$ in (13.1.14), we see that

$$G_4(\tau) = \frac{1}{\pi^2}\left[\wp\left(\frac{1}{2} \,\middle|\, \tau\right) - \wp\left(\frac{\tau}{2} \,\middle|\, \tau\right)\right] \tag{13.1.44}$$

(We note parenthetically that this means $G_4(\tau) = (e_1 - e_3)/\pi^2$ and (13.1.14) is just (10.5.99)of Part 2A! But we'll provide a proof by the means of this section.)

Returning to the definition of $G_4$, we see $G_4(\tau+2) = G_4(\tau)$ and that $G_4(\tau) = 1 + O(e^{-\pi \operatorname{Im}\tau})$, proving (a) and (c). For (b), we note that, by (10.4.5) of Part 2A,

$$\wp\left(x \,\middle|\, -\frac{1}{\tau}\right) = \tau^2 \wp(\tau x \mid \tau) \tag{13.1.45}$$

so (recall $\wp(x \mid \tau) = \wp(-x \mid \tau)$)

$$\wp\left(\frac{1}{2} \,\middle|\, -\frac{1}{\tau}\right) - \wp\left(\frac{1}{2\tau} \,\middle|\, -\frac{1}{\tau}\right) = \tau^2\left(\wp\left(\frac{\tau}{2} \,\middle|\, \tau\right) - \wp\left(\frac{1}{2} \,\middle|\, \tau\right)\right) \tag{13.1.46}$$

which, given (13.1.44), implies (13.1.40).

Finally, we turn to (13.1.42). Since $\wp(x \mid 1+\tau) = \wp(x \mid \tau)$ (since $(1,\tau)$ and $(1, 1+\tau)$ generate the same lattices)

$$\begin{aligned} G_4\left(1-\frac{1}{\tau}\right) &= \frac{1}{\pi^2}\left[\wp\left(\frac{1}{2} \,\middle|\, -\frac{1}{\tau}\right) - \wp\left(\frac{1}{2} - \frac{1}{2\tau} \,\middle|\, -\frac{1}{\tau}\right)\right] \\ &= \frac{\tau^2}{\pi^2}\left[\wp\left(\frac{\tau}{2} \,\middle|\, \tau\right) - \wp\left(\frac{\tau}{2} - \frac{1}{2} \,\middle|\, \tau\right)\right] \qquad (13.1.47) \\ &= \tau^2 \sum_k \left[\frac{1}{\sin^2(\pi(k-\frac{1}{2})\tau)} - \frac{1}{\cos^2(\pi(k-\frac{1}{2})\tau)}\right] \qquad (13.1.48) \end{aligned}$$

where (13.1.47) used (13.1.45) and (13.1.48) used (13.1.43). The dominant terms come from $k = 0, 1$ so there are four terms in all. Since the leading term for $\operatorname{Im}\alpha$ is large if $\sin^2(\alpha) = [\frac{1}{2i}(e^{-i\alpha})^2] = -\frac{1}{4}(e^{-i\alpha})^2$, each dominant term is $-4(e^{-\frac{\pi i}{2}\tau})^2$, so

$$G_4\left(1 - \frac{1}{\tau}\right) = -16\tau^2 e^{i\pi\tau} + O(e^{-3\pi|\operatorname{Im}\tau|}) \tag{13.1.49}$$

which implies (13.1.42). □

**Proof of Theorem 13.0.2.** Let

$$f(\tau) = \frac{G_4(\tau)}{\theta^\sharp(\tau)^4} \tag{13.1.50}$$

Now just follow the proof of Theorem 13.0.1 with Proposition 13.1.6 replacing Proposition 13.1.5. □

**Notes and Historical Remarks.** The two underlying results that all integers are the sum of four squares and that certain numbers are sums of two squares have a long history. The two-squares result is put together from three basic facts:

(1) $(a^2 + b^2)(c^2 + d^2) = A^2 + B^2$ (13.1.51)

where

$$A = ac \pm bd, \quad B = bc \mp bd \tag{13.1.52}$$

which is straightforward algebra. The modern view is that $|a+ib|^2|c\mp id|^2 = |A + iB|^2$.

(2) If $p = 4k + 1$ is prime, it can be written as $a^2 + b^2$.

(3) If $p = 4k + 3$ is prime, it cannot be written as a sum of two squares. This is immediate from the fact that, mod 4, any $a^2$ is conjugate to 0 or 1.

It has been conjectured that Diophantus knew (1) since he remarked that $65 = 8^2 + 1^2 = 4^2 + 7^2$ because $65 = 13 \times 5$, $13 = 3^2 + 2^2$, and $5 = 2^2 + 1^2$. It is also believed that he knew (3) above and that he conjectured the four-squares results.

Pierre de Fermat (1606–65) claimed results on both problems. In 1638, he wrote to Mersenne and years later to Pascal and Huygens that he had proven that every number is a sum of three triangular numbers, four squares, five pentagonal, etc. (triangular numbers are $1, 3, 6, 10, \dots$, the number of points in a triangle with side $1, 2, 3, \dots$, so $T_1 = 1$, $T_{n=1} = T_n + (n+1)$, so $T_n = n(n+1)/2$. Squares obey $S_1 = 1$, $S_{n+1} = S_n + 2n + 1$ (so $S_n = n^2$), and pentagonal $P_1 = 1$, $P_{n+1} = P_n + 3n + 1$) (see Figure 13.1.1), but there are doubts that he had a proof, especially since there is no indication that

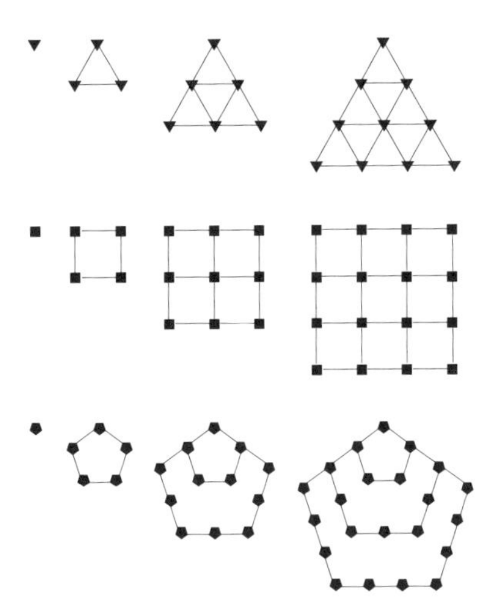

**Figure 13.1.1.** Polygonal numbers.

he knew the four-square analog of (13.1.51)/(13.1.52). He also said he had a proof of the two-square result for $4k+1$ primes and he said he could prove it by the method of descent which works, so it is believed he did, although no detailed proof exists in his known letters.

Euler looked into these questions starting in the 1740s. In 1747, he settled [**125**] the two-square questions and exploited the fact that a number of the form $4k+1$ was prime if and only if it had a unique representation $a^2+b^2$ with $0 < a < b$ and $a, b$ mutually prime. In particular, he determined $2,626,557$ was prime this way. He proved the three-triangular number result but was unable to prove the four-squares result, although he found the analog of (13.1.51)/(13.1.52) for four squares (something which, in modern language, comes from products of quaternions).

In 1770, the four-squares result was found by Joseph Lagrange (1736–1813) [**237**]. Euler wrote a paper [**129**] congratulating him, complaining his proof was involved, and finding a streamlined proof. Over a hundred years later, Hurwitz [**203, 205**] found a variant of Euler's proof using quaternions. The detailed counting results that appear in Theorems 13.0.1 and 13.0.2 are from Jacobi [**217, 218**] in 1828 and 1829.

The idea behind the proof via a theorem like Theorem 13.1.3 is due to Mordell [**280**] and Hardy [**173**]. We follow, in part, the approach of Stein–Shakarchi [**366**].

There have been numerous alternate or partially alternate approaches to the theorems of this section. We'll focus on the approaches to the full Jacobi result, except for mentioning Zagier's "one-sentence" proof [**417**] that a prime of the form $4k+1$ can be written $a^2+b^2$. For proofs of the full Jacobi results that rely on Jacobi's triple product formula, see [**192, 193, 194, 16**].

There is an intriguing proof of the two-squares theorem that goes back to a letter that Dirichlet sent Liouville. A modern exposition is by Cartier [**76**]. The idea relies on the zeta function of the Gaussian integers $\mathbb{Z}[i]$, that is, $\{a+bi \mid a,b \in \mathbb{Z}\}$. Its zeta function is defined by

$$Z(s) = \sum_{\substack{a+bi\in\mathbb{Z}(i) \\ a+ib\neq 0}} |a+ib|^{-2s} \tag{13.1.53}$$

which, of course, is given by

$$Z(s) = 4\sum_{n=1}^{\infty} r_2(n)n^{-s} \tag{13.1.54}$$

By analyzing the prime factorization in $\mathbb{Z}[i]$, one proves that (see Problems 2 and 3)

$$Z(s) = 4\zeta(s)L_\chi(s) \tag{13.1.55}$$

where $\zeta(s)$ is the Riemann zeta function (13.0.11) and $L_\chi$ is the Dirichlet $L$-function given by (13.0.22). Thus,

$$Z(s) = 4(1+2^{-s}+3^{-s}\dots)(1-3^{-s}+5^{-s}+\dots)$$

which implies (since $n^{-s}$ has terms from $(n/d)^{-s}$ in $\zeta$ and $d^{-s}$ in $L_\chi$)

$$r_2(n) = 4\sum_{\substack{d|n \\ d \text{ odd}}} (-1)^{(d-1)/2} = 4[d_1(n) - d_3(n)]$$

Many of the series of this section are of the form

$$a_0 + \sum_{n=1}^{\infty} \frac{a_n q^n}{1-q^n} \tag{13.1.56}$$

which are known as *Lambert series* after Johann Lambert (1728–77), who we've seen before in Problem 18 of Section 2.3 of Part 2A and the Notes to Section 7.5 of Part 2A. Indeed, that problem involved the simplest series of the form (13.1.56) where $a_0 = 0$, $a_n \equiv 1$ for $n \geq 1$. Often, as in that problem, simple Lambert series have natural boundaries. As we saw, some series that don't quite look like Lambert series are just slight rewritings, for example, by the proof of (13.1.1),

$$\sum_{n=1}^{\infty} \frac{q^n}{1+q^{2n}} = \sum a_n \frac{q^n}{1-q^n} \tag{13.1.57}$$

where $a_n = 1$ (respectively, $-1,0,0$) if $n \equiv 1$ (respectively, $3,0,2$) mod 4. For more on Lambert series, see [**55**, Ch. 3].

Formally,

$$\sum_{n=1}^{\infty} \frac{a_n q^n}{1-q^n} = \sum_{n=1}^{\infty} b_n q^n \tag{13.1.58}$$

$$b_n = \sum_{d|n} a_d \tag{13.1.59}$$

That means that in terms of the Möbius function, $\mu(j)$, of the next section (Example 13.2.11 and Problem 11 of that section),

$$a_n = \sum_{d|n} \mu\left(\frac{d}{n}\right) b_d \tag{13.1.60}$$

showing the family of Lambert series with polynomially bounded $a_n$ are exactly the family of Taylor series with polynomially bounded coefficients. Problem 19 of the next section has further discussion of Lambert series.

The themes of this section have been generalized. In 1770, Edward Waring (1736–98) conjectured that for each $k = 2, 3, \dots$, there is a minimal $g(k)$ so that every integer is a sum of at most $g(k)$ $k$-th powers. Lagrange's result is, of course, $g(2) = 4$. Waring conjectured that $g(3) = 9$, $g(4) = 19$. In 1909, Hilbert [**189**] finally proved Waring's conjecture on the existence of $g(k) < \infty$ for any $k$. For a comprehensive review of the extensive literature involving explicit bounds, see Ellison [**115**] and Vaughan–Wooley [**395**].

There is also considerable literature on $r_k(n)$ for values of $k$ other than 2 and 4. Jacobi's original paper had results on $k = 6$ and $k = 8$. This is a problem worked on by Stieltjes, Hurwitz, Hardy, Mordell, and others. Dickson [**102**] has an exhaustive summary on the pre-1920 work on these problems. For some other work on the subject, see [**202, 173, 90, 28, 55**].

## Problems

1. (a) By using a contour integral around the rectangle with vertices $\pm R$ and $\pm R + 2\pi i$ and taking $R \to \infty$, prove that for $k \neq 0$,

$$(1 - e^{2\pi k}) \int_{-\infty}^{\infty} \frac{e^{-ikx}}{\cosh x}\, dx = (2\pi)(e^{\pi k/2} - e^{3\pi k/2})$$

(b) Using $(b - b^3)/(1 - b^4) = 1/(b + b^{-1})$ with $b = e^{\pi k/2}$, prove that

$$\int_{-\infty}^{\infty} \frac{e^{-ikx}}{\cosh x}\, dx = \frac{\pi}{\cosh(\pi k/2)}$$

and thus, if $f_a(x) = [\cosh(ax)]^{-1}$ and $\widehat{g}(k) = (2\pi)^{-1/2} \int e^{-ikx} g(k)\, dk$, then

$$\widehat{f_a}(k) = \frac{\pi}{a}\, (2\pi)^{-1/2} f_{\pi/2a}(k)$$

(c) Using the Poisson summation formula and (b), prove (13.1.34).

Problems 2 and 3 prove (13.1.55) following Cartier [**76**]. The more algebraic part will be asserted with reference to Cartier. In $\mathbb{Z}[i]$, one defines the norm, $N(a+bi)$, by $N = a^2+b^2 \in \mathbb{Z}$ so $N(zw) = N(z)N(w)$. This is part of an alternate proof of the Jacobi two-square theorem, so you should not use that theorem but find another argument that primes of the form $4k+1$ are a sum of two squares.

2. (a) A *unit* is $u \in \mathbb{Z}[i]$ so there is $\tilde{u} \in \mathbb{Z}[i]$ with $u\tilde{u} = 1$. Prove that every unit has $N(u) = 1$ and then prove $\mathbb{Z}[i]$ has exactly four units.

   (b) A *prime* in $\mathbb{Z}[i]$ is $p$, not a unit, so that if $p = zw$, then either $z$ or $w$ is a unit. Show for every prime, $p$, there is a unit $u$ so $up = a+ib$ with $a > 0$. Show that for any prime in $\mathbb{Z}$, $p$ is not prime in $\mathbb{Z}[i]$ if $p = a^2+b^2$ has a solution in integers. Then show primes in $\mathbb{Z}[i]$ are up to a unit of three types: (1) $1+i$; (2) $p = 4k+3$, a prime in $\mathbb{Z}$; (3) $a \pm ib$ with $a^2+b^2 = 4k+1$, a prime in $\mathbb{Z}$ where $a > b > 0$. (*Hint*: Use Problem 6 of Section 13.0.)

For Problem 3, you should assume that you know that $\mathbb{Z}[i]$ has unique prime factorization, that is, any $z \in \mathbb{Z}[i]$ is uniquely a product $up_1^{k_1} \dots p_\ell^{k_\ell}$ where $u$ is a unit and $p_j = a_j + ib_j$, a prime with $a_j \geq |b_j| \geq 0$ and $1-i$ not allowed.

3. (a) Using prime factorization, prove that with $Z(s)$ given by (13.1.53) that
$$Z(s) = 4 \prod_{p \in \mathcal{P}[i]} (1 - N(p)^{-s})^{-1}$$
   where $\mathcal{P}[i]$ is the primes in $\mathbb{Z}[i]$ obeying $a \geq |b| \geq 0$ and $1-i$ not allowed.

   (b) Prove that
$$Z(s) = 4(1-2^{-s})^{-1} \prod_{p \in \mathcal{P}_1^{(4)}} (1-p^{-s})^{-2} \prod_{p \in \mathcal{P}_3^{(4)}} (1-p^{-2s})^{-1}$$

   (c) Prove (13.1.55).

## 13.2. Dirichlet Series

A *Dirichlet series* is a formal series of the form
$$f(s) = \sum_{n=1}^{\infty} a_n n^{-s} \tag{13.2.1}$$
thought of as a function of a complex variable $s = \sigma + it$ (this mixed notation, where $\sigma$ and $t$ are real and $s$ is complex, is weird but common!). We already

met two Dirichlet series, $\zeta(s)$ and $L_\chi(s)$, so we know they are central to the rest of this chapter.

A *generalized Dirichlet series* is of the form

$$f(s) = \sum_{n=1}^{\infty} a_n e^{-\lambda_n s} \tag{13.2.2}$$

where $\lambda_1 < \lambda_2 < \dots$ are all real. $\lambda_n = \log n$ corresponds to what we've called Dirichlet series, while $\lambda_n = (n-1)$ are ordinary power series in $e^{-s}$, so the generalized theory includes both cases of interest. That said, we'll mainly consider (13.2.1), leaving (13.2.2) to the Problems (see Problems 7–9). We note that many authors formally use "Dirichlet series" and "ordinary Dirichlet series" where we use "generalized Dirichlet series" and "Dirichlet series," but they then drop the word "ordinary" even when only discussing (13.2.1).

In some ways, the theory of Dirichlet series closely follows that of Taylor series, but there are two striking differences. The regions of absolute and conditional convergence of a Taylor series differ in a very thin set and the functions defined by Taylor series always have singularities on the border of their region of convergence. For Dirichlet series, neither of these is true. For example, the Dirichlet eta function,

$$\eta(s) = \sum_{n=1}^{\infty} \frac{(-1)^{n-1}}{n^s} \tag{13.2.3}$$

as we'll see below, has convergence absolutely in $\{s \mid \operatorname{Re} s > 1\}$, conditionally in $\{s \mid \operatorname{Re} s > 0\}$, and has an analytic continuation from its region of convergence to an entire function of $s$ with no singularities.

There is one aspect of singularities that does extend from Taylor series to Dirichlet: If $a_n \geq 0$, then $f(s)$ has a singularity at the point in $\mathbb{R}$ that is on the boundary of the region of convergence. This theorem of Landau, the subtlest of this section, will play a central role in determining regions without zeros for $L_\chi$ and $\zeta$, and that in turn will be critical in the proofs of the prime progression and prime number theorems. The analog for Taylor series is described in Problem 16 of Section 2.3 of Part 2A.

The key to the basic convergence theorems is a summation by parts formula, encoded in

**Proposition 13.2.1.** (a) *For complex numbers $\{x_j\}_{j=1}^{n+1} \cup \{y_j\}_{j=1}^{n}$, we have that*

$$\sum_{j=1}^{n} (x_{j+1} - x_j) y_j = \sum_{j=1}^{n-1} x_{j+1}(y_j - y_{j+1}) + x_{n+1} y_n - x_1 y_1 \tag{13.2.4}$$

(b) *Let* $\{c_j, b_j\}_{j=1}^n$ *be complex numbers,*

$$C_k = \sum_{j=1}^{k} c_j \tag{13.2.5}$$

*and* $C_0 = 0$. *Then for* $k = 1, \dots, n$,

$$\sum_{j=k}^{n} c_j b_j = \sum_{j=k}^{n-1} (C_j - C_{k-1})(b_j - b_{j+1}) + (C_n - C_{k-1}) b_n \tag{13.2.6}$$

(c) *Let* $\{c_j\}_{j=1}^n$ *be complex numbers and* $k = 2, \dots, n$. *Let* $C_k$ *be given by* (13.2.5). *Then for* $s = \sigma + it$ *with* $\sigma > 0$,

$$\left| \sum_{j=k}^{n} c_j j^{-s} \right| \le \max_{j=k,\dots,n} |C_j - C_{k-1}| \left[ k^{-\sigma} + \frac{|s|}{\sigma} (k-1)^{-\sigma} \right] \tag{13.2.7}$$

**Remark.** The basic summation by parts formula (13.2.4) is sometimes called Abel's formula after Abel [**1**].

**Proof.** (a) is elementary algebra.

(b) follows from (a) with $y_j = b_{k-1+j}$ and $x_j = C_{k-2+j} - C_{k-1}$, $j$ in sums replaced by $j + 1 - k$, and $n$ replaced by $n + 1 - k$.

(c) In (13.2.6), we take $b_j = j^{-s}$, so (13.2.7) follows from

$$|n^{-s}| \le |k^{-s}| = k^{-\sigma} \tag{13.2.8}$$

$$\sum_{j=k}^{n} |j^{-s} - (j+1)^{-s}| \le \frac{|s|}{\sigma} (k-1)^{-\sigma} \tag{13.2.9}$$

(13.2.8) is immediate from $n \ge k$ and $\operatorname{Re} s = \sigma$, while (13.2.9) comes from first using

$$\begin{aligned} |j^{-s} - (j+1)^{-s}| &= \left| \int_0^1 s(j+x)^{-s-1}\, dx \right| \\ &\le |s| j^{-\sigma-1} \end{aligned} \tag{13.2.10}$$

and then

$$\sum_{j=k}^{\infty} j^{-\sigma-1} \le \sum_{j=k}^{\infty} \int_{j-1}^{j} x^{-\sigma-1}\, dx = \sigma^{-1} (k-1)^{-\sigma} \tag{13.2.11}$$

□

**Theorem 13.2.2.** (a) *For any sequence* $\{a_n\}_{n=1}^\infty$, *if for* $s_0 \in \mathbb{C}$, $\sum_{j=1}^n a_j j^{-s_0}$ *converges, then* $\sum_{j=1}^n a_j j^{-s}$ *converges uniformly on each sector* $S_\theta(s_0) = \{s \mid \arg(s - s_0) \le \theta\}$ *for* $0 < \theta < \pi/2$.

(b) *There is* $\sigma_0 \in [-\infty, \infty]$ *so that* $\sum_{j=1}^{n} a_j j^{-s}$ *converges on* $\{s \mid \operatorname{Re} s > \sigma_0\}$ *uniformly on compacts and diverges on* $\{s \mid \operatorname{Re} s < \sigma_0\}$. *The limit,* $f(s)$, *is analytic on* $\{s \mid \operatorname{Re} s > \sigma_0\}$.

(c) *If* $\{\tilde{a}_n\}_{n=1}^{\infty}$ *is a second sequence and* $\sigma_0, \tilde{\sigma}_0$ *are both finite with* $f(s) = \widetilde{f}(s)$ *for* $s > \max(\sigma_0, \tilde{\sigma}_0)$, *then* $a_n = \tilde{a}_n$ *for all* $n$.

**Remark.** $\sigma_0$ is called the *abscissa of convergence* of the Dirichlet series.

**Proof.** (a) If $s \in S_\theta(s_0)$, then $|s - s_0|/|\sigma - \sigma_0| \le 1/\cos\theta$, so by (13.2.7) with $c_j = a_j j^{-s_0}$ and $s$ replaced by $s - s_0$,

$$\sup_n \left| \sum_{j=k}^{n} a_j j^{-s} \right| \le \sup_{j \ge k} |C_j - C_{k-1}| \, [1 + \cos(\theta)^{-1}] \tag{13.2.12}$$

since $\sigma - \sigma_0 > 0$. The convergence at $s_0$ says $C_j$ converges, so the sup goes to zero as $k \to \infty$, and we have the claimed uniform convergence.

(b) By taking a union over $0 \le \theta < \pi/2$, we see convergence at any $s_0$ implies convergence in $\{s \mid \operatorname{Re} s > \operatorname{Re} s_0\}$, so if $\sigma_0 = \inf\{s \in \mathbb{R} \mid \sum_{n=1}^{\infty} a_n n^{-\sigma}$ convergence$\}$, then the stated convergence/divergence holds. $\sigma_0 = \infty$ applies if the set is empty and $\sigma_0 = -\infty$ if it is $\mathbb{C}$. By the uniformity in (a) and Weierstrass' theorem, we have the claimed analyticity.

(c) So long as $\sigma_0 < \infty$, it is easy to see inductively that

$$a_j = \lim_{\substack{\sigma \to \infty \\ \sigma \in \mathbb{R}}} \left[ f(\sigma) - \sum_{k=1}^{j-1} a_k k^{-\sigma} \right] \tag{13.2.13}$$

which implies uniqueness. $\square$

Next is an analog of the Cauchy radius formula:

**Theorem 13.2.3.** *Suppose that* $\sum_{n=1}^{N} a_n$ *does not converge. Let*

$$A_N = \sum_{n=1}^{N} a_n \tag{13.2.14}$$

*Then*

$$\sigma_0 = \limsup_{n \to \infty} \frac{\log|A_n|}{\log n} \tag{13.2.15}$$

**Proof.** Let $\tilde{\sigma}$ be the lim sup. If $\tilde{\sigma} < 0$, $|A_n| \to 0$, so the assumed nonconvergence implies $\tilde{\sigma} \ge 0$.

Let $\sigma \equiv \operatorname{Re} s > \tilde{\sigma} + \varepsilon$ for some $\varepsilon > 0$. By (13.2.15), eventually, say for $n \ge N$, $|A_n| \le n^{\tilde{\sigma} + \varepsilon}$. By (13.2.6) with $c_j = a_j$, $b_j = j^{-s}$, and $k > N$, and by (13.2.10),

$$\left|\sum_{j=k}^{m} a_k j^{-s}\right| \leq 2\sum_{j=k}^{n-1} j^{\tilde{\sigma}+\varepsilon}\,|s|\,(j-1)^{-s-1} + 2n^{\tilde{\sigma}+\varepsilon}n^{-\sigma} \tag{13.2.16}$$

This goes to zero as $k \to \infty$ since $\sigma > \tilde{\sigma} + \varepsilon$. Thus, $\sum_{j=1}^{\infty} a_j j^{-s}$ converges, and so $\sigma_0 \leq \tilde{\sigma} + \varepsilon$ for all $\varepsilon$ and therefore, $\sigma_0 \leq \tilde{\sigma}$.

On the other hand, the nonconvergence assumption implies $\sigma_0 \geq 0$. Let $\sigma > \sigma_0$ (so $\sigma > 0$). Let $C \equiv \sup_n |\sum_{j=1}^{n} a_j n^{-\sigma}| < \infty$ by the assumed convergence (13.2.6) with $c_j = a_j n^{-\sigma}$ and $b_j = n^{\sigma}$, then by $\sigma > 0$ and (13.2.6),

$$\begin{aligned} |A_n| &\leq 2C\sum_{j=1}^{n-1}(b_{j+1} - b_j) + 2Cn^{\sigma} \\ &\leq 4C_n^{\sigma} \end{aligned}$$

so $\tilde{\sigma} \leq \sigma$, and thus, $\tilde{\sigma} \leq \sigma_0$. We conclude $\tilde{\sigma} = \sigma_0$. $\square$

**Corollary 13.2.4.** *If* $\limsup|a_n| > 0$ *and* $\limsup|A_n| < \infty$, *then* $\sigma_0 = 0$.

**Proof.** By the first condition, $\sum_{n=1}^{N} a_n$ is not convergent, so as in the proof, $\tilde{\sigma} \geq 0$. But clearly, since $\log n \to \infty$, $\tilde{\sigma} \leq 0$. Thus, $\sigma_0 = \tilde{\sigma} = 0$. $\square$

There is a final result on general convergence that we want to state:

**Theorem 13.2.5.** *There is* $\bar{\sigma} \in [-\infty, \infty]$ *so that* $\sum_{n=1}^{\infty}|a_n|n^{-\sigma} < \infty$ *if* $\sigma > \bar{\sigma}$ *and is* $= \infty$ *if* $\sigma < \bar{\sigma}$. $f(s) = \sum_{n=1}^{\infty} a_n n^{-s}$ *converges uniformly on each* $\{s \mid \operatorname{Re} s \geq \bar{\sigma} + \varepsilon\}$ *(for* $\varepsilon > 0$*). Moreover,*

$$\sigma_0 \leq \bar{\sigma} \leq \sigma_0 + 1 \tag{13.2.17}$$

**Remark.** $\bar{\sigma}$ is called the *abscissa of absolute convergence.*

**Proof.** The first sentence follows from Theorem 13.2.2(b) for $\sum_{n=1}^{\infty}|a_n|n^{-\sigma}$ and the second sentence is easy since $\sum_{n=1}^{\infty}|a_n|n^{-\bar{\sigma}-\varepsilon} < \infty$. That $\sigma_0 \leq \bar{\sigma}$ is obvious. Finally, if $s > \sigma_0 + 1$, then for some $\delta$, $\sup_n |a_n|n^{-s+1+\delta} < \infty$, so $\sum_{n=1}^{\infty}|a_n|n^{-s} = \sum_{n=1}^{\infty}(|a_n|n^{-s+1+\delta})n^{-1-\delta} < \infty$. Thus, $\bar{\sigma} \leq s$, that is, $\bar{\sigma} \leq \sigma_0 + 1$. $\square$

**Remark.** We'll see examples where $\sigma_0 = \bar{\sigma}$ (e.g., $a_n > 0$) and where $\sigma_0 = \bar{\sigma} - 1$ (e.g., $\sum_{n=1}^{\infty}(-1)^n n^{-s}$). If the Riemann hypothesis holds, the Dirichlet series associated to $\log(\sum_{n=1}^{\infty}(-1)^n n^{-s})$ (see Example 13.2.11) has $\bar{\sigma} = 1$ and $\sigma_0 = \frac{1}{2}$.

In the examples below, we'll sometimes write $f_{\{a_n\}}(s)$ for the sum of Dirichlet series associated to a sequence $\{a_n\}$.

**Example 13.2.6** (Riemann Zeta Function). $a_n \equiv 1$, so

$$\zeta(s) = \sum_{n=1}^{\infty} n^{-s} \tag{13.2.18}$$

This has $\sigma_0 = \bar{\sigma} = 1$, as is easy to see. We see shortly that $(s-1)\zeta(s)$ has an analytic continuation to $\operatorname{Re} s > 0$ and, in the next section, that it is entire. As we explained in the chapter introduction, $\lim_{s\downarrow 1}(s-1)\zeta(s) = 1$, so $\zeta(s)$ is entire meromorphic with a pole only at $s = 1$ with residue 1. Later in this section, we'll show $\zeta(s)$ is nonvanishing on $\{s \mid \operatorname{Re} s > 1\}$ and, in the next section, also on $\{s \mid \operatorname{Re} s \geq 1\}$. $\square$

**Example 13.2.7** (Hurwitz Zeta Function and Periodic Dirichlet Series). Let $x \in \mathbb{H}_+$. Define the *Hurwitz zeta function*,

$$\zeta(s, x) = \sum_{n=0}^{\infty} (n + x)^{-s} \tag{13.2.19}$$

If $x = 1$, this is the Riemann zeta function, and for $x = 2, 3, 4 \dots$, it is a Dirichlet series. For other $x$, it is only a generalized Dirichlet series (with $\lambda_n = \log(n+x)$), but it is closely related to both $\zeta(s)$ and to other Dirichlet series, as we'll see momentarily. In the next section, we'll prove it is entire meromorphic with a single pole at $s = 1$ with residue 1.

It is relevant to us here for the following reason. Let $p, q \in \mathbb{Z}_+$ and let

$$a_n^{p,q} = \begin{cases} 1 & \text{if } n = mq + p, \quad m = 0, 1, 2, \dots \\ 0, & \text{otherwise} \end{cases} \tag{13.2.20}$$

Then

$$\begin{aligned} f_{\{a_n^{p,q}\}}(s) &= \sum_{m=0}^{\infty} (mq + p)^{-s} \\ &= q^{-s} \sum_{m=0}^{\infty} \left( m + \frac{p}{q} \right)^{-s} \end{aligned}$$

so

$$\zeta\left( s; \frac{p}{q} \right) = q^s f_{\{a_n^{p,q}\}}(s) \tag{13.2.21}$$

Thus, if $a$ is a sequence so that for some $q \in \mathbb{Z}_+$, and all $n$,

$$a_{n+q} = a_n \tag{13.2.22}$$

then

$$f_{\{a\}}(s) = q^{-s} \sum_{p=1}^{q} a_p \zeta\left( s; \frac{p}{q} \right) \tag{13.2.23}$$

so our route to proving analyticity of $f$'s associated to periodic $a$'s will be via $\zeta(s; x)$ and (13.2.23). $\square$

The next two examples are important periodic cases.

**Example 13.2.8** (Dirichlet Eta Function). This is the Dirichlet series with $a_n = (-1)^{n+1}$

$$\eta(s) = \sum_{n=1}^{\infty} \frac{(-1)^{n+1}}{n^s} \tag{13.2.24}$$

Notice that $\zeta(s) - \eta(s) = \sum_{n=1}^{\infty} \frac{2}{(2n)^s} = 2^{1-s}\zeta(s)$, so

$$\eta(s) = (1 - 2^{1-s})\zeta(s) \tag{13.2.25}$$

By Corollary 13.2.4, for $\eta$, $\sigma_0 = 0$ (and $\bar{\sigma} = 1$). Thus, $\eta$ has an analytic continuation to $\{s \mid \operatorname{Re} s > 0\}$. By (13.2.25), $\zeta(s)$ has a meromorphic continuation to the same region with a pole only at $s = 1$ and, potentially, also at another solution of $2^{1-s} = 1$. Notice that since $(1 - 2^{1-s}) = (s - 1)\log 2 + O((s-1)^2)$, (13.2.25) also implies $\eta(1) = \log 2$ (which is the value of $1 - \frac{1}{2} + \frac{1}{3} - \frac{1}{4} \dots$ by Abel's summation theorem; see Problem 8 of Section 2.3 of Part 2A). $\square$

**Example 13.2.9** (Dirichlet $L$-Function). A *Dirichlet character* mod $m \in \mathbb{Z}_+$ is a function $\chi$ on $\mathbb{Z}_+$ with $\chi(n) = 0$ if $m$ and $n$ are not relatively prime, with

$$\chi(k + m) = \chi(k) \tag{13.2.26}$$

and

$$\chi(kn) = \chi(k)\chi(n) \tag{13.2.27}$$

and $\chi(1) = 1$. If $\chi(k) = 1$ for all $k$ relatively prime to $m$, $\chi$ is called *principal*, otherwise, *nonprincipal*.

The Dirichlet $L$-function is defined by

$$L_\chi(s) = \sum_{n=1}^{\infty} \frac{\chi(n)}{n^s} \tag{13.2.28}$$

In the next section (see Corollary 13.3.2), we'll prove that if $\chi^{(0)}$ is the principal character, then

$$L_{\chi^{(0)}}(s) = \zeta(s) \prod_{p \mid m} \left(1 - \frac{1}{p^s}\right) \tag{13.2.29}$$

where the product is over the finite set of all prime divisors of $m$. This shows $L_{\chi^{(0)}}$ has a meromorphic continuation to $\{s \mid \operatorname{Re} s > 0\}$ with a pole only at the poles of $\zeta$ and, in particular, a pole at $s = 1$ with residue $\prod_{p\mid m}(1 - \frac{1}{p})$.

On the other hand, if $\chi$ is a nonprincipal character, we'll prove in Section 13.4 (see (13.4.4)) that

$$\sum_{k=1}^{m} \chi(k) = 0 \tag{13.2.30}$$

which implies

$$\sup_n \left[\left|\sum_{k=1}^n \chi(n)\right|\right] = \sup_{n=1,\dots,m-1}\left|\sum_{k=1}^n \chi(n)\right| < \infty$$

so, by Corollary 13.2.4, $L_\chi$ is analytic in $\{s \mid \operatorname{Re} s > 0\}$. We'll eventually prove that $L_\chi$ is entire. □

Next, we want to consider products of Dirichlet series. Here's the key fact:

**Theorem 13.2.10.** *Let $f(s) = \sum_{n=1}^\infty a_n n^{-s}$ and $g(s) = \sum_{n=1}^\infty b_n n^{-s}$ be two functions defined by Dirichlet series with $\bar\sigma(a) < \infty$, $\bar\sigma(b) < \infty$. Define the* Dirichlet convolution, *$c_n$, by*

$$c_n = \sum_{d|n} a_d b_{n/d} \tag{13.2.31}$$

*then*

$$\bar\sigma(c) \le \max(\bar\sigma(a), \bar\sigma(b)) \tag{13.2.32}$$

*and if $h(s) = \sum_{n=1}^\infty c_n n^{-s}$, then for $\operatorname{Re} s > \max(\bar\sigma(a), \bar\sigma(b))$, $h(s) = f(s)g(s)$.*

**Proof.** If $s \ge \max(\bar\sigma(a), \bar\sigma(b))$, then both series are absolutely convergent, so their product can be rearranged and

$$\begin{aligned} f(s)g(s) &= \sum_{d,k} a_d d^{-s} b_k k^{-s} \\ &= \sum_{n=1}^\infty n^{-s}\left(\sum_{d,k \text{ so that } dk=n} a_d b_k\right) \\ &= \sum_{n=1}^\infty c_n n^{-s} \end{aligned} \tag{13.2.33}$$

where $c$ is given by (13.2.31). Since $|c_n| \le \sum_{d|n} |a_d|\,|b_d|$, we see $\sum_{n=1}^\infty |c_n| n^{-\sigma} = (\sum_{d=1}^\infty |a_d| d^{-\sigma})(\sum_{k=1}^\infty |b_k| k^{-\sigma})$, proving (13.2.32). □

**Remark.** One can show that the series for $c$ is convergent in a region where one of the $a$ or $b$ series is absolutely convergent and the other only convergent.

We'll write $a*b$ in *this chapter only* for Dirichlet convolution of sequences.

**Example 13.2.11** (Möbius and Mertens Functions). The *Möbius function*, $\mu(n)$, is defined in terms of the prime factorization of $n = p_1^{k_1} \dots p_\ell^{k_\ell}$,

$k_j \geq 1$, by ($\mu(1) = 1$)

$$\mu(n) = \begin{cases} 0, & \text{some } k_j \geq 2 \\ 1, & \ell \text{ even, all } k_j = 1 \\ -1, & \ell \text{ odd, all } k_j = 1 \end{cases} \tag{13.2.34}$$

The reason $\mu$ is important is that it obeys

$$\sum_{d|n} \mu(d) = \begin{cases} 1, & n = 1 \\ 0, & n > 1 \end{cases} \tag{13.2.35}$$

The reader will prove this in Problem 4(b), but we note here that it is true because $\sum_{j=0}^{L} \binom{L}{j}(-1)^j = 0$ (aka the number of even subsets of $\{1, \dots, L\}$ and the number of odd subsets is the same). (13.2.35) can be written as

$$\mu * 1 = \delta \tag{13.2.36}$$

where $1_n \equiv 1$ and $\delta_n \equiv \delta_{n1}$. Thus, if $f_\mu$ is the Dirichlet series for $\mu$, (13.2.36) says $f_\mu(s)\zeta(s) = 1$, or

$$\zeta(s)^{-1} = \sum_{n=1}^{\infty} \mu(n) n^{-s} \tag{13.2.37}$$

at least for $\operatorname{Re} s > 1$. See Problem 6 for another proof of (13.2.37). More generally, if $\chi$ is a Dirichlet character, then (Problem 12)

$$L_\chi(s)^{-1} = \sum_{n=1}^{\infty} \mu(n)\chi(n) n^{-s} \tag{13.2.38}$$

for $\operatorname{Re} s > 1$.

The Möbius function is also connected with the Möbius inversion formula (Problem 11), which inverts $f(n) = \sum_{d|n} g(d)$.

The *Mertens function* is defined by

$$M(n) = \sum_{k=1}^{n} \mu(k) \tag{13.2.39}$$

By Theorem 13.2.3, if

$$\limsup \frac{\log|M(n)|}{\log n} = \frac{1}{2} \tag{13.2.40}$$

(aka $|M(n)| \leq C_\varepsilon n^{\frac{1}{2}+\varepsilon}$ for any $\varepsilon > 0$), then $\zeta(s)^{-1}$ is analytic and so $\zeta(s)$ has no zero in $\{s \mid \operatorname{Re} s > \frac{1}{2}\}$, which implies the Riemann hypothesis. In fact, it is known that (13.2.40) is equivalent to the Riemann hypothesis. (13.2.40), which can be stated with only reference to prime factorization of integers, is the most "elementary" expression of the Riemann hypothesis. □

**Example 13.2.12** (Totient and Divisor Functions). The *Euler totient function*, $\varphi(m)$, is the number of elements among $\{1,\dots,m\}$ relatively prime to $m$, that is, the order of $\mathbb{Z}_m^\times$, the invertible elements of the ring $\mathbb{Z}_m$ ($\varphi(1)\equiv 1$). The *divisor function*, $d(n)$, is the total number of divisors of $m$.

Euler found an explicit formula for $\varphi(n)$, namely,

$$\varphi(p_1^{k_1}\dots p_\ell^{k_\ell}) = \prod_{j=1}^{\ell} p_j^{k_j-1}(p_j-1) \tag{13.2.41}$$

where $p_1,\dots,p_\ell$ are distinct primes; equivalently,

$$\varphi(n) = n\prod_{\substack{p\in P\\ p|n}}\left(1-\frac{1}{p}\right) \tag{13.2.42}$$

Problems 13, 14, and 15 have proofs of this, and Problem 6 of Section 13.4 has another proof.

Clearly, $\sum_{d|n} 1 = d(n)$, which says $1*1 = d$ where $d_n = d(n)$. Thus,

$$\sum_{n=1}^\infty d(n)n^{-s} = \zeta(s)^2 \tag{13.2.43}$$

In Problem 2, the reader will prove $\sum_{d|n}\varphi(d) = n$, so $\varphi * 1 = \tilde n$, where $\tilde n$ is the sequence $\tilde n_n = n$. Since $\sum_{n=1}^\infty nn^{-s} = \zeta(s-1)$, this immediately implies

$$\sum_{n=1}^\infty \varphi(n)n^{-s} = \frac{\zeta(s-1)}{\zeta(s)} \tag{13.2.44}$$

□

**Example 13.2.13** (Derivatives). In the region of convergence

$$f_a'(s) = -\sum_{n=1}^\infty a_n(\log n)n^{-s} \tag{13.2.45}$$

(this is obvious in the region of absolute convergence and true in the region of convergence; see Problem 3). Thus, $f_a^{(k)}(s)$ is the Dirichlet series with sequence $(-1)^k(\log n)^k a_n$. □

**Example 13.2.14** (von Mangoldt Function). The *von Mangoldt function* is defined by

$$\Lambda(n) = \begin{cases} \log p & \text{if } n=p^k,\ k=1,2,\dots\ p=\text{prime} \\ 0, & \text{otherwise} \end{cases} \tag{13.2.46}$$

It is easy to see (Problem 5) that

$$\sum_{d|n} \Lambda(d) = \log n \tag{13.2.47}$$

so $\Lambda * 1 = \widetilde{\log n}$ (where $\widetilde{\log n}_n = n$). By (13.2.45), this says that

$$\sum_{n=1}^{\infty} \Lambda(n) n^{-s} = -\frac{\zeta'(s)}{\zeta(s)} \tag{13.2.48}$$

Many proofs of the prime number theorem use $\Lambda(n)$. We'll use a close relative. Note one can rewrite (13.2.48) as

$$\begin{aligned} -\frac{\zeta'(s)}{\zeta(s)} &= \sum_{p\in\mathcal{P}} \sum_{k=1}^{\infty} (\log p) p^{-ks} \\ &= \sum_{p\in\mathcal{P}} \frac{\log p}{p^s - 1} \end{aligned} \tag{13.2.49}$$

Problem 6 has a direct proof of this, and so an alternate proof of (13.2.48). □

Finally, we turn to the theorem of Landau mentioned earlier:

**Theorem 13.2.15** (Landau). *Let $\{a_n\}_{n=1}^{\infty}$ be a sequence of nonnegative numbers with a finite abscissa of convergence, $\sigma_0$. Then $f(s) = \sum_{n=1}^{\infty} a_n n^{-s}$ has a singularity at $s = \sigma_0$ in the sense that $f(s)$ defined in $\{s \mid \operatorname{Re} s > \sigma_0\}$ cannot be analytically continued to all of any $\mathbb{D}_\delta(\sigma_0)$ with $\delta > 0$.*

**Remark.** This is an analog of Pringsheim's theorem for Taylor series discussed in Problem 16 of Section 2.3 of Part 2A. Many discussions appeal to that theorem—we instead just follow its proof!

**Proof.** Suppose $f$ can be analytically continued throughout $\mathbb{D}_\delta(\sigma_0)$. Let

$$g(s) = f(-s + \sigma_0 + 1) \tag{13.2.50}$$

analytic in $\{s \mid \operatorname{Re} s < 1\} \cup \mathbb{D}_\delta(1)$, so

$$g(s) = \sum_{n=1}^{\infty} a_n n^{-\sigma_0 - 1} e^{s \log n} \tag{13.2.51}$$

Thus, by the positivity of $a_n$ and the monotone convergence theorem,

$$g^{(k)}(1) = \lim_{\varepsilon \downarrow 0} g^{(k)}(1-\varepsilon) = \sum_{n=1}^{\infty} a_n n^{-\sigma_0} (\log n)^k \tag{13.2.52}$$

By the supposed analyticity of $g$ in $\mathbb{D}_\delta(1)$,

$$\infty > \sum_{k=0}^{\infty} \frac{g^{(k)}(1)}{k!}\left(\frac{\delta}{2}\right)^k = \sum_{n=1}^{\infty} a_n n^{-\sigma_0} \sum_{k=0}^{\infty} \frac{(\log n)^k}{k!}\left(\frac{\delta}{2}\right)^k = \sum_{n=1}^{\infty} a_n n^{-\sigma_0+\delta/2} \tag{13.2.53}$$

where we used positivity to interchange sums in (13.2.53). Thus, the region of convergence of $f$ is bigger than $\{s \mid \operatorname{Re} s \geq \sigma_0\}$. This contradiction shows $f$ must have a singularity at $\sigma_0$. □

**Notes and Historical Remarks.** Dirichlet series played a key role in his work on number theory and a systematic review appeared in his lectures [**105**], published posthumously by Dedekind, who took lecture notes and expanded on them. Dirichlet and Dedekind worked with $s$ as a real variable (and, indeed, our proof in Section 13.4 only depends on functions for real $s$; that said, our proof that $L_\chi(1) \neq 0$ in Section 13.3 relies on complex variables, but there are other proofs, including Dirichlet's, that do not). It was Jensen [**221, 222**] who first explored complex values and determined half-planes of convergence. Systematic early books on Dirichlet series were Landau [**242**] and Hardy–Riesz [**178**]. Recent books are Helson [**188**] and Mandelbrojt [**269**].

One might ask why it is Dirichlet, not Taylor, series that enter in number theory. The answer is in part the multiplicative structure in Theorem 13.2.10 and in part the multiplicative structure in the Euler factorization formula.

Theorem 13.2.15 is from Landau [**241**]; see also [**242**].

Hurwitz [**201**] introduced his zeta function, and for $0 < x \leq 1$ and $0 < \operatorname{Re} s < 1$ proved

$$\zeta(1-s,x) = \frac{\Gamma(s)}{(2\pi)^s}\{e^{-i\pi s/2}\beta(x,s) + e^{i\pi s/2}\beta(-x,s)\} \tag{13.2.54}$$

where $\beta$ is the Dirichlet series

$$\beta(x,s) = \sum_{n=1}^{\infty} \frac{e^{2\pi i n x}}{n^s} \tag{13.2.55}$$

In particular, since $\beta$ has an analytic continuation to $\{s \mid \operatorname{Re} s > 0\}$, this formula can be used to analytically continue $\zeta(s,x)$ to $\mathbb{C}\backslash\{1\}$. Since $\beta(1,s) = \zeta(s)$, this also leads directly to a proof of the functional equation for $\zeta$ (see Section 13.3 and Problem 6 of Section 14.7) and, with more work, to functional equations for $\zeta(s,x)$. A nice presentation of the proofs of (13.2.54) and its consequences is in Chapter 12 of Apostol [**18**]; see also Problem 7 of Section 14.7.

The Möbius function was introduced by Möbius in 1832 [**279**] and Mertens introduced his function in 1897 [**272**]. The von Mangoldt function is from [**397**], also in 1897. Section 7.3 of Part 2A has a capsule biography of Möbius.

As mentioned, $|M(n)| \leq C_\varepsilon n^{\frac{1}{2}+\varepsilon}$ is equivalent to the Riemann hypothesis. In 1885, Stieltjes announced in a letter to Hermite that he had proven $|M(n)| \leq C\sqrt{n}$ but the details never appeared—and the current belief is that this is false (see below). Mertens conjectured that $|M(n)| \leq \sqrt{n}$ holds and it is actually true for $n \leq 10^{14}$. But in 1985, Odlyzko–Riele [**304**] proved it false, although an explicit $n$ for which it fails is not known. Graphs of $\mu(n)$ for $n$ large look kind of random and sums of random $\pm 1$ have $\sup_{m\leq n} S(m) \sim \sqrt{n}\log(\log n)$ (law of the iterated logarithm; see Section 7.2 of Part 1), so current beliefs are that is the large $n$ behavior of $M(n)$.

## Problems

1. The simplest $L$-function is the principal one for $\mathbb{Z}_2$:
$$L_{\chi_2^{(0)}}(s) = \sum_{m=0}^{\infty} \frac{1}{(2m+1)^s}$$
By relating this to $\zeta$ and $\eta$, prove $L_{\chi_2^{(0)}}(s) = \zeta(s)(1-\frac{1}{2^s})$, which is a special case of (13.2.29).

2. You'll prove
$$\sum_{d|n} \varphi(d) = n \tag{13.2.56}$$
in this problem.

(a) A *primitive $k$-th-root of unity* is $\omega \in \mathbb{C}$ so $\omega^k = 1$ and $\omega^m \neq 1$ for $m = 1, 2, \dots, k-1$. Prove that $\omega$ is a prime $k$-th root if and only if $\omega = e^{2\pi i\ell/k}$ where $0 < \ell < k$ and with $\ell$ and $k$ relatively prime. Thus, the number of primitive $k$-th roots of unity is $\varphi(k)$.

(b) Prove that any $n$-th root of unity is a primitive $d$-th root for a unique $d \mid n$.

(c) Conclude (13.2.56) from the fact that there are exactly $n$ $n$-th roots of unity.

3. Prove that in the region of convergence, (13.2.45) holds. (*Hint*: First show the sum from $n = 1$ to $N$ has the claimed derivative and then use uniformity of convergence.)

4. (a) If $n = p_1^{k_1} \dots p_\ell^{k_\ell}$ with $k_j \geq 1$, prove $\mu(d) \neq 0$ for $d \mid n$ if and only if $d = p_1^{\alpha_1} \dots p_\ell^{\alpha_\ell}$ where $\alpha_\ell = 0$ or 1, with $\mu(d) = (-1)^{\alpha_1 + \cdots \alpha_\ell}$.

(b) Prove that $\sum_{d|n} \mu(d) = (1-1)^\ell = 0$.

(c) Prove that $\sum_{d|n} |\mu(d)| = 2^\ell$.

5. If $n = p_1^{k_1} \dots p_\ell^{k_\ell}$, prove the sum in (13.2.47) only has nonzero terms if $d \in \{p_j^{m_j}\}_{m_j \le k_j,\, j=1,\dots,\ell}$ and conclude (13.2.47).

6. (a) Assuming the Euler factorization theorem (see Theorem 13.3.1), $\zeta(s) = \prod_{p\in\mathcal{P}}(1 - \frac{1}{p^s})^{-1}$, prove directly that $-\zeta'(s)/\zeta(s)$ is given by (13.2.49) for $s > 1$.

   (b) By expanding the product $\zeta(s)^{-1} = \prod_{p\in\mathcal{P}}(1 - \frac{1}{p^s})$, find another proof of (13.2.37).

7. Prove the analogs of Proposition 13.2.1 and Theorem 13.2.2(a),(b) for generalized Dirichlet series, where $n^{-s}$ is replaced by $e^{-\lambda_n s}$ with $\lambda_1 < \lambda_2 < \dots$.

8. Prove the analog of Theorem 13.2.3 for general Dirichlet series where $\log n$ in (13.2.15) is replaced by $\lambda_n$.

9. Show that the generalized Dirichlet series $\sum_{n=1}^\infty \frac{(-1)^n (\log n)^s}{\sqrt{n}}$ has $\bar\sigma = \infty$, $\sigma_0 = -\infty$.

10. Prove that $\sum_{m|n} \mu(m) d(\frac{n}{m}) = 1$. (*Hint*: $1 * 1 = d$ and $\mu * 1 = \delta$.)

11. Suppose $f, g$ are related by

$$f(n) = \sum_{d|n} g(d) \tag{13.2.57}$$

   Prove that (*Möbius inversion*)

$$g(n) = \sum_{d|n} \mu\left(\frac{n}{d}\right) f(d) \tag{13.2.58}$$

12. (a) Let $f$ be a *completely multiplicative* function on $\mathbb{Z}_+$, that is, $f(nm) = f(n)f(m)$ and $f(1) = 1$. Prove that

$$f(g * h) = (fg) * (fh) \tag{13.2.59}$$

   (b) If $f$ is completely multiplicative, prove that

$$(f\mu) * f = \delta \tag{13.2.60}$$

   (c) Let $\chi$ be a Dirichlet character. Prove that

$$L_\chi(s)^{-1} = \sum_{n=1}^\infty \frac{\chi(n)\mu(n)}{n^s} \tag{13.2.61}$$

   for $\operatorname{Re} s > 1$.

13. (a) Using (13.2.56) and Möbius inversion, prove that the Euler totient function, $\varphi$, obeys

$$\varphi(n) = \sum_{d|n} \frac{n}{d}\, \mu(d) \tag{13.2.62}$$

(b) Prove that if $p_1, \dots, p_\ell$ are all the prime divisors of $n$, then

$$\begin{aligned}\left(1 - \frac{1}{p_1}\right)\left(1 - \frac{1}{p_2}\right)\cdots\left(1 - \frac{1}{p_\ell}\right) &= 1 - \sum_{j=1}^{\ell} \frac{1}{p_j} + \sum_{1 \le j_1 < j_2 \le \ell} \frac{1}{p_{j_1} p_{j_2}} - \cdots \\ &= \sum_{d|n} \mu(d)\, \frac{1}{d}\end{aligned}$$

which is therefore $\varphi(n)/n$, and so prove (13.2.42).

14. (a) $f$ on $\mathbb{Z}_+$ is called *multiplicative* if $f(nm) = f(n)f(m)$ for $n, m$ relatively prime. Prove that the Dirichlet convolution of multiplicative functions is multiplicative. (*Hint*: First show that if $n$ and $m$ are relatively prime and $d \mid nm$, then $d = d_1 d_2$ with $d_1 \mid n$, $d_2 \mid m$ in a unique way.)

    (b) Prove that $\varphi$ is multiplicative. (*Hint*: (13.2.62).)

    (c) Prove that $\varphi(p^k) = p^k - p^{k-1}$ for $p$ prime. (*Hint*: Find those $m$ not relatively prime to $p^k$ and at most $p^k$.)

    (d) Prove that $\varphi(p_1^{k_1} \dots p_\ell^{k_\ell}) = \prod_{j=1}^{\ell} p_j^{k_j - 1}(p_j - 1)$.

15. This provides another proof of the key to (13.2.41) that if $n$ and $m$ are relatively prime, then $\varphi(nm) = \varphi(n)\varphi(m)$. Below, suppose $n$ and $m$ are relatively prime.

    (a) Prove that $k$ is relatively prime to $nm$ if and only if it is relatively prime to $n$ and to $m$.

    (b) For any $k$, prove that $k$ is relatively prime to $m$ if and only if $k = \ell + bm$, where $0 \le \ell < m$ and $\ell$ is relatively prime to $m$.

    (c) Let $0 \le \ell < m$ fixed. Prove that the mod $n$ conjugacy classes of $\ell, \ell + m, \dots, \ell + m(n-1)$ are all the $n$ elements of $\mathbb{Z}_n$.

    (d) Prove that if $0 \le \ell < m$ and $\ell$ is relatively prime to $m$, then exactly $\varphi(n)$ among $\ell, \ell + m, \dots, \ell + m(n-1)$ are relatively prime to $mn$.

    (e) Conclude that $\varphi(nm) = \varphi(n)\varphi(m)$.

16. The *Liouville function* is defined by

$$\lambda(p_1^{k_1} \dots p_\ell^{k_\ell}) = (-1)^{k_1 + \cdots + k_\ell} \tag{13.2.63}$$

    (a) Prove that $\sum_{d|n} \lambda(d) = 1$ (respectively, 0) if $n$ is a perfect square (respectively, $n$ is not).

(b) Prove that

$$\sum_n \frac{\lambda_n}{n^s} = \frac{\zeta(2s)}{\zeta(s)} \qquad (13.2.64)$$

17. (a) Prove that $\sum_{n=0}^k (2m+1)(-1)^{k+m} = (k+1)$. (*Hint*: Induction in $k$, starting from $k=0$.)

(b) Prove that $d(p_1^{k_1}\dots p_\ell^{k_\ell}) = \prod_{j=1}^\ell (k_j+1)$.

(c) Prove that $\sum_{m|n} d(m^2)\lambda(\frac{n}{m}) = d(n)$ where $\lambda$ is the Liouville function of (13.2.63). (*Hint*: Use (a).)

(d) Prove that

$$\sum_{n=1}^\infty \frac{d(n^2)}{n^s} = \frac{\zeta(s)^3}{\zeta(2s)}$$

(*Hint*: (13.2.43) and (13.2.64).)

(e) Prove that $\sum_{m=0}^k (2m+1) = (k+1)^2$. (*Hint*: Induction in $k$.)

(f) Prove that $\sum_{m|n} d(m^2) = d(n)^2$.

(g) Prove that

$$\sum_{n=1}^\infty \frac{d(n)^2}{n^s} = \frac{\zeta(s)^4}{\zeta(2s)}$$

18. Define $d_k(n)$ inductively by $d_1(n) \equiv d(n)$,

$$d_k(n) = \sum_{m|n} d_{k-1}(m) d\left(\frac{n}{m}\right)$$

(a) Prove $d_k(n) = \#\{(n_1,\dots,n_{k+1}) \in \mathbb{Z}_+^{k+1} \mid n_1\dots n_{k+1} = n\}$.

(b) Prove that $\sum_{n=1}^\infty d_k(n) n^{-s} = \zeta(s)^{k+1}$.

19. The relation between Lambert series $(\sum_{n=1}^\infty a_n q^n/(1-q^n)$ and Taylor series $(\sum_{n=1}^\infty b_n q^n)$ is given by (13.1.59). Prove that

(a) $$\sum_{n=1}^\infty \frac{\mu(n)q^n}{1-q^n} = q$$

(b) $$\sum_{n=1}^\infty \frac{\phi(n)q^n}{1-q^n} = \frac{q}{(1-q)^2}$$

(c) $$\sum_{n=1}^\infty \frac{\lambda(n)q^n}{1-q^n} = \sum_{n=1}^\infty q^{n^2}$$

(d) $$\sum_{n=1}^\infty \frac{nq^n}{1-q^n} = \sum_{n=1}^\infty \sigma(n) q^n$$

## 13.3. The Riemann Zeta and Dirichlet $L$-Function

We've already defined the Riemann zeta function (Example 13.2.6), $\zeta(s)$, and Dirichlet $L$-function associated to a Dirichlet character $\chi$ (Example 13.2.9). In the next section, we will explore what $\chi$'s exist. We've already proven these functions have a meromorphic continuation to $\{s \mid \operatorname{Re} s > 0\}$ with the only possible pole at $s = 1$. For our applications, these regions of analyticity suffice, but it so easy with our results in Section 9.7 of Part 2A to get analyticity in $\mathbb{C} \setminus \{1\}$, we will. There are two other results we'll focus on in this section: the Euler factorization formulae we've already mentioned several times and the proof that $\zeta(s)$ is nonvanishing on $\{s \mid \operatorname{Re} s = 1\}$.

Because we'll take logs, regions with no zeros are important to the proofs of both the prime progression and prime number theorems. That $L_\chi(1) \neq 0$ for nonprincipal characters will be proven in the next section. We'll leave the functional equation for $\zeta$, which we won't need later, to the Notes and Problems.

We turn first to Euler factorization. A function $f$ on $\mathbb{Z}_+$ is called *completely multiplicative* if and only if $f(nm) = f(n)f(m)$ for all $n$ and $m$ and $f(1) = 1$. Examples include $f(n) \equiv 1$, Dirichlet characters, and the Liouville function of (13.2.63).

**Theorem 13.3.1** (Euler Factorization Formula). *Let $f$ be a bounded completely multiplicative function. Then for* $\operatorname{Re} s > 1$,

$$\sum_{n=1}^{\infty} f(n) n^{-s} = \prod_{p \in \mathcal{P}} (1 - f(p) p^{-s})^{-1} \tag{13.3.1}$$

*where, in the indicated region, the sum on the left and the product on the right are absolutely convergent.*

**Remark.** This is more often called the *Euler product formula*, a name we used instead for the product representation of $\sin(\pi x)$. The name is also sometimes restricted to the case $f(n) \equiv 1$ of the zeta function.

**Proof.** Since $f$ is bounded, for the Dirichlet series on the left, $\bar{\sigma} \leq 1$ and we have uniform and absolute convergence in $\operatorname{Re} s > 1$. Similarly, since $\sum_{p \in \mathcal{P}} |p^{-s}| \leq \sum_{n=1}^{\infty} |n^{-s}|$, the product on the right (before the inverse) is absolutely convergent on $\operatorname{Re} s > 1$.

Since $f(n^k) = f(n)^k$, $f$ bounded implies $|f(n)| \leq 1$. Thus, no factor in $\prod_{p \in \mathcal{P}} (1 - f(p) p^{-s})$ is zero, and the product is invertible.

Since $f$ is completely multiplicative,

$$f(p_1^{k_1} \dots p_\ell^{k_\ell})(p_1^{k_1} \dots p_\ell^{k_\ell})^{-s} = \prod_{j=1}^{\ell} [f(p_j)^{k_j} (p_j^{k_j})^{-s}] \tag{13.3.2}$$

and so, by prime factorization,

$$\begin{aligned}\sum_{n=1}^{\infty} f(n)n^{-s} &= 1+\sum_{\ell=1}^{\infty}\sum_{p_1<\cdots<p_\ell\in\mathcal{P}}\sum_{k_1,\dots,k_\ell=1}\prod_{j=1}^{\ell}(f(p_j)p_j)^{-sk_j} \\ &= \prod_{p\in\mathcal{P}}(1+f(p)p^{-s}+f(p)^2p^{-2s}+\dots) \\ &= \prod_{p\in\mathcal{P}}(1-f(p)p^{-s})^{-1} \qquad (13.3.3)\end{aligned}$$

where the rearrangement is justified by the absolute convergence. □

This implies, once again, that $\zeta$ and $L$-functions do not vanish in $\{s \mid \operatorname{Re} s > 1\}$. It can also be used to prove $\zeta(s)^{-1} = \sum_{n=1}^{\infty} \mu(n)n^{-s}$ (see Problem 6 of Section 13.2). It also implies that:

**Corollary 13.3.2.** *Let $\chi_m^{(0)}$ be the principal Dirichlet character for $\mathbb{Z}_m$. Let $p_1,\dots,p_\ell$ be the prime factors of $m$. Then*

$$L_{\chi_m^{(0)}}(s) = \zeta(s)\prod_{j=1}^{\ell}(1-p_j^{-s}) \qquad (13.3.4)$$

**Proof.** For any prime $p$, $\chi_m^{(0)}(p) = 1$ (respectively, 0) if $p \notin \{p_1,\dots,p_\ell\}$ (respectively, $p \in \{p_1,\dots,p_\ell\}$). Thus, (13.3.4) is immediate from comparing the Euler factorization of $\zeta(s)$ and $L_{\chi_m^{(0)}}(s)$. □

**Theorem 13.3.3.** *The Hurwitz zeta function, $\zeta(s;x)$, for $x \in \mathbb{H}_+$ given by* (13.2.19) *has a meromorphic continuation to $\mathbb{C}$ with $\zeta(s;x)-(s-1)^{-1}$ entire analytic.*

**Proof.** Let $g(y) = (y+x)^{-s}$ with $\operatorname{Re} s > 1$. Then (9.7.24) of Part 2A (Euler–Maclaurin expansion) with $k = 2m$ and with $n \to \infty$ says

$$\begin{aligned}\zeta(s;x) &= \tfrac{1}{2}(x+1)^{-s} + (s-1)^{-1}(x+1)^{1-s} \\ &\quad -\sum_{\ell=1}^{m}\frac{B_{2\ell}}{(2\ell)!}P_{2\ell-2}(s)(x+1)^{-s-2\ell+2} \qquad (13.3.5) \\ &\quad -\frac{1}{(2m)!}\int_1^{\infty} B_{2m}(\{y\})P_{2m}(s)(x+y)^{-s-2m}\,dy\end{aligned}$$

where $P_n(s) = \prod_{j=0}^{n}(-s-j)$. The terms before the integral are entire meromorphic in $s$ with a pole only at $s-1$ with residue 1, and the integral converges to an analytic function in $\{s \mid \operatorname{Re} s \geq 1-2m\}$. This allows a continuation to bigger and bigger regions whose union is $\mathbb{C}$. □

**Remark.** Problem 7 of Section 14.7 has another proof that $\zeta(s;x)$ has a meromorphic continuation.

**Corollary 13.3.4.** *Let $a$ be a function on $\mathbb{Z}_+$ with $a_{n+m} = a_n$ for all $n$ and a fixed $m$. Then $(s-1)f_a(s)$ has an analytic continuation from* $\operatorname{Re} s > 1$ *to all of $\mathbb{C}$. If*

$$c = \sum_{n=1}^{m} a_n \tag{13.3.6}$$

*then $f_a$ has a pole at $s = 1$ with residue $c$ if $c \neq 0$ and a removable singularity at $s = 1$ if $c = 0$.*

**Proof.** This is immediate from Theorem 13.3.3 and (13.2.23). □

We will eventually see that nonprincipal Dirichlet characters have $\sum_{n=1}^{m} \chi(n) = 0$, so such $L_\chi(s)$ are entire functions. Finally, we turn to proving $\zeta(s)$ has no zeros on $\{s \mid \operatorname{Re} s = 1\}$ and its consequences:

**Theorem 13.3.5.** *$\zeta(s)$ has no zeros in $\{s \mid \operatorname{Re} s \geq 1\}$.*

**Proof.** We already know $\zeta$ is nonvanishing on $\operatorname{Re} s > 1$ (and we'll reprove it below), so suppose for $a \in \mathbb{R}$, $a \neq 0$, that $\zeta(1+ia) = 0$. Then $\zeta(1-ia) = 0$ also since $\zeta(\bar{s}) = \overline{\zeta(s)}$. Define

$$f(s) = \zeta(s)^2 \zeta(s+ia)\zeta(s-ia), \qquad g(s) = \log(f(s)) \tag{13.3.7}$$

The proof will exploit the fact that, under our assumption, $\zeta$ has at least two zeros on the line $\operatorname{Re} s = 1$ and only one pole, so products can cancel poles and leave zeros left over. Indeed, tracking zeros and poles, $f(s)$ is regular at $s = 1$ and has zeros at $s = 1 \pm ia$, $1 \pm 2ia$.

By the Euler factorization formula and

$$-\log(1-y) = \sum_{k=1}^{\infty} \frac{y^k}{k} \tag{13.3.8}$$

for $|y| < 1$, we see, for $\operatorname{Re} s > 1$,

$$g(s) = \sum_{\substack{p \in \mathcal{P} \\ k=1,2,\dots}} k^{-1} p^{-sk} [2 + p^{-iak} + p^{iak}] \tag{13.3.9}$$

Since

$$[2 + p^{-iak} + p^{iak}] = 2 + 2\cos(ak\log(p)) \geq 0 \tag{13.3.10}$$

we conclude that Landau's theorem applies. We also see that $f(1) = \lim_{\varepsilon \downarrow 0} \exp(g(1+\varepsilon)) \geq 1$.

Since $f$ is nonvanishing on $[1, \infty)$ and analytic in a neighborhood of it, $g$ has no singularities on that set. Since Landau's theorem applies, the $\bar{\sigma}$ for $g$ has $\bar{\sigma} < 1$. Thus, $g$ is analytic in $\{s \mid \operatorname{Re} s > \bar{\sigma}\}$. But clearly, $g$ is not analytic at $1 \pm ia$, $1 \pm 2ia$ since $f$ vanishes there. Thus, there is a contradiction. □

The following exploits the main results of this section and will be critical in the proof of the prime number theorem in Section 13.5:

**Theorem 13.3.6.** *Let*

$$\Phi(s) = \sum_{p\in\mathcal{P}} \frac{\log p}{p^s} \tag{13.3.11}$$

*Then $\Phi(s) - 1/(s-1)$ has an analytic continuation to a neighborhood of $\{s \mid \operatorname{Re} s \geq 1\}$.*

**Proof.** We first claim that

$$\Phi(s) + \frac{\zeta'(s)}{\zeta(s)} \tag{13.3.12}$$

has an analytic continuation to $\{s \mid \operatorname{Re} s > \frac{1}{2}\}$. For by the Euler factorization theorem,

$$(13.3.12) = \sum_{p\in\mathcal{P}} \frac{\log p}{p^s} - \sum_{p\in\mathcal{P}} \frac{\log p}{p^s - 1} \tag{13.3.13}$$

$$= -\sum_{p\in\mathcal{P}} \frac{\log p}{p^s(p^s-1)} \tag{13.3.14}$$

is analytic there since $\sum_{p\in\mathcal{P}} p^{-2\sigma} \leq \sum_{n=1}^{p} n^{-2\sigma} < \infty$ if $\sigma > \frac{1}{2}$. (13.3.13) comes from $\log(\zeta(s)) = -\sum_{p\in\mathcal{P}} \log(1-p^{-s})$ and

$$\frac{d}{ds}\log(1-p^{-s}) = \frac{p^{-s}(\log p)}{1-p^{-s}} = \frac{\log p}{p^s - 1} \tag{13.3.15}$$

On the other hand, since $\zeta(s) - 1/(s-1)$ is analytic at $s = 1$, $\zeta'(s)/\zeta(s) + 1/(s-1)$ is analytic near $s = 1$ and, by Theorem 13.3.5, in a neighborhood of $\{s \mid \operatorname{Re} s \geq 1\}$. Subtracting this function from (13.3.12) yields the claimed result. $\square$

**Notes and Historical Remarks.** The word "completely" appears before "multiplicative" because a function on $\mathbb{Z}_+$ is called *multiplicative* if $f(nm) = f(n)f(m)$ for relatively prime $n$ and $m$. Examples of multiplicative functions which are not completely multiplicative are the Möbius function and the Euler totient function (see Problem 1). Euler factorization requires complete multiplicativity, although a multiplication function obeys

$$\sum_{n=1}^{\infty} f(n)n^{-s} = \prod_{p\in\mathcal{P}} (1 + f(p)p^{-s} + f(p^2)p^{-2s} + \dots) \tag{13.3.16}$$

The factor in the product is a special case of a *Bell series* (Problem 2)

$$f_p(x) = \sum_{n=0}^{\infty} f(p^n)x^n \tag{13.3.17}$$

Euler factorization is from Euler [**124**] in 1737 and was used to prove $\sum_{p\in\mathcal{P}} p^{-1} = \infty$, and indeed that the divergence should be roughly as $\log(\log n)$.

It was Riemann in 1859 [**327**] who first considered $\zeta(s)$ for complex $s$ and proved the meromorphic continuation in Theorem 13.3.3 (for $x = 1$). His first proof of this relied on the integral representation (9.6.121) of Part 2A, which also shows $\zeta(-2n) = 0$, $n = 1, 2, \dots$ (Problem 3). These are called the trivial zeros of $\zeta$ and are listed explicitly by Riemann. The formula also shows

$$\zeta(-n) = (-1)^n \frac{B_{n+1}}{n+1}, \qquad n = 0, 1, \dots \tag{13.3.18}$$

where $B_n$ are the Bernoulli numbers (Problem 3). Problem 5 of Section 14.7 will have another proof that $\zeta$ has a continuation and of (13.3.18).

Riemann's paper was a mere eight pages (!) and has evoked great interest and admiration. See [**225, 312**] for books on the mathematics of the Riemann zeta function and [**100, 337**] for books intended for nonmathematicians (the history chapters of [**100**] are readable and lively; the chapters on the math are not recommended for mathematicians). In particular, Edwards [**113**] discusses Riemann's paper in detail. In places, the paper is lacking details although Riemann claims to have proven them (an example is a Hadamard factorization for $\zeta$). In others, he makes what are clearly conjectures. Some of what he claims to have a proof of (but not presented) are open today! His notes were kept in the Göttingen library. Remarkably in 1932, Siegel [**350**] published a formula he found in these notes and then proved what is now known as the Riemann–Siegel formula.

Riemann also proved what is called the *functional equation for* $\zeta$:

$$\xi(s) = \xi(1-s) \tag{13.3.19}$$

where

$$\xi(s) = \pi^{-s/2} s(1-s)\Gamma\left(\frac{s}{2}\right)\zeta(s) \tag{13.3.20}$$

Notice that the $\Gamma(\frac{s}{2})$ poles are exactly cancelled by the trivial zeros of $\zeta$ and the poles of $\zeta$ at $s = 1$ and $\Gamma(\frac{s}{2})$ at $s = 0$ by the $s(1-s)$ factor. Thus, $\xi(s)$ is an entire function.

Notice that, by (13.3.19) and the fact that $\zeta$ is nonvanishing on $\{s \mid \operatorname{Re} s \geq 1\}$, we see $\xi(s)$ has all its zeros in the *critical strip* $\{s \mid 0 < \operatorname{Re} s < 1\}$. In Problem 4, the reader will prove $\xi(s)$ is of order 1 (in the sense of Section 9.10 of Part 2A and then, by the symmetry (13.3.19), that it has an infinity of zeros.

Much more is known than merely that $\zeta(s)$ has an infinity of zeros in the critical strip $\{s \mid 0 \leq \operatorname{Re} s \leq 1\}$. In his paper, Riemann claims that if $N(T)$ is the number of zeros in the strip with $0 < \operatorname{Im} s < T$, then as

$T \to \infty$, $N(T) = \frac{T}{2\pi}\log(\frac{T}{2\pi}) - \frac{T}{2\pi} + O(\log T)$. Riemann says this follows from a contour integral which he doesn't explain how to do (!); it was only proven in 1905 by von Mangoldt [**398**] using other methods. The zeros are important because Riemann found a wonderful formula for $\pi(x)$ in terms of the zeros—it is discussed, for example, in Edwards [**113**, Ch. 1]. It can be expressed most simply in terms of

$$\psi(x) = \sum_{n \le x} \Lambda(n)$$

where $\Lambda$ is given by (13.2.46). It says that for $x > 0$, $x \neq p^n$ for any prime, one has

$$\psi(x) = x - \lim_{T\to\infty} \sum_{|\operatorname{Im}\rho| \le T} \frac{x^\rho}{\rho} - \frac{\zeta'(0)}{\zeta(0)} - \tfrac{1}{2}\log(1 - x^{-2})$$

where the sum is over all zeros, $\rho$, of $\zeta$. See Ivić [**211**, Ch. 12].

Riemann's original proof of the functional equation relied on the symmetry of the Jacobi theta function given in (10.5.48) of Part 2A. In Problem 5, the reader will follow this proof. Clearly, the functional equation, proven initially in the critical strip, provides another proof of the meromorphic continuation.

There are, of course, other proofs of the functional equation. One, going back to Hurwitz [**201**] in the paper where he introduced his zeta function, relies on a contour integral representation; see Apostol [**18**, Ch. 12] and Problems 5 and 6 of Section 14.7. It leads to functional equations for $\zeta(s;x)$ and for Dirichlet $L$-functions. An interesting simple proof that doesn't rely on $\theta$ functions is in Ivić [**211**, §1.2].

Interestingly enough, in 1749 Euler [**127**] found some special cases of the functional equation, and conjectured the general real case. Specifically, he defined $\sum_{n=1}^\infty (-1)^{n+1} n^s$ for $s$ a positive integer by what we'd now call Abel summation $\lim_{x\uparrow 1} \sum_{n=1}^\infty (-1)^{n+1} x^n n^s$. Since (see (13.2.25)), $\sum_{n=1}^\infty (-1)^{n+1} n^{-s} = (1-2^{1-s})\zeta(s)$, the resulting formula implies one for $\zeta$, but abelian summation doesn't work without the $(-1)^{n+1}$. Euler proved the relation between the sum above and $\sum_{n=1}^\infty (-1)^{n+1} n^{-(s-1)}$ for $s = 1, \dots, 12$ and wrote etc. to indicate that he believed he had proven it for arbitrary positive integral $s$. One $s$-dependent sequence was $1, 0, -1, 0, \dots$, which he conjectures is interpolated by $\cos((s+1)\pi/2)$. He then made the conjecture of the functional equation for all rational $s$, checking it explicitly for $s = \frac{1}{2}$, where he replaced a $(-\frac{1}{2})!$ by $\Gamma(\frac{1}{2}) = \sqrt{\pi}$, for $s$ a negative integer and for $s = \frac{3}{2}$, where he relied on a numeric evaluation via Euler–Maclaurin summation! Euler also had what we would call a functional equation for the Dirichlet $L$-function $\sum_{n=1}^\infty (-1)^{n+1}(2n-1)^{-s}$. Euler also wrote down what derivatives of his relation would mean!

Both de la Vallée Poussin [**97**] and Hadamard [**171**] in their proofs of the prime number theorem realized it depends on the nonvanishing of $\zeta(s)$ on $\{s \mid \operatorname{Re} s \geq 1\}$. Hadamard's proof is fairly direct but not so simple. The standard proof for many years (and even now) relies on an inequality Mertens [**273**] proved in 1898 to show that $\zeta(s)$ is nonvanishing on $\operatorname{Re} s = 1$, namely,

$$3 + 4\cos\theta + \cos(2\theta) \geq 0 \tag{13.3.21}$$

which the reader proves and applies in Problem 6. It has the advantage of providing lower bounds on $|\varphi(1+it)|$. The proof we use that relies on looking at $\zeta(s)^2\zeta(s+ia)\zeta(s-ia)$ and using Landau's theorem is from Narasimhan [**285**].

De la Vallée Poussin's proof that $\zeta(1+it)$ has no zeros for $t$ real runs about fifteen pages! Both authors actually obtain explicit zero-free regions of the form no zeros in $\{s \mid \operatorname{Re} s > 1-\varepsilon(|\operatorname{Im} s|)\}$, where the function $\varepsilon(|\operatorname{Im} s|) \to 0$ as $|\operatorname{Im} s| \to \infty$ as $1/\log(|\operatorname{Im} s|)$.

Charles Jean Gustave Nicolas Baron de la Vallée Poussin (1866–1962) was a Belgian mathematician made a baron by the King on the 35th anniversary of his professorship at Louvain, where his father was a professor of geology and where he spent his entire career. Besides the work on the PNT, he made significant contributions to measure theory (see Section 2.5 of Part 3), Fourier series (see Section 3.5 of Part 1), and potential theory.

Both proofs used the functional equation for $\zeta$. It was Landau who found a proof only requiring analyticity of $\zeta$ in $\operatorname{Re} s > 0$.

### Problems

1. Prove that the following functions on $\mathbb{Z}_+$ are multiplicative but not completely multiplicative: (a) the Möbius function, (13.2.34); (b) the Euler totient function, $\varphi(m)$ = the number of invertible elements in $\mathbb{Z}_m$; (c) $d(n)$, the number of divisors of $n$; (d) $\sigma_k(n)$, the sum of $k$-th powers of all divisors of $n$.

2. For each of the functions in Problem 1, compute the Bell series, (13.3.17), and so the analog of Euler factorization of $\sum_{n=1}^\infty f(n)n^{-s}$. Confirm directly from this that (13.2.34) holds.

3. (a) Suppose $f(x)$ is bounded and measurable on $[1,\infty)$ and obeys $|f(x)| \leq C_N(1+|x|)^{-N}$ for all $N$. Prove that $F(z) = \int_1^\infty x^z f(x)\,dx$ is an entire function of $z$.

   (b) Suppose $f(x)$ is bounded and measurable on $[0,1]$ and obeys $|f(x)| \leq C_N x^N$ for some fixed $N$. Prove that $F(z) = \int_0^1 x^z f(x)\,dx$ is analytic in $\{z \mid \operatorname{Re} z > -N-1\}$.

(c) Prove $F(z) = \int_0^1 x^{z+k}\, dz$ defined originally in $\{z \mid \operatorname{Re} z > -k-1\}$ has a meromorphic continuation to $\mathbb{C} \setminus \{-k-1\}$ with a pole at $-k-1$ with residue 1.

(d) Using the representation (9.6.121) of Part 2A, prove that $F(z) = \zeta(z)\Gamma(z)$, originally defined in $\{z \mid \operatorname{Re} z > 1\}$, has an analytic continuation to $\mathbb{C} \setminus \{1, 0, -1, -2, \dots\}$ with removable singularities or poles at the excluded points and with

$$\lim_{z \to 1-n} (z - 1 + n)\zeta(z)\Gamma(z) = \frac{B_n}{n!}, \qquad n = 0, 1, 2, \dots$$

where $B_n$ are the Bernoulli numbers given by (3.1.48) of Part 2A.

(e) Prove (13.3.18). (*Hint*: Use (9.6.24) of Part 2A.)

**Remarks.** 1. This is from Riemann's paper [**327**].

2. See Problem 5 of Section 14.7 for another way to go from (9.6.121) of Part 2A to an analytic continuation of $\zeta$ and to (d) above.

4. Let $\xi(x)$ be given by (13.3.19). This problem will show that $\xi(s)$ has order 1 and use that plus the functional equation to conclude infinitely many zeros.

   (a) Prove that in the region $\{s \mid \operatorname{Re} s \geq \frac{1}{2}\}$, we have $|(s-1)\zeta(s)| \leq C|s|^2$. (*Hint*: Use the idea behind (13.3.5) but with the first-order Euler–Maclaurin series.)

   (b) Prove that in the region $\{s \mid \operatorname{Re} s \geq \frac{1}{2}\}$, $\log|\xi(s)| \leq C|s|(\log|s| + 1)$. (*Hint*: Stirling's formula.)

   (c) Prove that $\xi(s)$ has order 1.

   (d) Prove that there is an entire function $G$ of order $\frac{1}{2}$ so that $\xi(s) = G((s - \frac{1}{2})^2)$.

   (e) Conclude $\xi$ has an infinite number of zeros. (*Hint*: Theorem 9.10.11 of Part 2A.)

5. This problem will lead you through Riemann's proof of the functional equation (13.3.19), which we write in the form: $\Xi(s) = \Xi(1-s)$ where $\Xi(s) = \pi^{-s/2}\Gamma(\frac{s}{2})\zeta(s)$.

   (a) Prove that with $\theta_0$ given by (10.5.47) of Part 2A, we have for $\operatorname{Re} s > 1$ that

$$\Xi(s) = \int_0^\infty \psi(x) x^{\frac{s}{2}-1}\, dx$$

   where $\psi = \frac{1}{2}\theta_0 - 1$ and $\theta_0$ is given by (10.5.47) of Part 2A. (*Hint*: Prove first that $n^{-s}\pi^{-s/2}\Gamma(\frac{s}{2}) = \int_0^\infty e^{-xn^2\pi} x^{\frac{s}{2}-1}\, dx$.)

(b) Using (10.5.48) of Part 2A, prove that

$$\Xi(s) = \int_1^\infty \psi(x)[x^{s/2} + x^{(1-s)/2}]\,\frac{dx}{x} - \frac{1}{s(1-s)}$$

(*Hint*: Use (10.5.48) of Part 2A to turn $\int_0^1$ into an integral $\int_1^\infty$.)

(c) Using the decay of $\psi(x)$, prove that the integral defines an entire function invariant under $s$ to $1-s$ and conclude that $\Xi(s)$ has a continuation to a meromorphic function with poles at 0 and 1 invariant under $s$ to $1-s$.

6. This problem will lead you through Hadamard's original proof that $\zeta(s)$ has no zeros on $\{s \mid \operatorname{Re} s = 1\}$.

(a) Prove (13.3.21). (*Hint*: Look at $(1+\cos\theta)^2$.)

(b) Prove that Euler's formula implies $\log(\zeta(s)) = \sum_{n=1}^\infty a_n s^{-n}$ for $\operatorname{Re} s > 1$ with $a_n \geq 0$.

(c) Fix $t \in \mathbb{R}$. Let $\phi(s) = \zeta^3(s)\zeta^4(s+it)\zeta(s+2it)$. Prove that for $s$ real with $s > 1$,

$$\log|\phi(s)| = \sum_{n=1}^\infty a_n n^{-s}(3 + 4\cos(t\log n) + \cos(2t\log n)$$

(d) Conclude for any $t \in \mathbb{R}$, $t \neq 0$,

$$\lim_{s\downarrow 1}|\zeta(s)^3\zeta^4(s+it)\zeta(s+2it)| \geq 1$$

(e) Show $\zeta$ cannot have a zero at $1+it$ for $t \neq 0$.

## 13.4. Dirichlet's Prime Progression Theorem

> The total number of Dirichlet's publications is not large: jewels are not weighed on a grocery scale.
>
> —*C. F. Gauss*

Our goal in this section is to prove Theorem 13.0.5. We'll first prove the existence and properties of Dirichlet characters, then prove the critical fact that for nonprincipal characters, $L_\chi(1) \neq 0$, and finally put it all together.

Recall we care about functions modulo $m$, that is, on $\mathbb{Z}_m = \mathbb{Z}/m\mathbb{Z}$. We are especially interested in $\mathbb{Z}_m^\times$, the group of invertible elements in the field $\mathbb{Z}_m$; equivalently, those $k$ relatively prime to $m$, so $\#(\mathbb{Z}_m^\times) = \varphi(m)$. A *Dirichlet character*, $\chi$, vanishes on divisors of $m$ other than 1 and is nonzero exactly on $\mathbb{Z}_m^\times$ and obeys

$$\chi(k\ell) = \chi(k)\chi(\ell) \tag{13.4.1}$$

for $k \in \mathbb{Z}_m^\times$. $\chi(k) \equiv 1$ (and $\chi(k) = 0$ if $k$ and $m$ are not relatively prime) is the principal character and all other characters are nonprincipal. We'll use $\widehat{\mathbb{Z}_m^\times}$ for the set of all characters. The key fact is that there are $\varphi(m)$ distinct Dirichlet characters and they obey the *orthogonality relations*:

**Theorem 13.4.1.** (a) *There are $\varphi(m)$ distinct Dirichlet characters.*

(b) (First orthogonality relation) *For $\chi_1, \chi_2 \in \widehat{\mathbb{Z}_m^\times}$,*

$$\frac{1}{\varphi(m)} \sum_{k \in \mathbb{Z}_m^\times} \overline{\chi_1(k)}\, \chi_2(k) = \delta_{\chi_1, \chi_2} \tag{13.4.2}$$

(c) (Second orthogonality relation) *For $k, \ell \in \mathbb{Z}_m^\times$,*

$$\frac{1}{\varphi(m)} \sum_{\chi \in \widehat{\mathbb{Z}_m^\times}} \overline{\chi(k)}\, \chi(\ell) = \delta_{k\ell} \tag{13.4.3}$$

**Remarks.** 1. In particular, if $\chi$ is a nonprincipal character,

$$\sum_{k \in \mathbb{Z}^m} \chi(k) = 0 \tag{13.4.4}$$

proving $L_\chi(s)$ is nonsingular at $s = 1$, and thus, an entire function.

2. The proof depends on elementary facts about unitary matrices; see Section 1.3 of Part 2A.

**Proof.** Consider $\ell^2(\mathbb{Z}_m^\times)$, the functions on $\mathbb{Z}_m^\times$ with the inner product,

$$\langle f, g \rangle = \frac{1}{\varphi(m)} \sum_{k \in \mathbb{Z}_m^\times} \overline{f(k)}\, g(k) \tag{13.4.5}$$

Define unitary maps, $\{U_k\}_{k \in \mathbb{Z}_m^\times}$, on $\ell^2(\mathbb{Z}_m^\times)$ by

$$(U_k f)(\ell) = f(k\ell) \tag{13.4.6}$$

Since $\mathbb{Z}_m^\times$ is an abelian group, $U_k U_\ell = U_\ell U_k$, so (by Theorem 1.3.5 of Part 2A), there exists an orthonormal basis, $\{\psi_j\}_{j=1}^{\varphi(m)}$, of common eigenvectors.

If $\chi_j(k)$ is the eigenvalue, we have

$$\psi_j(k\ell) = \chi_j(k) \psi_j(\ell) \tag{13.4.7}$$

Since $U$ is unitary, $|\chi_j(k)| = 1$, and thus, $|\psi_j(k)| = |\psi_j(1)| = 1$ since $\langle \psi_j, \psi_j \rangle = 1$. We can thus normalize $\psi_j$ so $\psi_j(1) = 1$. Taking $\ell = 1$ in (13.4.6), we see that

$$\psi_j(k) = \chi_j(k) \tag{13.4.8}$$

The fact that the $\psi_j$'s are orthonormal is thus (13.4.2). The orthogonality implies $\psi_j \neq \psi_k$ for $j \neq k$, so the $\chi$'s are distinct and thus, there are $\varphi(m)$ distinct $\chi$'s.

That the $\chi_j$ are an orthonormal basis for functions implies

$$\delta_{k\ell} = \sum_{j=1}^{\varphi(m)} \langle \chi_j, \delta_k \rangle \chi_j(\ell)$$
$$= \frac{1}{\varphi(m)} \sum \overline{\chi_j(k)}\, \chi_j(\ell) \tag{13.4.9}$$

which is (13.4.3). □

Next, we are heading towards a proof that for any nonprincipal character that $L_\chi(1) \neq 0$. We'll do this by considering

$$\alpha(s) = \prod_{\chi \in \widehat{\mathbb{Z}_m^\times}} L_\chi(s) \tag{13.4.10}$$

To apply Landau's theorem, both parts of the following are critical:

**Proposition 13.4.2.** *The Dirichlet series associated to the function, $\alpha$, of (13.4.10) has nonnegative terms and diverges at $s = 1/\varphi(m)$.*

**Proof.** Let $\beta(s) = \log(\alpha(s))$. By Theorem 13.3.1 and $-\log(1-x) = \sum_{k=1}^\infty x^k/k$,

$$\beta(s) = \sum_{\chi \in \mathbb{Z}_m^\times} \sum_{k=1}^\infty \sum_{p \in \mathcal{P}} \chi(p)^k \frac{p^{-ks}}{k} \tag{13.4.11}$$

Since the sums converge absolutely if $s > 1$, we can rearrange freely. Since $\chi(p)^k = \chi(p^k)$ and $\sum_{\chi \in \mathbb{Z}_m^\times} \chi(p^k) = \varphi(m)\tilde{\delta}_{p^k,1}$ (where $\tilde{\delta}$ is, mod $m$, a Kronecker delta), we get

$$\beta(s) = \sum_{p \in \mathcal{P}} \sum_{k \text{ s.t. } p^k \equiv 1, \text{ mod } m} \frac{\varphi(m) p^{-ks}}{k} \tag{13.4.12}$$

Clearly, the coefficients of this Dirichlet series, $\tilde{\delta}_{p^k,1}\varphi(m)/k$, are nonnegative. Moreover (Problem 1), for any $p$ not dividing $m$, $p^{\varphi(m)} \equiv 1 \pmod m$, so $p^{-s\varphi(m)}$ terms occur for all $p$, with $p \nmid m$.

Since $\alpha(s) = 1 + \beta(s) + \beta(s)^2/2! + \dots$, $\alpha$ also has nonnegative Dirichlet coefficients and they are larger than those of $\beta$. Thus,

$$\alpha\left(\frac{1}{\varphi(m)}\right) \geq \sum_{\substack{p \in \mathcal{P} \\ p \nmid m}} p^{-1} = \infty \tag{13.4.13}$$

by (13.0.15) since $\sum_{p \mid m} p^{-1} < \infty$. □

**Theorem 13.4.3.** *For every nonprincipal character,*

$$L_\chi(1) \neq 0 \tag{13.4.14}$$

**Remark.** One can show (Problem 2) that for every character and $t \neq 0$, $t \in \mathbb{R}$, $L_\chi(1+it) \neq 0$ by mimicking the proof of Theorem 13.3.5.

**Proof.** Each $L_\chi$ is either entire or analytic on $\mathbb{C} \setminus \{1\}$ with a simple pole at $s=1$. Thus, $\alpha(s)$ is either entire or has a pole at $s=1$. If some $L_\chi(1)=0$, since only one $L_{\chi^{(0)}}$ has a pole, the product would be regular at $s=1$. But then, $\alpha$ is entire, so by Landau's theorem (Theorem 13.2.15), the Dirichlet series for $\alpha$ is convergent at all $s$.

Since we proved the series is divergent at $s = 1/\varphi(m)$, we see there must be a pole at $s=1$, that is, no $L_\chi(1)=0$. □

Let $\mathcal{P}_k^{(m)} = \{p \in \mathcal{P} \mid p \equiv k \bmod m\}$. The following implies that $\mathcal{P}_k^{(m)}$ is infinite, and so proves Theorem 13.0.5:

**Theorem 13.4.4.** *Let $k \in \mathbb{Z}_m^\times$. Then*

$$\lim_{s \downarrow 1} \Bigl[ \sum_{p \in \mathcal{P}_k^{(m)}} p^{-s} \Bigr] = \infty \tag{13.4.15}$$

*In fact,*

$$\sup_{s \in (1,2]} \Bigl| \sum_{p \in \mathcal{P}_k^{(m)}} p^{-s} - \varphi(m)^{-1} \log(s-1) \Bigr| < \infty \tag{13.4.16}$$

**Proof.** Fix $m$. Let $f_k(s)$ be the function whose limit is in (13.4.15), and for any character $\chi$, let

$$g_\chi(s) = \sum_{p \in \mathcal{P}} \chi(p) p^{-s} \tag{13.4.17}$$

By (13.4.3),

$$f_k(s) = \frac{1}{\varphi(m)} \sum_{\chi \in \widehat{\mathbb{Z}_m^\times}} \overline{\chi(k)}\, g_\chi(s) \tag{13.4.18}$$

Let $\chi^{(0)}$ be the principal character. Since (13.0.15) holds, since $\chi^{(0)}(p)=1$ (respectively, 0) for $p \nmid m$ (respectively, $p \mid m$), and since only finitely many $p \mid m$, we have

$$\sup_{s \in (1,2]} |\chi^{(0)}(k) g_{\chi^{(0)}}(s) - \log(s-1)| < \infty \tag{13.4.19}$$

On the other hand, by Theorem 13.3.1, with $f = \chi$,

$$\log(L_\chi(s)) - g_\chi(s) = \sum_{p \in \mathcal{P}} \Bigl[ \sum_{k=2}^\infty \frac{\chi(p)^k p^{-ks}}{k} \Bigr] \tag{13.4.20}$$

defines an analytic function in $\{s \mid \operatorname{Re} s > \frac{1}{2}\}$ since $\sum_{p\in\mathcal{P}} p^{-2\sigma} < \infty$ for $\sigma > \frac{1}{2}$. Thus, by (13.4.14),

$$\sup_{s\in(1,2]} |\overline{\chi(k)}\, g_\chi(s)| < \infty \tag{13.4.21}$$

for any nonprincipal character. (13.4.18), (13.4.19), and (13.4.20) prove (13.4.16). □

In Problem 3, the reader will prove an analog of Theorem 13.3.6 for $\mathcal{P}_k^{(m)}$.

**Notes and Historical Remarks.** Theorem 13.0.5 was proven by Dirichlet [**104**] in 1837. This landmark paper not only led to an extensive theory of other number theoretic and geometric $L$-functions, but it was a precursor to the theory of representations of groups, developed initially by Frobenius and his student, Schur. In 1785, Legendre stated what we now call Dirichlet's prime progression theorem, but he gave no proof!

The proof we give of Theorem 13.4.1 is group representation theoretic. For any finite group, one defines the *regular representation*, $(U_k f)(\ell) = f(k^{-1}\ell)$ on $L^2(G)$. Its direct sum decomposition yields any irreducible representation of dimension $d$, $d$-times. When $G$ is abelian, $d = 1$ for all representations. For textbook treatments of representations of finite groups, see Serre [**345**] and Simon [**354**].

The set of characters is a group, $\widehat{G}$, since $\chi_1\chi_2$, the pointwise product, is a character with the principal character as identity and $\chi_1^{-1} = \overline{\chi}_1$. In Problem 4, the reader will show for finite abelian groups that $\widehat{G} \cong G$ (the proof will assume the fundamental structure theorem for finite abelian groups).

For $\mathbb{Z}_m^\times$, one has (Problem 5) that $\mathbb{Z}_{mn}^\times = \mathbb{Z}_m^\times \times \mathbb{Z}_n^\times$ if $m$ and $n$ are relatively prime—this implies (13.2.41).

Clearly, the key to the proof of the prime progression theorem is the fact that $L_\chi(1) \neq 0$ for nonprincipal characters. It has a variety of proofs, some depending on algebraic number theory. Others avoid Landau's theorem by separately treating real and complex characters, where real characters are those with $\chi(k) = \pm 1$ for all $k \in \mathbb{Z}_m^\times$ and complex characters are those with some $\chi(k) \notin \mathbb{R}$. If $\chi$ is complex, $\chi \neq \overline{\chi}$. Since $\overline{L_\chi(s)} = L_{\overline{\chi}}(\bar{s})$, if $L_\chi(1)$ were 0, then $\lim_{s\downarrow 1} \prod_{\chi\in\widehat{\mathbb{Z}_m^\times}} L_\chi(s) = 0$ since there is one pole and two zeros. This violates the positivity of the Dirichlet series. The proofs that $L_\chi(1) \neq 0$ for real characters are often quite involved—see Apostol [**18**] and Stein–Shakarchi [**365**] for two approaches.

Dirichlet's proof that $L_\chi(1) \neq 0$ relied on the theory of quadratic field extensions of $\mathbb{Q}$. There are many other proofs relying on algebraic number theory. Chapter 25 of [**227**] has an exposition that relies on Kummer's 1847 formula for the zeta function of the field obtained by adding a primitive $m$-th root of unity to $\mathbb{Q}$.

In Problems 5–7, the reader will explore when $\mathbb{Z}_m^\times$ has nonprincipal real characters and, in particular, will prove all characters are real if and only if $m$ is a divisor of 24 (i.e., for $m = 2, 3, 4, 6, 8, 24$).

Johann Peter Gustav Lejeune Dirichlet (1805–59) was a German mathematician of Belgian extraction (Le jeune de Richelet means the young one from Richelet in French). Because he was concerned about the poor level of mathematics education in Germany, he studied under Poisson and Fourier in Paris. Because he did not have a German degree, there were difficulties in appointing him to a professorship, solved by arranging an honorary degree for him. In turn, while a professor at Berlin, his students included Eisenstein, Kronecker, and Lipschitz, and essentially also Riemann. He married the sister of the composer Felix Mendelssohn and was very close friends with Jacobi. He died of heart disease at age 54, shortly after going to Göttingen as Gauss' successor.

Besides his work on primes in arithmetic progressions, Dirichlet did fundamental work on convergence of Fourier series (see Section 3.5 of Part 1) and in algebraic number theory.

**Problems**

1. (a) Prove Euler's theorem that if $k \in \mathbb{Z}_m^\times$, then $k^{\varphi(m)} \equiv 1 \pmod{m}$.

   (b) Prove that if $p$ is prime and $p \nmid m$, then $p^{\varphi(m)} \equiv 1 \pmod{m}$.

2. (a) Suppose $\alpha(1 + it_0) = 0$ for some $t_0 \in (0, \infty)$ where $\alpha$ is given by (13.4.10). Show $\gamma(s) = \alpha(s + it_0)\alpha(s)^2\alpha(s - it_0)$ is analytic in a neighborhood of $\{1 + it \mid t \in \mathbb{R}\}$ including at $t = 0$.

   (b) Prove $\log \gamma$ has a positive Dirichlet series.

   (c) Prove $\gamma$ has a zero at $s = 1 + it_0$ and deduce a contradiction.

   (d) Prove all $L_\chi$ are nonvanishing on $\{1 + it \mid t \in \mathbb{R}\}$.

3. (a) Prove that for any character, $\chi$,

$$\frac{L'_\chi(s)}{L_\chi(s)} + \sum_{p \in \mathcal{P}} \frac{\chi(p) \log p}{p^s}$$

   is analytic in a neighborhood of $\{s \mid \operatorname{Re} s \geq 1\}$.

   (b) Prove that for any $m$ and $k \in \mathbb{Z}_m^\times$,

$$\frac{1}{\varphi(m)} \sum_{\chi \in \widehat{\mathbb{Z}_m^\times}} \overline{\chi(k)} \frac{L'_\chi(s)}{L_\chi(s)} + \sum_{p \in \mathcal{P}_k^{(m)}} \frac{\log p}{p^s}$$

   is analytic in a neighborhood of $\{s \mid \operatorname{Re} s \geq 1\}$.

(c) Prove that for any $m$ and $k \in \mathbb{Z}_m^\times$,

$$\sum_{p \in \mathcal{P}_k^{(m)}} \frac{\log p}{p^s} - \frac{1}{\varphi(m)} \frac{1}{s-1}$$

has an analytic continuation to a neighborhood of $\{s \mid \operatorname{Re} s \geq 1\}$.

**Remark.** This problem requires the result proven in Problem 2.

4. This problem will assume the theorem (for a proof, see [**354**, §4.1]) that any finite abelian group, $G$, is isomorphic to a product of cyclic groups $\mathbb{Z}_m$ (in fact, it is a product of such groups with $m = p^k$ for some prime $p$). For a general finite abelian group, $G$, a character is a group homomorphism to $\partial\mathbb{D}$ as a multiplicative group. It will also use the fact that the set of characters is a group under the product $\chi_1\chi_2(g) \equiv \chi_1(g)\chi_2(g)$. Let $\widehat{G}$ denote the group of all characters on $G$.

   (a) Prove $\widehat{G_1 \times G_2} = \widehat{G}_1 \times \widehat{G}_2$ under $(\chi_1, \chi_2)(g_1, g_2) = \chi_1(g_1)\chi_2(g_2)$.

   (b) If $\mathbb{Z}_m$ is the cyclic group, with $[0]$ the additive identity, and $\chi$ is a character, prove $\chi([0])$ determines $\chi$ and that $\chi([0])$ is an arbitrary $m$-th root of 1. Conclude that $\widehat{\mathbb{Z}}_m \cong \mathbb{Z}_m$.

   (c) Prove $\widehat{G} = G$ for any finite abelian group.

5. The main point of this problem is to prove that if $m$ and $n$ are relatively prime, then $\mathbb{Z}_{mn}^\times$ is isomorphic as a group to $\mathbb{Z}_m^\times \times \mathbb{Z}_n^\times$, a fact we'll need in the analysis of real characters in Problem 7 below. A key will be the result that there exist $a, b \in \mathbb{Z}$ so

$$am + bn = 1 \qquad (13.4.22)$$

   (see Section 1.3 of Part 2A). In essence, this problem proves a variant of the *Chinese remainder theorem.*

   (a) For $k, \ell \in \mathbb{Z}_m \times \mathbb{Z}_n$, define $f(k, \ell) = bnk + am\ell$. Prove that $f$ is a well-defined additive group morphism of $\mathbb{Z}_m \times \mathbb{Z}_n$ into $\mathbb{Z}_{mn}$.

   (b) Prove if $f(k, \ell) = 0$, then $m \mid k$ and $n \mid \ell$, so $\ker(f) = \{0\} \times \{0\}$ and so $f$ is an additive group isomorphism. (*Hint*: Show $m$ is relatively prime to $bn$ and $n$ to $am$.)

   (c) Prove $f(k, 1)f(1, \ell) \equiv f(k, \ell) \pmod{mn}$ and $f(k_1k_2, 1) = f(k_1, 1)f(k_2, 1)$, and $f(1, \ell_1\ell_2) = f(1, \ell_1)f(1, \ell_2)$, and conclude that $f$ is a ring isomorphism of $\mathbb{Z}_m \times \mathbb{Z}_n$ and $\mathbb{Z}_{mn}$. (*Hint*: Repeatedly use (13.4.22).)

   (d) Prove $f$ is an isomorphism of the multiplicative groups $\mathbb{Z}_m^\times \times \mathbb{Z}_n^\times$ and $\mathbb{Z}_{mn}^\times$. (*Hint*: These groups are the invertible elements of the rings.)

6. Let $p$ be a prime.

   (a) Prove $\mathbb{Z}_p^\times$ is isomorphic to the cyclic group $\mathbb{Z}_{p-1}$. (*Hint*: $\mathbb{Z}_p$ is a field, so prove $x^k - 1$ has at most $k$ solutions for any $k$.)

   (b) Prove $\#(\mathbb{Z}_{p^\ell}^\times) = p^{\ell-1}(p-1)$.

   (c) Prove again Euler's formula that $\varphi(p_1^{k_1} \dots p_\ell^{k_\ell}) = \prod p_j^{k_j - 1}(p_j - 1)$. (*Hint*: You also need Problem 5.)

   (d) Prove that $\{1, 1+p^{k-1}, 1+2p^{k-1}, \dots, 1+(p-1)p^{k-1}\}$ is a subgroup of $\mathbb{Z}_{p^k}^\times$ and the quotient is $\mathbb{Z}_{p^{k-1}}^\times$.

   (e) Prove that if $p, k, \ell$ are such that there is $x$ in $\mathbb{Z}_{p^{k-1}}^\times$ of order $\ell$, then there are $y$ in $\mathbb{Z}_{p^k}^\times$ of order at least $\ell$.

7. (a) Let $\chi$ be a character on $\mathbb{Z}_m^\times$. Prove $\chi$ is real if and only if $\chi^2 = \chi^{(0)}$ where $\chi^{(0)}$ is the principal character and $\chi^2$ is the group product in $\widehat{\mathbb{Z}_m^\times}$, that is, pointwise multiplication.

   (b) Prove that for $m \geq 3$, $\varphi(m)$ is even and use this to prove that every $\mathbb{Z}_m^\times$ for $m \geq 3$ has elements of order 2 and so, since $\widehat{\mathbb{Z}_m^\times} \cong \mathbb{Z}_m^\times$, there are some real characters.

   (c) Let $G$ be an abelian group in which every element other than 1 has order 2. Prove $o(G) = 2^k$ for some $k$, and conclude if $\mathbb{Z}_m^\times$ has this property, and $m = p_1^{k_1} \dots p_\ell^{k_\ell}$, then $k_j = 1$ unless $p_j = 2$.

   (d) If $\mathbb{Z}_m^\times$ is a group with every element of order 2, prove $m = 2^b$ or $m = 2^b 3$ for some $b$. (*Hint*: Use Problem 6(b).)

   (e) Prove $\mathbb{Z}_8^\times = \mathbb{Z}_2 \times \mathbb{Z}_2$ and $\mathbb{Z}_{16}^\times = \mathbb{Z}_4 \times \mathbb{Z}_2$. Conclude $\mathbb{Z}_m^\times$ has all real characters if and only if $m = 2^b$ or $m = 2^b 3$ with $b \leq 3$, that is, if and only if $m \mid 24$.

## 13.5. The Prime Number Theorem

> Mathematicians have tried in vain to this day to discover some order in the sequence of prime numbers, and we have reason to believe that it is a mystery into which the human mind will never penetrate. To convince ourselves, we have only to cast a glance at tables of primes, which some have taken the trouble to compute beyond a hundred thousand, and we should perceive at once that there reigns neither order nor rule.
>
> —*L. Euler* (translated by Pólya) quoted in [**415**]

Here we'll prove Theorem 13.0.6, one of the most famous theorems in mathematics. Our proof is based on ideas of Newman. As we'll explain in the Notes, there were important precursors in Wiener and Ingham and elucidation of Newman's ideas by Korevaar and Zagier. The key is the following Tauberian theorem of Newman that we'll prove at the conclusion of this section.

**Theorem 13.5.1** (Newman's Tauberian Theorem). *Let $f$ be a bounded measurable function on $(0,\infty)$. For $\operatorname{Re} z > 0$, define*

$$g(z) = \int_0^\infty e^{-zt} f(t)\, dt \tag{13.5.1}$$

*Suppose $g$ has an analytic continuation to a neighborhood of $\{z \mid \operatorname{Re} z \geq 0\}$. Then $\lim_{T\to\infty} \int_0^T f(t)\, dt$ exists and*

$$\lim_{T\to\infty} \int_0^T f(t)\, dt = g(0) \tag{13.5.2}$$

Changing variables to a form closer to a Dirichlet series, let $u = e^t$, $s = z+1$, so $G(s) = g(z) = g(s-1)$ and $F(u) = f(t) = f(\log u)$. Then $e^{-tz}\, dt = u^{-z}(\frac{du}{u}) = u^{-s}\, du$, so

**Corollary 13.5.2.** *Let $F$ be a bounded measurable function on $(1,\infty)$. For $\operatorname{Re} s > 1$, define*

$$G(s) = \int_1^\infty F(u) u^{-s}\, du \tag{13.5.3}$$

*Suppose $G$ has an analytic continuation to a neighborhood of $\{s \mid \operatorname{Re} s \geq 1\}$. Then $\lim_{U\to\infty} \int_1^U F(u) u^{-1}\, du$ exists and*

$$\lim_{U\to\infty} \int_1^U F(u) u^{-1}\, du = G(1) \tag{13.5.4}$$

We will apply this to an $F$ related to the *Chebyshev $\theta$-function* (to be distinguished from the Jacobi theta function!) defined for $x > 0$ by

$$\theta(x) = \sum_{\substack{p\leq x \\ p\in\mathcal{P}}} \log p \tag{13.5.5}$$

Boundedness of $F$ will come from:

**Proposition 13.5.3** (Chebyshev's Lemma). *We have that*

$$\theta(x) \leq (4\log 2) x \tag{13.5.6}$$

**Proof.** Let $p \in \mathcal{P}$, $n < p \leq 2n$. Then $p \mid \binom{2n}{n}$, since $p$ is a factor of $2n!$ that cannot be cancelled by $(n!)^2$. Since $\sum_{j=0}^{2n} \binom{2n}{j} = 2^{2n}$, we have

$$2^{2n} \geq \prod_{\substack{p\in\mathcal{P} \\ n<p\leq 2n}} p = \exp(\theta(2n) - \theta(n))$$

so

$$\theta(2n) - \theta(n) \leq 2n \log 2 \tag{13.5.7}$$

which implies

$$\theta(2^k) - \theta(2^{k-1}) \le 2^k \log 2 \tag{13.5.8}$$

Using $\theta(1) = 0$ and summing $\theta(2^{k-j}) - \theta(2^{k-j+1})$, $j = 0, \dots, k-1$, implies that

$$\begin{aligned} \theta(2^k) &\le (2^k + 2^{k-1} + \dots) \log 2 \\ &\le 2^{k+1} \log 2 \end{aligned} \tag{13.5.9}$$

Now let $x \in \mathbb{R}$, $x \ge 1$. Pick an integer $k \ge 0$ so that $2^k \le x \le 2^{k+1}$. Then

$$\theta(x) \le \theta(2^{k+1}) \le 4\, 2^k \log 2 \le x(4 \log 2)$$

which is (13.5.6). □

**Proposition 13.5.4.** *Let*

$$\pi(x) = \#\{p \in \mathcal{P} \mid p \le x\} \tag{13.5.10}$$

*Then*

$$\lim_{x\to\infty} \frac{\pi(x)}{x/\log x} = 1 \Leftrightarrow \lim_{x\to\infty} \frac{\theta(x)}{x} = 1 \tag{13.5.11}$$

**Proof.** If $p \le x$, then $\log p \le \log x$, so

$$\theta(x) \le \pi(x) \log x \tag{13.5.12}$$

On the other hand, if $\alpha < 1$ and $p \ge x^\alpha$, then $\log p \ge \alpha \log x$, so

$$\pi(x) \log x \le \pi(x^\alpha) \log x + \alpha^{-1} \theta(x) \tag{13.5.13}$$

Since $\pi(x^\alpha) \le x^\alpha$, we see that $\frac{\pi(x^\alpha)}{x} \log x \to 0$. Thus, if $x_j$ is a sequence for which $\lim_{j\to\infty} \theta(x_j)/x_j$ and $\lim_{j\to\infty} \pi(x_j) \log x_j / x_j$ exist (either can be 0 or $\infty$), then

$$\lim_{j\to\infty} \frac{\pi(x_j) \log x_j}{x} \le \alpha^{-1} \lim_{j\to\infty} \frac{\theta(x_j)}{x_j} \tag{13.5.14}$$

Since $\alpha$ is arbitrary in $(0,1)$, (13.5.12) and (13.5.14) imply (13.5.11). □

**Proposition 13.5.5.** *If*

$$\lim_{U\to\infty} \int_1^U \left[ \frac{\theta(u) - u}{u^2} \right] du \tag{13.5.15}$$

*exists and is finite, then* $\lim_{x\to\infty} \theta(x)/x = 1$.

**Remark.** Normally, convergence of an integral doesn't imply an integrand goes pointwise to zero because of narrow peaks (see Problem 1). Since $\theta$ is monotone, formally $(\theta(x) - x)' \ge -1$, and that prevents the narrow peaks.

**Proof.** Suppose for some $\alpha > 1$, there is $x_j \to \infty$ with $\theta(x_j) > \alpha x_j$. Then, by monotonicity of $\theta$, for $x_j < x < \alpha x_j$, $\theta(x) - x \geq \alpha x_j - x > 0$, so

$$\begin{aligned}\int_{x_j}^{\alpha x_j} \frac{\theta(x) - x}{x^2}\, dx &\geq \int_{x_j}^{\alpha x_j} \frac{\alpha x_j - x}{x^2}\, dx \\ &= \int_1^\alpha \left(\frac{\alpha - y}{y^2}\right) dy \equiv \eta(\alpha) > 0 \qquad (13.5.16)\end{aligned}$$

using $y = x/x_j$. Thus, if $Q(U)$ is the integral in (13.5.15), we have $\liminf_{j\to\infty}(Q(\alpha x_j) - Q(x_j)) \geq \eta(\alpha) > 0$, violating the hypothesis that the limit exists. It follows that, eventually, $\theta(x) < \alpha x$, that is, $\limsup \theta(x)/x \leq 1$.

Similarly, if $\alpha < 1$ and $\theta(x_j) < \alpha x_j$, we look at the integral from $\alpha x_j$ to $x_j$, and see it is less than some $\eta_1(\alpha) < 0$, so $\liminf \theta(x)/x \geq 1$. □

**Proof of Theorem 13.0.6.** Define

$$F(u) = \frac{\theta(u) - u}{u} \qquad (13.5.17)$$

and define $G$ for $\operatorname{Re} s > 1$ by (13.5.3). By (13.5.6), $\|F\|_\infty \leq 4\log 2$. For $\operatorname{Re} s > 1$,

$$\int_1^\infty u^{-s}\, du = (s-1)^{-1} \qquad (13.5.18)$$

so the $-u/u$ part of (13.5.17) contributes $-(s-1)^{-1}$ to $G$. Next, by the absolute convergence of sum and integral if $\operatorname{Re} s > 1$,

$$\begin{aligned}\int_1^\infty \theta(u) u^{-s-1}\, du &= \int_1^\infty \Bigl(\sum_{\substack{p<u \\ p\in\mathcal{P}}} (\log p) u^{-s-1}\Bigr)\, du \\ &= \sum_{p\in\mathcal{P}} \int_p^\infty (\log p) u^{-s-1}\, du \\ &= s^{-1} \sum_p \frac{\log p}{p^s}\end{aligned}$$

Thus, in terms of the function $\Phi$ of (13.3.11),

$$G(s) = s^{-1}\Phi(s) - \frac{1}{s-1} = s^{-1}\left[\Phi(s) - \frac{1}{s-1}\right] + s^{-1}$$

so, by Theorem 13.3.6, $G(s)$ has an analytic continuation to a neighborhood of $\{s \mid \operatorname{Re} s \geq 1\}$. By Corollary 13.5.2,

$$\lim_{U\to\infty} \int_1^U \frac{\theta(u) - u}{u^2}\, du$$

exists and is finite. So, by Proposition 13.5.5, $\lim_{x\to\infty} \theta(x)/x = 1$. By Proposition 13.5.4, (13.0.4) holds. □

All that remains is

**Proof of Theorem 13.5.1.** For $0 < T < \infty$, define

$$g_T(z) = \int_0^T e^{-zt} f(t)\, dt \tag{13.5.19}$$

which is an entire function of $z$, and let $\|f\|_\infty = \sup_{0 \le t < \infty} |f(t)| < \infty$.

For each $R$, by compactness, there is $\delta_R$ so $g$ is analytic in $\{z \mid |\mathrm{Im}\, z| < R+1,\ \mathrm{Re}\, z > -2\delta_R\}$. We want to show that

$$\lim_{T\to\infty} |g_T(0) - g(0)| = 0 \tag{13.5.20}$$

Let $C_R$ be the closed contour shown in Figure 13.5.1, that is, the part of the circle $\{z \mid |z| = R\}$ with $\mathrm{Re}\, z \ge -\delta_R$ and the straight line with $\mathrm{Re}\, z = -\delta_R$ that closes the contour.

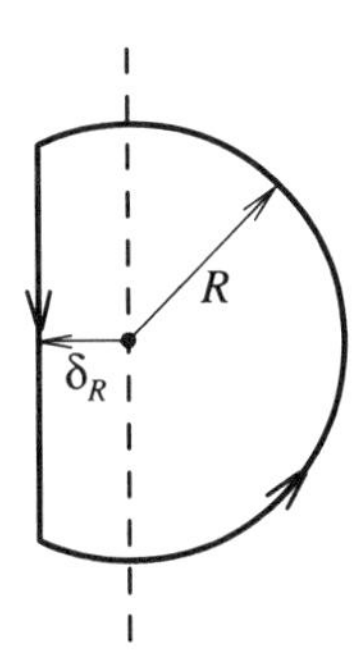


**Figure 13.5.1.** Newman's contour.

We'll use the following consequence of the Cauchy formula:

$$g_T(0) - g(0) = (2\pi i)^{-1} \oint_{C_R} (g_T(z) - g(z)) e^{zT} \left(1 + \frac{z^2}{R^2}\right) \frac{dz}{z} \tag{13.5.21}$$

We begin by noting that

$$\begin{aligned} \mathrm{Re}\, z > 0 \Rightarrow |g_T(z) - g(z)| &\le \|f\|_\infty \int_T^\infty e^{-(\mathrm{Re}\, z)t}\, dt \\ &= \|f\|_\infty e^{-T(\mathrm{Re}\, z)} (\mathrm{Re}\, z)^{-1} \end{aligned} \tag{13.5.22}$$

$$\begin{aligned} \mathrm{Re}\, z < 0 \Rightarrow |g_T(z)| &\le \|f\|_\infty \int_{-\infty}^T e^{-(\mathrm{Re}\, z)t}\, dt \\ &= \|f\|_\infty e^{-T(\mathrm{Re}\, z)} |\mathrm{Re}\, z|^{-1} \end{aligned} \tag{13.5.23}$$

Moreover, if $|z| = R$ so $z = Re^{i\theta}$, then $|z^{-1} + z/R^2| = 2R^{-1}|\cos\theta|$, so

$$z = Re^{i\theta} \Rightarrow \left| z^{-1} + \frac{z}{R^2} \right| = \frac{2|\mathrm{Re}\, z|}{R^2} \tag{13.5.24}$$

Since $|e^{zT}| = e^{\operatorname{Re} zT}$, on $C_R^+ = C_R \cap \{z \mid \operatorname{Re} z > 0\}$, we get a bound on the integrand of (13.5.21) by $2\|f\|_\infty R^{-2}$ (using (13.5.22) and (13.5.24)), so

$$\begin{aligned}|\text{Contribution of } C_R^+ \text{ to (13.5.21)}| &\leq (2\pi)^{-1} 2\|f\|_\infty R^{-2}(\pi R) \\ &= \|f\|_\infty R^{-1} \qquad (13.5.25)\end{aligned}$$

Since $g_T$ is entire, we can shift the contour $C_R^- = C_R \cap \{z \mid \operatorname{Re} z < 0\}$ to $\{z \mid |z| = R,\ \operatorname{Re} z < 0\}$, and then use (13.5.23) as above to see

$$|\text{Contribution of } C_R^- \text{ from } g_T(z)| \leq \|f\|_\infty R^{-1} \qquad (13.5.26)$$

Fix $R$. As for the contribution of $g$, notice $C_R^-$ is bounded, as is $|g(z)||z^{-1} + zR^{-2}|$ and $|e^{zT}| \to 0$ monotonically as $T \to \infty$ since $\operatorname{Re} z < 0$, so

$$\limsup_{T\to\infty} |\text{Contribution of } C_R^- \text{ from } g(z)| = 0 \qquad (13.5.27)$$

Thus, for each fixed $R$,

$$\limsup_{T\to\infty} |g_T(0) - g(0)| \leq 2\|f\|_\infty R^{-1}$$

Taking $R \to \infty$ proves (13.5.20). □

**Notes and Historical Remarks.** That $\pi(x) \sim x/\log x$ was conjectured around 1795 by Legendre [**262**] and Gauss. Gauss published nothing but noted in a letter fifty years later that when he was about fifteen years old, he reached this belief but was unable to prove it.

Chebyshev [**82, 83**] in 1852 studied the problem with elementary methods and proved that if the ratio has a limit, it must be 1 and proved $.9R \leq x/\log x \leq 1.11$ for all large $x$. In particular, Proposition 13.5.4 and its proof are from his work.

It was de la Vallée Poussin [**97**] and Hadamard [**171**] who finally proved the result in 1896 (Diamond [**101**] and Bateman–Diamond [**29**] summarize the earlier and later history). Proofs by Tauberian theorems that relied on deep Fourier analysis were associated with the work of Wiener [**412**], Ikehara [**206**], and Ingham [**209, 210**]. We will present Ingham's proof in Section 6.12 of Part 4 and the needed Tauberian theorem in Section 6.11.

Our proof here is based on ideas of Newman [**296**], as fleshed out by Korevaar [**230**] and Zagier [**418**]. Newman also discusses his proof in his book [**298**]. Newman works with sums; the integral form is from the later papers, but the key idea of using $(1 + z^2/R^2)$ is Newman's. Theorem 13.5.1 is already in Ingham's 1935 paper, but not central and with a proof that depends on Wiener's arguments. Newman's book [**296**] is sometimes opaque, but its point of view emphasizing Landau's theorem has strongly influenced our presentation in Sections 13.2–13.5.

Diamond [**101**] and Goldfeld [**153**] discuss proofs of the prime number theorem, going back to Erdős and Selberg, which do not use analytic functions.

There is a lot to say about explicit error estimates on $\pi(x) - Li(x)$ where $Li(x) = \text{pv} \int_0^x \frac{dt}{\log t}$, a function asymptotic to $x/\log x$ and used already by Riemann. We'll settle for noting that the Riemann hypothesis is equivalent to $\pi(x) - Li(x) = 0(x^{1/2} \log x)$ and that the best error current estimates are still quite far from that.

**Problems**

1. Let $|\varepsilon_n| < \frac{1}{2}$ and let $f$ be the function which is $a_n$ at $n = 1, 2, \dots$, zero on $\mathbb{R} \setminus \bigcup_{n=1}^\infty (n - \varepsilon_n, n + \varepsilon_n)$ and linear from $n$ to $n \pm \varepsilon_n$. Prove that $\int_1^\infty |f(u)u^{-1}|\, du \le \sum_{n=1}^\infty (n - \frac{1}{2})^{-1} |a_n| \varepsilon_n$ can converge even though $a_n \to \infty$ very fast.

2. Let $k$ and $m$ be relatively prime. Prove $\lim_{x\to\infty}(\#\{p \le x \mid p \in \mathcal{P}_k^{(m)}\}/(x/\log x)) = \varphi(m)^{-1}$. (*Hint*: Mimic the proof of this section using Problem 3 of Section 13.4.)

   **Note.** This result was in the original papers of de la Vallée Poussin [**97**] and Hadamard [**171**] that proved the prime number theorem.

3. In this problem, you will prove an Abelian analog of Newman's theorem. Let $f$ be bounded, $g$ given by (13.5.1), and $S(T) = \int_0^T f(t)\, dt$.

   (a) For any $a$ and $s > 0$, prove that
   $$a - g(s) = s \int_0^\infty e^{-st}(a - S(t))\, dt$$

   (b) For any $T$, prove that
   $$|a - g(s)| \le sT(a + T\|f\|_\infty) + \sup_{t \ge T} |a - S(t)|$$

   (c) Show that if $\lim_{T\to\infty} S(T) = a$, then $\lim_{s\downarrow 0} g(s) = a$.

*Chapter 14*

# Ordinary Differential Equations in the Complex Domain

> A friend recently wrote: "Mathematics is the creation of flesh and blood, not just novelty-curious and wise automatons. We ought to devote some part of our efforts to increasing understanding of the observable universe." This book is a contribution to these efforts. We praise famous men, men who created beautiful structures and directed the course of mathematics... The structures that the masters built are not just beautiful to the eye; they are also eminently useful... it is fitting to recall that few domains of mathematics are so widely applicable as the theory of ordinary differential equations...
>
> —*E. Hille* [**190**], *1976*

**Big Notions and Theorems:** Hypergeometric Equation and Functions, Bessel Equation and Functions, Airy Equation and Functions, Monodromy and ODEs, Circuit Matrix, Regular Singular Point, Fuchsian ODEs, Euler–Cauchy Equation, Indicial Equation, Riemann $P$-Functions, Papperitz's Theorem, Papperitz–Riemann Equation, Euler Integral, Pfaff Identity, Kummer's Solutions, Gauss Contiguous Relations, Rodrigues' Formula, Jacobi Polynomials, Legendre Equation, $Y$, $I$, and $K$ Functions, Hankel Functions, Movable Singularity, Painlevé Functions, Integral Representations, Hankel Contour, Sommerfeld Contour, Schläfli's Formula

The next two chapters focus on two closely linked subjects: ordinary differential equations (ODEs) with analytic or meromorphic coefficients, and asymptotics, especially of integrals. The link will be via integral representations of solutions of certain ODEs, the subject of Section 14.7.

It is no coincidence that Airy and Bessel, whose functions play a starring role, and Debye, the father of the method of steepest descent, were astronomers and physicists—these subjects are central to applications of analysis while also being significant in wide swaths of pure mathematics.

General analytic ODEs are given by

$$\frac{d^n f}{dz^n} = F(z; f(z), f'(z), \dots, f^{(n-1)}(z)) \tag{14.0.1}$$

where $F$ is an analytic function of $n+1$ variables. While Section 14.6 will have some small number of remarks about this general nonlinear case, most of the chapter will focus on the linear case,

$$\sum_{k=0}^{n} b_k(z) f^{(k)}(z) = 0 \tag{14.0.2}$$

with $\{b_k\}_{k=0}^n$ analytic so that for each $k = 0, 1, \dots, n-1$, $b_k$ and $b_n$ have no common zero. Equivalently,

$$f^{(n)}(z) = \sum_{k=0}^{n-1} a_k(z) f^{(k)}(z) \tag{14.0.3}$$

where

$$a_k(z) = -\frac{b_k(z)}{b_n(z)} \tag{14.0.4}$$

are meromorphic with no common zero. Points, $z_0$, where $b_n(z_0) \neq 0$ (equivalently, all $a_k(z_0)$ are analytic) are called *regular points* for the ODE. Points, $z_0$, where $b_n(z_0) = 0$ or where some $a_k(z_0)$ has a pole, are called *singular points.*

We'll be particularly interested in the case where the $b$'s are all polynomials ($a$'s rational), especially three classical cases. First, the *hypergeometric equation* (in Gauss form), where $a, b, c$ are parameters,

$$z(1-z)f'' + [c - (a+b+1)z]f' - abf = 0 \tag{14.0.5}$$

whose singular points are 0, 1, and $\infty$ (the meaning of singularity at infinity will require a separate analysis found in Section 14.3) and whose solution analytic at $z = 0$ with $f(0) = 1$ is given by (for $|z| < 1$)

$$F(a, b, c; z) = 1 + \sum_{n=1}^{\infty} \frac{(a)_n (b)_n}{n!(c)_n} z^n \tag{14.0.6}$$

where

$$(a)_n = a(a+1)\dots(a+n-1) \tag{14.0.7}$$

is the *shifted factorial* or *Pochhammer symbol* (so $(1)_n = n!$). This is the *hypergeometric function*, ${}_2F_1$.

Second is *Bessel's equation* ($\alpha$ is a parameter)

$$z^2 f'' + zf' + (z^2 - \alpha^2)f = 0 \tag{14.0.8}$$

whose only finite singular point is 0 ($\infty$ is also singular). There are solutions near $z = 0$ given by

$$J_\alpha(z) = \sum_{k=0}^\infty \frac{(-1)^k}{\Gamma(k+1)\Gamma(\alpha+k+1)} \left(\frac{z}{2}\right)^{2k+\alpha} \tag{14.0.9}$$

If $\alpha^2 \neq n^2$ ($n = 0, 1, 2, \dots$), $J_\alpha$ and $J_{-\alpha}$ are two linearly independent solutions. $J_\alpha$ is the *Bessel function* of order $\alpha$.

The third equation is the *Airy equation*

$$f'' - zf = 0 \tag{14.0.10}$$

which is singular only at $\infty$. A distinguished solution, the *Airy function*, is given for $z$ real by

$$\mathrm{Ai}(z) = \frac{1}{\pi} \int_0^\infty \cos(\tfrac{1}{3}\, t^2 + zt)\, dt \tag{14.0.11}$$

defined as an improper integral (Problem 1). This has the power series expansion

$$\mathrm{Ai}(z) = \frac{1}{3^{2/3}\pi} \sum_{n=0}^\infty \frac{3^{n/3}\Gamma(\frac{1}{3}(n+1))}{n!} \sin\left(\frac{2(n+1)}{3}\, \pi\right) z^n \tag{14.0.12}$$

(so coefficients of $z^2, z^5, z^8, \dots$ are all zero). Airy functions can be expressed in terms of Bessel functions of order $\frac{1}{3}$ but are important in their own right.

In Theorem 11.2.3 of Part 2A and the discussion following it, we showed that if $\Omega$ is any region, with $\{a_k(z)\}_{k=1}^{n-1}$ analytic in $\Omega$, then any local solution of (14.0.3) can be analytically continued along any curve, yielding an $n$-dimensional set of germs about any point, so that at $z_0 \in \Omega$, $f \to (f(z_0), f'(z_0), \dots, f^{(n-1)}(z_0))$ is onto $\mathbb{C}^n$. In Section 14.1, we'll prove a converse of this and prove that monodromies with this extra nondegeneracy condition always come from analytic ODEs.

This will lead, in Section 14.2, to focusing on isolated singularities, that is, where $\Omega$ is analytic in $\mathbb{D}_\delta(z_0) \setminus \{z_0\}$ for some $z_0$ and $\delta > 0$. We prove generically in that case, there is a basis of solutions of the form

$$(z - z_0)^{\lambda_j} \sum_{n=-\infty}^\infty a_n (z - z_0)^n \tag{14.0.13}$$

and, in general, one only need add some powers of $\log(z - z_0)$. Section 14.3 will explore when an ODE has a basis where the series part of (14.0.13) only has a removable or a polar singularity, that is, the sum in (14.0.13) doesn't run to $-\infty$. If that happens, the singularity is called a *regular singular point.* To avoid some messy technicalities, we'll restrict parts of our discussion to the case where only first powers of logs are needed—something automatic in the case of second-order equations, which are the main focus of our interest.

If all singular points, including infinity, are regular singular points, the ODE is called *Fuchsian.* The simplest nontrivial Fuchsian ODE has second order and three singular points. Their solutions are given, up to an FLT and multiplication by $z^\lambda$, by hypergeometric functions, the subject of Section 14.4. Section 14.5 is on Bessel and Airy functions, and is mainly a glossary of their differing forms, such as Hankel and spherical Bessel functions.

Section 14.6 has a very few brief remarks on the nonlinear case, focusing on an important difference from the linear case—the existence of movable singularities. Section 14.7 presents some integral representations for the special functions discussed earlier in this chapter. These representations will be used in the next chapter to study asymptotics of the functions they represent.

**Notes and Historical Remarks.** This chapter is a mere introduction to a vast subject on which there are many books, of which we mention [**190, 207, 208, 349, 401, 402**]. For a masterful presentation of the classical theory, see Hille's book [**190**]—the title of this chapter is an homage to this book. For some of the recent literature, see Ilyashenko–Yakovenko [**207**].

Our approach in this book is fairly classical. There are more recent approaches, especially in the algebraic geometry literature, that view the basic objects on which we study monodromy as vector bundles, where the unique continuation by analyticity defines what a differential geometer would call a connection. The ODE written in matrix form is then a covariant derivative. For a review of some of the literature, see Varadarajan [**390**]; and for two early works on this modern approach, see Manin [**271**] and Deligne [**98**].

The second half of this chapter studies some special functions. Two masterful monographs on special functions are [**15, 31**]. More elementary but also useful are [**72, 196**].

## Problems

1. (a) Prove that for any real $z$, $\lim_{R\to\infty} \int_1^R \exp(\pm\frac{i}{3}t^3 + izt)\,dt$ exists. (*Hint*: Use $\frac{1}{it^2}\frac{d}{dt}e^{it^3/3} = e^{it^3/3}$ and integrate by parts.)

   (b) Prove that for any real $z$, $\lim_{R\to\infty} \int_0^R \cos(\frac{1}{3}t^3 + zt)\,dt$ exists.

## 14.1. Monodromy and Linear ODEs

We are going to focus on finite-dimensional families of analytic functions that can be continued along any curve and which remain within the family upon a closed loop. The following notion is frequently used, but doesn't seem to have a standard name.

**Definition.** A *monodromic family of dimension* $n$ on a domain $\Omega$ is a dimension-$n$ vector space, $V$, of germs of analytic functions at some point $z_0 \in \Omega$ so that

(i) Any $\mathfrak{g} \in V$ can be analytically continued along any curve $\gamma$ in $\Omega$ with $\gamma(0) = z_0$.
(ii) If $\gamma$ is a closed curve with $\gamma(0) = \gamma(1) = z_0$ and $\mathfrak{g} \in V$, then $\tau_\gamma(\mathfrak{g}) \in V$.

For any closed $\gamma$, $\tau_\gamma$ defines a map of $V$ to $V$ which is invertible since, by the monodromy theorem (Theorem 11.2.1 of Part 2A), $\gamma \mapsto \tau_\gamma$ is a representation of the fundamental group $\pi(\Omega, z_0)$, so $\tau_\gamma^{-1} = \tau_{\gamma^{-1}}$. In addition, if $\gamma$ is a curve from $z_0$ to $z_1 \in \Omega$, $\mathfrak{g} \mapsto \tau_\gamma(\mathfrak{g})$ shows the set of germs at $z_1$ obtained from $V$ by analytic continuation is also $n$-dimensional. We call this set $\sigma_{z_1}(V)$, the germs of $V$ at $z_1$.

**Definition.** Let $A_n^w$ map the germ at $w$ to $\mathbb{C}^n$ by mapping the germ $\mathfrak{g}$ to $(g(w), g'(w), \dots, g^{(n-1)}(w))$ for a representative $g$ of $\mathfrak{g}$. We say a monodromic family, $V$, of dimension $n$ is *strongly nondegenerate* if for all $z_1 \in \Omega$, $A_n^{z_1}$ maps $\sigma_{z_1}(V)$ bijectively to $\mathbb{C}^n$.

**Example 14.1.1.** Let $\Omega = \mathbb{C}$, $z_0 = 0$, and $V$ the germs of all polynomials, $P$, of degree at most $n+1$ with $P(0) = 0$. It is easy to see $A_n^{z_1}$ is a bijection for $z_1 \neq 0$, but not for $z_1 = 0$, so $V$ is not strongly nondegenerate. This shows that not all monodromic families are strongly nondegenerate. $\Box$

Theorem 11.2.4 of Part 2A says that for any linear differential equation of the form (14.0.3) with analytic coefficients, the germs of solutions near any $z_0$ are a monodromic family. Uniqueness of solutions then implies each $A_n^{z_1}$ is one-one, and so a bijection. Thus,

**Theorem 14.1.2.** *Let* $\{a_j\}_{j=0}^{n-1}$ *be analytic functions on a region* $\Omega$. *The germs of all solutions to the linear ODE* (14.0.3) *at some* $z_0 \in \Omega$ *form a strongly nondegenerate monodromic family on* $\Omega$.

The main result of this section is a converse:

**Theorem 14.1.3.** *Let* $z_0 \in \Omega$, *a region in* $\mathbb{C}$. *Let* $V$ *be a set of germs at* $z_0$ *that form a strongly nondegenerate monodromic family. Then there exist analytic functions* $\{a_j\}_{j=0}^{n-1}$ *on* $\Omega$, *so that* $V$ *is precisely the set of germs of solutions of* (14.0.3).

**Proof.** Let $f_1(z), \dots, f_n(z)$ be a basis of germs at $z_0$ for $V$. Fix $f \in V$. Consider the $(n+1) \times (n+1)$ matrix, $M_{n+1}(z)$, whose first row is $(f(z), f'(z), \dots, f^{(n)}(z))$ and whose $(j+1)$-st row, $j = 1, \dots, n$, is $(f_j(z), \dots, f_j^{(n)}(z))$. Let $D(z)$ be its determinant which is identically zero near $z_0$ since the rows are linearly dependent.

All elements of $M_{n+1}(z)$ can be analytically continued along any curve, $\gamma$, in $\Omega$, starting at $z_0$. Expand $D(z)$ in minors in the top row and see that

$$\sum_{k=0}^{n} b_k(z) f^{(k)}(z) = 0 \tag{14.1.1}$$

where, for example,

$$b_n(z) = (-1)^n \det((f_j^{(k-1)}(z))_{j,k=1,\dots,n}) \tag{14.1.2}$$

Let $\mathcal{M}_{[\gamma]}$ be the monodromy matrix written in $f_j$ basis. Then each minor transforms under continuation along $\gamma$ by multiplying on the left by $\mathcal{M}_{[\gamma]}$, so each $b_j$ is multiplied by $\det(\mathcal{M}_{[\gamma]})$ so ratios are invariant, that is,

$$a_j \equiv -\frac{b_j}{b_n}, \qquad j = 0, 1, \dots, n-1 \tag{14.1.3}$$

are analytic single-valued functions on $\Omega$. Since $V$ is strongly nondegenerate, $a_j$ has no polar singularities. Thus, any $f \in V$ obeys (14.0.3), for $a_j$ given by (14.1.3), so by dimension counting, $V$ is the monodromic family of an analytic ODE. □

We note next that for general monodromic families without an assumption of strong nondegeneracy, (14.1.1) holds, and one can show (Problem 1) that $b_n$ has only isolated zeros, so the family is the set of all solutions of a differential equation of the form (14.0.3), but with coefficients meromorphic on $\Omega$. Finally, it is illuminating and useful to know that the $a_j$ are unique:

**Theorem 14.1.4.** *Let $V$ be a strongly nondegenerate monodromic family. Then the set of functions $\{a_j\}_{j=0}^{n-1}$ for which* (14.0.3) *holds for all germs in $V$ is unique.*

**Proof.** Let $W = \{(f(z_0), \dots, f^{(n)}(z_0)) \mid f \in V\} \subset \mathbb{C}^{n+1}$. By the strong nondegeneracy, $W$ has dimension $n$. Thus, $W^\perp$ is one-dimensional, implying that if $(\bar{a}_0, \dots, \bar{a}_n, 1)$ and $(\bar{c}_0, \dots \bar{c}_n, 1)$ are both in $W$, then $a_j = c_j$. □

**Notes and Historical Remarks.** For $n = 2$, $b_n$ of (14.1.2) is the famous *Wronskian* $f_1 f_2' - f_1' f_2$. For general $n$, the leading $b_n$ is a generalized Wronskian. Related to the question of this section is which monodromy groups arise from monodromic families, a question answered by Grauert [**158**]. This much more subtle issue is discussed in the book of Sibuya [**349**].

Józef Maria Hoëné, (1778–1853) was born of Czech parents in Poland and took the name Wronski around 1815. He spent most of his adult life in France pursuing crackpot schemes and making overblown claims in areas as diverse as philosophy, physics, economics, and law. His checkered legacy is aptly summarized in the Wikipedia biography[1] as: "Though during his lifetime nearly all his work was dismissed as nonsense, some of it has come in later years to be seen in a more favorable light. Although nearly all his grandiose claims were in fact unfounded, his mathematical work contains flashes of deep insight and many important intermediary results."

**Problems**

1. Let $V$ be a monodromic family of dimension $n$ on a region $\Omega$. Let $f_1, \dots, f_n$ be a local basis near $z_0$. Let $g_k(z)$ for $k = 1, 2, \dots, n$ be the determinant of the $k \times k$ matrix $\{f_j^{(\ell-1)}(z)\}_{\ell=1,\dots,k;\, j=1,\dots,k}$.

   (a) If $g_{k-1}(z_1) \neq 0$ but $g_k(z) \equiv 0$ near $z_1$, prove that $f_1, \dots, f_k$ solve an ODE of the form $f^{(k-1)}(z) = \sum_{j=0}^{k-2} a_j(z) f^{(j)}(z)$ with $a_j$ analytic near $z_1$, and conclude that $f_1, \dots, f_k$ cannot be linearly independent germs at $z_1$.

   (b) Prove inductively that $g_k(z)$ has isolated zeros.

   (c) Show $V$ is the set of solutions of an ODE of the form (14.0.3) with coefficients meromorphic in $\Omega$.

## 14.2. Monodromy in Punctured Disks

Here we'll study finite-dimensional monodromic families on $\mathbb{D}_\delta(z_0) \setminus \{z_0\}$ for some $\delta > 0$, $z_0 \in \mathbb{C}$. The paradigms are seen in

**Example 14.2.1.** Fix $\lambda \in \mathbb{C}$. For $z$ near 1, define

$$z^\lambda \equiv \exp(\lambda \log(z)) \tag{14.2.1}$$

where the branch of log is taken with $\log(1) = 0$ (so $1^\lambda = 1$). Then $f(z) = z^\lambda$ can be continued along any curve in $\Omega = \mathbb{C}^\times$. If $\gamma$ is a curve in $\Omega$ with $\gamma(0) = \gamma(1) = 1$ and $\gamma$ has winding number $m$ around 0, then

$$\tau_\gamma(f) = e^{2\pi i \lambda m} f \tag{14.2.2}$$

so multiples of $f$ form a one-dimensional monodromic family. Notice that the multiplicative factor is the same if any integer is added to $\lambda$. Equivalently, if $g$ is analytic in $\mathbb{C}^\times$ and $f(z) = z^\lambda g(z)$, (14.2.2) still holds.

Now, let

$$L(z) = \frac{1}{2\pi i} \log(z) \tag{14.2.3}$$

so that, for the same $\gamma$,

$$\tau_\gamma(L) = m + L \tag{14.2.4}$$

[1] http://en.wikipedia.org/wiki/Jozef_Maria_Hoene-Wronski

Thus, if $g$ is analytic in $\mathbb{C}^\times$, the span of $f_1(z) = z^\lambda g(z)$ and $f_2(z) = z^\lambda L(z)g(z)$ is a two-dimensional monodromic family with

$$\tau_\gamma(f_2) = e^{2\pi i\lambda m}[f_2 + mf_1] \tag{14.2.5}$$

that is, in $f_2, f_1$ basis, $\tau_\gamma$ has the form

$$\tau_\gamma = e^{2\pi i m\lambda}\begin{pmatrix} 1 & m \\ 0 & 1 \end{pmatrix} \tag{14.2.6}$$

□

Our main goal in this section is to show these paradigms are general. In $\mathbb{D}_\delta(z_0)$, the fundamental group is generated by $\gamma_0$, the loop $\gamma_0(t) = z_0 + \frac{1}{2}\delta e^{2\pi i t}$. The corresponding monodromy matrix, $C \equiv \tau_{\gamma_0}$, is called the *circuit matrix.* Things are especially simple if $C$ has no Jordan anomalies.

**Theorem 14.2.2.** *Let $V$ be an $n$-dimensional monodromic family on $\Omega = \mathbb{D}_\delta(z_0) \setminus \{z_0\}$. Suppose the circuit matrix, $C$, is diagonalizable so that its eigenvalues are $\{e^{2\pi i\lambda_j}\}_{j=1}^n$ with each $\lambda_j \in \mathbb{C}$. Then there are functions, $g_j$, analytic on $\Omega$ so that*

$$f_j(z) = (z - z_0)^{\lambda_j} g_j(z) \tag{14.2.7}$$

*is a basis for $V$.*

**Proof.** Let $\{f_j\}_{j=1}^n$ be a basis of eigenvectors for $C$ with

$$Cf_j = e^{2\pi i\lambda_j} f_j \tag{14.2.8}$$

and let

$$g_j(z) = (z - z_0)^{-\lambda_j} f_j(z) \tag{14.2.9}$$

Then $g_j$ is invariant under continuation around $\gamma_0$, so around $(\gamma_0)^m$, so around any curve by the monodromy theorem. □

In general, we just need powers of logs. For notational simplicity, we take $\delta = 1$ and $z_0 = 0$.

**Theorem 14.2.3.** *Let $V$ be an $n$-dimensional monodromic family on $\mathbb{D}^\times$. Then there exist $m_1, \dots, m_q$ so $m_1 + \cdots + m_q = n$, and $\lambda_1, \dots, \lambda_q$ in $\mathbb{C}$, and functions $\{f_{jk}\}_{j=1,\,k=1,\dots,q}^{m_k}$ analytic in $\mathbb{D}^\times$ so that (with $L(z)$ given by (14.2.3))*

(i) *The $f$'s are a basis for $V$.*
(ii) *The $f_{jk}$ obey*

$$(C - e^{2\pi i\lambda_k})f_{(j+1)k} = \begin{cases} e^{2\pi i\lambda_k} f_{jk}, & j \geq 1 \\ 0, & j = 0 \end{cases} \tag{14.2.10}$$

$$\text{(iii) } f_{jk}(z) = \sum_{\ell=1}^{j} z^{\lambda_k} L(z)^{\ell-1} g_{jk\ell}(z) \tag{14.2.11}$$

*where* $\{g_{jk\ell}\}_{k=1,\,j=1,\,\ell=1}^{q\quad m_k\quad j}$ *are analytic in* $\mathbb{D}^\times$.

In the case $m_k = 2$, we have two functions $g_{1k}$ and $g_{2k}$ (instead of $g_{1k1}$, $g_{2k1}$, and $g_{2k2}$) so that

$$f_{1k}(z) = z^{\lambda_k} g_1(z), \qquad f_{2k}(z) = f_{1k}(z)L(z) + z^{\lambda_k} g_2(z) \tag{14.2.12}$$

**Remarks.** 1. (14.2.12) is true for $f_{1k}, f_{2k}$ whenever $m_k \geq 2$, that is, $g_{2k2}$ can be taken equal to $g_{1k1}$.

2. In general, there are linear relations between the $g_{jk\ell}$ so that there are only $m_k$ independent $g$'s for a fixed $k$. Indeed, for $\ell \geq 2$, $g_{jk\ell}$ is a linear combination of $\{g_{pk1}\}_{p=1}^{j-1}$; see, for example, the discussion in Hille [**190**].

We begin the proof of Theorem 14.2.3 with a lemma that is pure algebra. Let $M$ be defined on germs of functions that can be continued along any curve of $\mathbb{D}^\times$ by

$$(Mf)(z) = L(z)f(z) \tag{14.2.13}$$

so, by (14.2.4), $CL = L + 1$, and so, $CMf = (CL)(Cf) = MCf + Cf$, that is, with $[A, B] = AB - BA$,

$$[C, M] = C \tag{14.2.14}$$

**Lemma 14.2.4.** *Let $C$ and $M$ be linear transformations on some complex vector space, $X$, obeying* (14.2.14). *Let $\mu \in \mathbb{C}^\times$. Then*

$$\mathbb{K}_n \equiv \ker((C-\mu)^n) = \{f_1 + Mf_2 + \cdots + M^{n-1} f_n \mid f_j \in \ker(C-\mu)\} \tag{14.2.15}$$

*Moreover, if $f \in \mathbb{K}_{n+1}$, then*

$$f - (\mu n)^{-1} M(C-\mu)f \in \mathbb{K}_n \tag{14.2.16}$$

**Proof.** We begin by noting that, by a simple inductive argument (Problem 1), for $n = 1, 2, \dots,$

$$[(C-\mu)^n, M] = nC(C-\mu)^{n-1} \tag{14.2.17}$$

Thus, if

$$g = (C-\mu)f \tag{14.2.18}$$

and $f \in \mathbb{K}_{n+1}$, then

$$\begin{aligned}(C-\mu)^n Mg &= M(C-\mu)^{n+1} f + nC(C-\mu)^n f \\ &= n\mu(C-\mu)^n f\end{aligned} \tag{14.2.19}$$

since $n(C-\mu)^{n+1} f = 0$. (14.2.19)/(14.2.18) imply (14.2.16).

Now we can prove (14.2.15) inductively. $n = 1$ is trivial. Suppose (14.2.15) holds for $n$. If $f \in \mathbb{K}_{n+1}$, then $(C - \mu)f \in \mathbb{K}_n$, so

$$(C - \mu)f = h_1 + \cdots + M^{n-1}h_n \tag{14.2.20}$$

By (14.2.16) and the inductive hypothesis,

$$\begin{aligned} f &= (\mu n)^{-1}[Mh_1 + M^2h_2 + \cdots + M^n h_n] + g_1 + \cdots + M^{n-1}g_n \qquad (14.2.21) \\ &= f_1 + Mf_2 + \cdots + M^n f_{n+1} \end{aligned}$$

where $f_1 = g_1$, $f_2 = g_2 + (\mu n)^{-1}h_1, \dots, f_n = g_n + (\mu n)^{-1}h_{n-1}$, $f_{n+1} = (\mu n)^{-1}h_n$. □

**Proof of Theorem 14.2.3.** The circuit matrix, $C$, can be written as a sum of $q$ Jordan blocks of size $m_1, \dots, m_q$ with $m_1 + \cdots + m_q = n$. This allows us to find a basis for $V$ so that $V$ is a direct sum of $q$ pieces invariant under analytic continuation. Thus, we can suppose $C$ has a single Jordan block with (14.2.10). Drop the $k$ index and set $\mu = e^{2\pi i\lambda}$.

By the argument in the proof of Theorem 14.2.2, every function, $f$, that has $f \in \ker(C - \mu)$ has the form

$$f(z) = z^\lambda g(z) \tag{14.2.22}$$

with $g$ analytic in $\mathbb{D}^\times$. This plus Lemma 14.2.4 implies the Jordan basis elements obey (14.2.11).

In case $m = 2$, we have $(C - \mu)f_2 = \mu f_1$ and $f_2 \in \mathbb{K}_2$, so by (14.2.16), $f_2 - Mf_1 \in \mathbb{K}_1$. Thus, (14.2.12) holds by (14.2.22). □

An interesting subclass of monodromies are those where none of the functions $g_{jk\ell}$ have essential singularities at $z_0$. The following is thus important:

**Theorem 14.2.5.** *Let $V$ be a finite-dimensional monodromic family on $\mathbb{D}^\times$. Then the following are equivalent:*

(1) *$V$ has a basis in which all the functions $g_{jk\ell}$ of* (14.2.11) *have poles or removable singularities at $z = 0$.*
(2) *Place a norm on $V$ so that it has a unit ball, $B$. Given a germ at* 1 *of $f$ in $V$, consider its unique continuation to $\mathbb{D} \setminus (-1, 0]$, also denoted by $f$. Then there exist $C > 0$ and $K$ so that*

$$\sup_{\substack{z \in \mathbb{D}\setminus(-\infty,0] \\ f \in B}} |f(z)| \le C|z|^{-K} \tag{14.2.23}$$

**Proof.** (1) ⇒ (2) is trivial. For the converse, note that since (14.2.23) also holds on the closure of $\mathbb{D} \setminus (-\infty, 0]$ on a multisheeted Riemann surface for $L(z)$, and since the monodromy matrix is bounded, we get a bound like (14.2.23) on a finite number of sheets over $\mathbb{D} \setminus (-\infty, 0]$. Since the $g_{jk\ell}$ can be determined from finitely many continuations of $f$ around 0, we get a bound like (14.2.23) for each $g_{jk\ell}$, and so that $g_{jk\ell}$ has a nonessential singularity. This proves (1). □

**Definition.** We say that a monodromic family, $V$, on $\mathbb{D}_\delta(z_0) \setminus \{z_0\}$ has $z_0$ as a *regular singular point* if and only if the equivalent conditions of Theorem 14.2.5 hold for $V$.

**Definition.** Let $V$ be a monodromic family with $z_0$ a regular singular point. Let $f \in V$ obey $(C - \mu)^k f = 0$ for some $k$, where $C$ is the circuit matrix about $z_0$. The *index*, $\lambda$, of $f$ is defined by

$$\text{(a)}\ |\mu| = e^{-2\pi \operatorname{Im} \lambda} \tag{14.2.24}$$

$$\text{(b)}\ -\operatorname{Re} \lambda = \inf\{K \mid (14.2.23) \text{ holds}\} \tag{14.2.25}$$

**Example 14.2.6.** If $f(z) = z^{\lambda_0} g(z)$, where $g$ is analytic at 0 with $g(0) \neq 0$, then $\lambda_0$ is the index of $f$. More generally, if

$$f(z) = z^{\lambda_0} \sum_{\ell=1}^{j} L(z)^{\ell-1} z^{N_\ell} g_\ell(z) \tag{14.2.26}$$

where each $g_\ell$ is analytic at $z$ with $g_\ell(0) \neq 0$, then the index, $\lambda$, of $f$ is given by

$$\lambda = \lambda_0 + \min_{\ell=1,\dots,j} (N_\ell) \tag{14.2.27}$$

In particular, if $f(z) = z^{\lambda_0}(g_0(z) + z^N g_1(z) L(z))$, with $g_2(0) \neq 0$, $g_3(0) \neq 0$, then the index, $\lambda$, of $f$ has $\lambda = \lambda_0$ if $N \geq 0$ and $\lambda = \lambda_0 + N$ if $N < 0$. □

**Definition.** Let $V$ be a monodromic family with $z_0$ a regular singular point. The *indices* of $V$ are the set of indices of a Jordan basis for $C$, the circuit matrix about $z_0$.

**Remark.** In the special case where $C$ has degenerate geometric eigenvalues (i.e., multiple Jordan blocks with the same eigenvalue), there are extra subtleties, and the indices can depend on choices made. For example, if a basis is $f_1(z) = z^{1/2}$ and $f_2(z) = z^{3/2}$, we have indices $\frac{1}{2}$ and $\frac{3}{2}$; but if $f_3 = f_1 + f_2$, we could take a basis of $f_1$ and $f_3$, both with index $\frac{1}{2}$. One can always make the choice in this case so that the indices have real parts as large as possible (i.e., $\frac{1}{2}$ and $\frac{3}{2}$ in this case)—and that is what we have in mind. With this choice, doubled indices are always associated with log terms. Because this is rare, we'll avoid explaining the associated contortions.

### Problems

1. Prove that (14.2.14) implies (14.2.17).

2. Let $V$ be a monodromic family with $z_0$ a regular singular point. Let $\lambda_1, \dots, \lambda_n$ be the indices of $V$ with $\operatorname{Re}\lambda_1 \le \operatorname{Re}\lambda_2 \le \cdots \le \operatorname{Re}\lambda_n$.

   (a) Show there is a basis of $V$ for which the inf over $N$ for which (14.2.23) is $-\operatorname{Re}\lambda_1$ for all basis elements.

   (b) If $f_1, \dots, f_n$ is any basis with the $a_j$ real, so $-a_j = \inf\{K \mid$ (14.2.23) holds for $f_2\}$ ordered by $a_1 \le a_2 \le \dots$, prove that $a_j \le \operatorname{Re}\lambda_j$.

## 14.3. ODEs in Punctured Disks

We've seen that a strongly nondegenerate monodromic family on $\Omega \subset \mathbb{C}$ is all the solutions of an ODE of the form (14.0.3) with $a_k(z)$ analytic in $z$. If $\Omega$ is a punctured disk which, by scaling and translation, we can take to be $\mathbb{D}^\times$, it is natural to ask when 0 is a regular singular point of the monodromy as defined in the last section. The most important idea in this section is that this is true if and only if $z^{n-k}a_k(z)$ has a removable singularity at $z = 0$. We'll prove:

**Theorem 14.3.1.** *Let $\{a_k(z)\}_{k=0}^{n-1}$ be a family of functions analytic in $\mathbb{D}^\times$. Suppose each $a_k$ has a removable singularity at $z = 0$ or a pole of order at most $n - k$ (and at least one has a pole). Then 0 is a regular singular point of the monodromy associated to* (14.0.3).

**Theorem 14.3.2.** *Let $V$ be a strongly nondegenerate monodromic family on $\mathbb{D}^\times$ of dimension $n$. Let $V$ have a regular singular point at $z = 0$. Then $V$ is the set of all solutions of a differential equation of the form* (14.0.3), *where each $a_k$ has a removable singularity or pole of order at most $n - k$ at $z = 0$.*

Besides this theme, we determine exactly what the leading behavior of the solutions is in terms of a polynomial equation (the indicial equation) depending on $\lim_{z\to 0,\, z\ne 0} z^{n-k}a_k(z)$, and we'll analyze in the second-order case what happens if the singular point is at infinity. We begin with a lemma needed for the proof of Theorem 14.3.1.

**Lemma 14.3.3.** *Let $\boldsymbol{f}$ be a $\mathbb{C}^n$-valued analytic function in a neighborhood, $N$, of $\{z \mid \alpha \le \arg z \le \beta,\ 0 < |z| < 1\}$ that obeys*

$$\boldsymbol{f}'(z) = A(z)\boldsymbol{f}(z) \tag{14.3.1}$$

*where $A(z)$ is a matrix obeying*

$$\|A(z)\| \le \frac{C}{|z|} \tag{14.3.2}$$

*for $|z| \le \frac{1}{2}$ and some $C > 0$. Then for $|z| \le \frac{1}{2}$ and some $D$,*

$$\|\boldsymbol{f}(z)\| \le D|z|^{-C} \tag{14.3.3}$$

**Proof.** A simple use of *Gronwall's inequality*. For $r < \frac{1}{2}$,

$$\|\boldsymbol{f}(re^{i\theta})\| \le \|\boldsymbol{f}(\tfrac{1}{2}\,e^{i\theta})\| + \int_r^{1/2} \left\| \frac{d\boldsymbol{f}}{dz}\,(\rho e^{i\theta}) \right\| d\rho \tag{14.3.4}$$

$$\le \|\boldsymbol{f}(\tfrac{1}{2}\,e^{i\theta})\| + C \int_r^{1/2} \frac{\|\boldsymbol{f}(\rho e^{i\theta})\|}{\rho}\, d\rho \tag{14.3.5}$$

by (14.3.1) and (14.3.2).

Fix $\theta$. Let $g(r)$ be the right side of (14.3.5), so

$$\frac{dg}{dr} = -\frac{C\|\boldsymbol{f}(re^{i\theta})\|}{r} \ge -\frac{Cg(r)}{r} \tag{14.3.6}$$

by (14.3.4). Thus,

$$\frac{d\log(g)}{dr} \ge -\frac{C}{r} \tag{14.3.7}$$

so

$$\log(g(r)) = \log(g(\tfrac{1}{2})) - \int_r^{1/2} \frac{d}{d\rho}\log(g(\rho))\,d\rho$$

$$\le \log(g(\tfrac{1}{2})) + C\log(2r) \tag{14.3.8}$$

by (14.3.4), which implies (14.3.3) with

$$D = 2^C \sup_{\alpha\le\theta\le\rho} \|\boldsymbol{f}(\tfrac{1}{2}\,e^{i\theta})\| \tag{14.3.9}$$

□

At first sight, this lemma seems useless for what we want! The standard way to turn an $n$-th-order (14.0.3) to a first-order vector system is to use the vector

$$\boldsymbol{f}(z) = \begin{pmatrix} f(z) \\ f'(z) \\ \vdots \\ f^{(n-1)}(z) \end{pmatrix} \tag{14.3.10}$$

which obeys (14.3.1) with

$$A(z) = \begin{pmatrix} 0 & 1 & 0 & \dots & 0 \\ 0 & 0 & 1 & \dots & 0 \\ \vdots & \vdots & \vdots & & \vdots \\ a_0(z) & a_1(z) & a_2(z) & \dots & a_{n-1}(z) \end{pmatrix} \tag{14.3.11}$$

which has $\|A(z)\| = O(z^{-n})$, not $O(z^{-1})$. The key is to note that $f^{(j)}(z)$ have different orders at $z = 0$. If $|f(z)| = O(|z|^{-\lambda})$, then $f^{(j)}(z) = O(|z|^{-\lambda-j})$. This suggests we define $\boldsymbol{g}(z)$ by

$$g_j(z) = z^{j-1} f^{(j-1)}(z) \tag{14.3.12}$$

This then obeys

$$\boldsymbol{g}'(z) = B(z)\boldsymbol{g}(z) \tag{14.3.13}$$

where (since $g_j'(z) = z^{-1}g_{j+1}(z) - (j-1)z^{-1}g_j(z)$, $j = 1, \dots, n-2$, and $g_{n-1}'(z) = -(n-1)z^{-1}g_{n-1} + \sum_{k=0}^{n-1} z^{n-k-1}a_k(z)g_{k+1}(z)$)

$$B(z) = \begin{pmatrix} 0 & z^{-1} & 0 & \dots & 0 \\ 0 & -z^{-1} & z^{-1} & \dots & 0 \\ 0 & 0 & -2z^{-1} & \dots & 0 \\ \vdots & \vdots & \vdots & & \vdots \\ z^{n-1}a_0(z) & z^{n-2}a_1(z) & z^{n-3}a_2(z) & \dots & (a_{n-1}(z) - (n-1)z^{-1}) \end{pmatrix} \tag{14.3.14}$$

Thus, if $a_k(z) = O(z^{-(n-k)})$, we have

$$\|B(z)\| = O(|z|^{-1}) \tag{14.3.15}$$

**Proof of Theorem 14.3.1.** By Theorem 14.2.5, we need only prove a bound of the form (14.2.23) and we can do that in $S_\pm \equiv \{z \mid 0 < |z| < \frac{1}{2}, \arg z \in \pm[-\frac{\pi}{2}, \pi]\}$. By Lemma 14.3.3 and the transformation (14.3.12), each component of each $\boldsymbol{g}$ obeys such a bound, and so each $f = g_1$ does. $\square$

**Lemma 14.3.4.** *Let $V_0$ be a one-dimensional monodromic family on $\mathbb{D}^\times$ and let $f \in V_0$ be nonvanishing on $\mathbb{D}^\times$. Then*

(a)

$$h(z) = \frac{f'(z)}{f(z)} \tag{14.3.16}$$

*is analytic on $\mathbb{D}^\times$.*

(b) *If*

$$D_f = \frac{d}{dz} - h(z) \tag{14.3.17}$$

*then $D_f g = 0$ if and only if $g$ is a constant multiple of $f$.*

(c) *$V_0$ has a regular singularity at $z = 0$ or is regular there if and only if $h(z)$ has a simple pole or removable singularity at $z = 0$.*

**Remark.** (c) is an extension of Theorem 3.8.3 of Part 2A.

**Proof.** (a) Since $V_0$ is one-dimensional, for any curve $\gamma$ in $\mathbb{D}^\times$, $\tau_\gamma(f) = \lambda_\gamma f$, so $\tau_\gamma(f)' = \lambda_\gamma f'$, and thus, $\tau_\gamma(h) = h$, that is, $h$ is globally analytic on $\mathbb{D}^\times$.

(b) If $D_f g = 0$, then $(g/f)' = (fg' - gf')/f^2 = (D_f g)/f = 0$, so $g/f$ is constant.

(c) If $f(z) = z^\lambda g(z)$ with $g(z)$ analytic in $\mathbb{D}^\times$ with $g(0) \neq 0$, then $h(z) = \lambda z^{-1} + g'(z)/g(z)$ has a simple pole at zero or is analytic there if $\lambda = 0$. Thus, if $V_0$ is regular or has a regular singularity at zero, then $h$ has a simple pole or removable singularity.

Conversely, if $h(z) = \lambda z^{-1} + p(z)$ with $p$ analytic at $z = 0$, then $f(z)z^{-\lambda}\exp(-\int_0^z p(u)\,du)$ has zero derivative, so $f$ has a regular singular point or regular point (if $\lambda = 0$) at $z = 0$. $\square$

We also need some algebra in the noncommutative ring of analytic linear differential operators. Given a region, $\Omega$, $\mathcal{D}(\Omega)$ is the set of finite sums of products of $D \equiv \frac{d}{dz}$ and $M_a$'s, the operators of multiplication by $a$, a function in $\mathfrak{A}(\Omega)$, that is, $M_a f = af$ for a germ, $f$. An operator of the form $\mathcal{O} \equiv \sum_{j=0}^n M_{a_j} D^j$, with $a_n \not\equiv 0$, is said to be in *normal form* and be of *degree* $n$, written $\deg(\mathcal{O}) = n$. By $[D, M_a] = M_{a'}$, every differential operator can be written in normal form. If $a_n \equiv 1$, we say $\mathcal{O}$ is *monic*.

**Proposition 14.3.5** (Division Algorithm for Differential Operators). *Let $h \in \mathfrak{A}(\Omega)$ and let $\mathcal{O} \in \mathcal{D}(\Omega)$ with $\deg(\mathcal{O}) = n \geq 1$. Then there exists $\widetilde{\mathcal{O}} \in \mathcal{D}(\Omega)$ with $\deg(\widetilde{\mathcal{O}}) = n - 1$ and $q \in \mathfrak{A}(\Omega)$ so that*

$$\mathcal{O} = \widetilde{\mathcal{O}}(D - M_h) + M_q \tag{14.3.18}$$

*If $\mathcal{O}$ is monic, so is $\widetilde{\mathcal{O}}$. Moreover, if there is a nonzero germ, $f$, at $z_0$, so $\mathcal{O}f = (D - M_h)f = 0$, then $q = 0$.*

**Proof.** We use induction in $n$. For $n = 1$, if $\mathcal{O} = M_{a_1}D + M_{a_0}$, then $\widetilde{\mathcal{O}} = M_{a_1}$ and $q = M_{a_0 + a_1 h}$. Since $M_{a_n}D^n = M_{a_n}D^{n-1}(D - M_h) + \mathcal{O}^\sharp$, with $\mathcal{O}^\sharp = M_{a_n}D^{n-1}M_h$ of degree $n - 1$, by induction we get (14.3.18).

Applying (14.3.18) to $f$, we see that if $(D - M_h)f = \mathcal{O}f = 0$, then $qf = 0$. So $q$ vanishes near $z_0$ and so is 0. $\square$

**Proof of Theorem 14.3.2.** The proof is by induction in the dimension, $n$, of $V$. Since the $a_j$'s are determined by $V$ (by Theorem 14.1.3) and are analytic away from $z = 0$ (also by Theorem 14.1.3), we need only prove the $a_j$'s have poles of order at most $n - j$ at $z = 0$, that is, control $a_j$ near $z = 0$.

If $n = 1$, pick a nonzero $f$ in $V$. Then $h$, given by (14.3.16), has a simple pole by Lemma 14.3.4, and $V = \{g \mid D_f g = 0\}$ by the same lemma. Thus, $a_0(z) = h(z)$ has a pole of order $1 - 0 = 1$ as required.

Suppose we have the result when $\widetilde{V}$ has dimension $n - 1$ and let $V$ be a dimension $n$ monodromic family. By Theorem 14.2.3, there exists $f \in V$ which is an eigenfunction of $C$, the circuit matrix. Since 0 is a regular singular point, $f$ is nonzero near 0 and so $h = f'/f$ is analytic in some $(\mathbb{D}_0(\delta))^\times$. Let $\mathcal{O}$ be the differential operator of the form (14.0.3) (i.e., $\mathcal{O} = D^n - \sum_{k=0}^{n-1} M_{a_k} D^k$) guaranteed by Theorem 14.1.3. By Proposition 14.3.5, we can write $\mathcal{O}$ in the form (14.3.18), and since $\mathcal{O}f = (D - M_h)f = 0$, $q = 0$.

Let $\widetilde{V} = (D - M_h)[V] = \{g' - hg \mid g \in V\}$. Since analytic continuation of $g$ yields a continuation of $g' - hg$, $\widetilde{V}$ is a monodromic family. By Lemma 14.3.4, $\dim(\widetilde{V}) = n - 1$ and, by (14.3.18), any germ, $\eta \in \widetilde{V}$, obeys $\widetilde{\mathcal{O}}\eta = 0$, so $\widetilde{V}$ is the set of all solutions of $\widetilde{\mathcal{O}}\eta = 0$. Thus, $\widetilde{V}$ is an $(n-1)$-dimensional, strongly nondegenerate monodromic family. Since, by a Cauchy estimate, $g$ polynomially bounded in $|z|^{-1}$ in a sector implies the same for $g' - hg$ in any smaller sector, we conclude, by the induction hypothesis, that $\widetilde{\mathcal{O}}$ has the form

$$\widetilde{\mathcal{O}} = \frac{d^{n-1}}{dz^{n-1}} - \sum_{k=0}^{n-2} \tilde{a}_k \left(\frac{d}{dz}\right)^k \tag{14.3.19}$$

with

$$|\tilde{a}_k(z)| \le \widetilde{C}_k |z|^{(-n-1-k)} \tag{14.3.20}$$

By (14.3.18), if

$$\mathcal{O} = \frac{d^n}{dz} - \sum_{k=0}^{n-1} a_k \left(\frac{d}{dz}\right)^k \tag{14.3.21}$$

then, by Leibniz's rule (Problem 1(a))

$$a_k = \tilde{a}_{k-1} + \binom{n-1}{k} h^{(n-1-k)} - \sum_{j=k}^{n-1} \binom{j}{k} \tilde{a}_j h^{(j-k)} \tag{14.3.22}$$

Since $|h^{(\ell)}| \le C|z|^{-\ell-1}$ near $z = 0$, we see that (Problem 1(b))

$$|a_k(z)| \le C_k |z|^{-(n-k)} \tag{14.3.23}$$

Thus, the theorem is proven inductively. □

A differential equation of the form (14.0.3) is thus said to have a *regular singular point* at $z = 0$ if (14.3.23) holds.

Having completed the most significant results of this section, we turn next to the issue of the indices of the monodromy as defined at the end of Section 14.2. An *Euler differential equation* (also called *Euler–Cauchy equation*) is one of the form

$$f^{(n)}(z) = \sum_{k=0}^{n-1} \frac{c_k}{z^{n-k}} f^{(k)}(z) \tag{14.3.24}$$

for constants, $c_k$. If $f(z) = z^\lambda$, it satisfies (14.3.24) if and only if

$$\lambda(\lambda - 1)\dots(\lambda - n + 1) - \sum_{k=0}^{n-1} c_k \lambda(\lambda - 1)\dots(\lambda - k + 1) = 0 \tag{14.3.25}$$

(where $\lambda(\lambda-1)\dots(\lambda-k+1)$ for $k = 0$ is interpreted as 1). (14.3.25) is called the *indicial equation* for (14.3.24). If the roots of (14.3.25) are distinct, say

$\lambda_1, \dots \lambda_n$, then a basis of solutions of (14.3.24) are $z^{\lambda_1}, \dots, z^{\lambda_n}$. If some $\lambda_j$ is a multiple root of multiplicity $\ell > 1$, we need $z^{\lambda_j}, z^{\lambda_j} L(z), \dots, z^{\lambda_j} L(z)^{\ell-1}$.

If the differential equation (14.0.3) has a regular singular point at $z_0 \in \mathbb{C}$, we set $c_k = \lim_{z \to z_0, z \neq z_0} (z - z_0)^{n-k} a_k(z)$ and call (14.3.25) the *indicial equation at* $z_0$ of the ODE. Here is the main point:

**Theorem 14.3.6.** *The indices of a strongly nondegenerate monodromic family, $V$, at a regular singular point are precisely the solutions of the indicial equation of its associated differential equation.*

**Proof.** Without loss, suppose $z_0 = 0$. We use induction in $n$. If $n = 1$ and $f$ solves the differential operator with $f(z) = z^{\lambda_0} g(z)$, where $g$ is regular at $z = 0$ with $g(0) \neq 0$, then the differential equation is $f' - hf = 0$, where $h = f'/f = \lambda_0/z + g'/g$. The indicial equation is $\lambda - \lambda_0 = 0$ and the index is $\lambda_0$.

Suppose we know the result for $\widetilde{V}$ of dimension $n - 1$ and that $V$ has dimension $n$. Pick an eigenfunction of the circuit matrix of the form $f(z) = z^{\lambda_0} g(z)$ and let

$$h(z) = \frac{f'(z)}{f(z)} = \frac{\lambda_0}{z} + \frac{g'(z)}{g(z)} \tag{14.3.26}$$

Let $q(z) \in V$ be another Jordan basis element. Let $\eta(z) = q'(z) - h(z)q(z)$.

We claim the indices $\mu_q$ and $\mu_\eta$ of $q$ and $\eta$ are related by

$$\mu_\eta = \mu_q - 1 \tag{14.3.27}$$

If $\mu_q \neq \lambda_0$, this is immediate from $(\frac{d}{dz} - \frac{\lambda_0}{z}) z^{\mu_q} = (\mu_q - \lambda_0) z^{\mu_q - 1}$. If $\mu_q = \lambda_0$, it relies on the remark at the end of Section 14.2 about the presence of logs when there are equal index solutions by our choice of basis.

The same analysis works on the Euler equation. Since the roots of the indicial equation are the powers in the solution of the Euler equation, we have the induction step. □

We are thus left with the following algorithm for a differential equation at a regular singular point, say $z_0 = 0$. Let $\lambda$ be a root of the indicial equation. If $\lambda - \mu \neq \mathbb{Z}$ for all the other roots $\mu$, then $e^{2\pi i \lambda}$ is a simple eigenvalue of the circuit matrix, so there is a solution of the form

$$\sum_{m=0}^{\infty} c_m z^{\lambda + m} \tag{14.3.28}$$

with $c_0 \neq 0$. If $\lambda - \mu \in \mathbb{Z}$ for some other root $\mu$, there may then be some log terms. Finer analysis shows if there is no $\mu$ with $\mu - \lambda \in \{1, 2, \dots\}$, then one still has a solution of the form (14.3.28).

As a final general subject, we turn to the question of what happens at infinity. We'll do this only in the second-order case. For comparison, it pays to write out the indicial equation at finite orders in this second-order case. For

$$\mathcal{O} = \frac{d^2}{dz^2} - a_1(z)\,\frac{d}{dz} - a_0(z) \tag{14.3.29}$$

if

$$a_1(z) = \frac{c_1}{z - z_0} + O(1), \qquad a_0(z) = \frac{c_2}{(z-z_0)^2} + O\biggl(\frac{1}{z-z_0}\biggr) \tag{14.3.30}$$

the indicial equation is

$$\lambda(\lambda-1) - c_1\lambda - c_0 = 0 \tag{14.3.31}$$

To look at infinity, let $w = z^{-1}$ so $z = w^{-1}$, and by the chain rule,

$$\frac{d}{dz} = \frac{dw}{dz}\,\frac{d}{dw} = -z^{-2}\,\frac{d}{dw} = -w^2\,\frac{d}{dw} \tag{14.3.32}$$

and

$$\biggl(\frac{d}{dz}\biggr)^2 = w^2\,\frac{d}{dw}\biggl(w^2\,\frac{d}{dw}\biggr) = w^4\,\frac{d^2}{dw^2} + 2w^3\,\frac{d}{dw} \tag{14.3.33}$$

Thus, the operator $\mathcal{O}$ of (14.3.29) becomes

$$\mathcal{O} = w^4\biggl[\frac{d}{dw^2} - b_1(w)\,\frac{d}{dw} - b_0(w)\biggr] \tag{14.3.34}$$

where

$$b_1(w) = -2w^{-1} - w^{-2}a_1\biggl(\frac{1}{w}\biggr) \tag{14.3.35}$$

$$b_0(w) = w^{-4}a_0\biggl(\frac{1}{w}\biggr) \tag{14.3.36}$$

For $w = 0$, to be a regular point, we need $b_1(w) = O(w^{-1})$ and $b_0(w) = O(w^{-2})$. Equivalently, $w^{-2}a_1(\frac{1}{w}) = O(\frac{1}{w})$ (or $a_1(\frac{1}{w}) = O(w)$) and $w^{-4}a_0(\frac{1}{w}) = O(\frac{1}{w^2})$ (or $a_0(\frac{1}{w}) = O(w^2)$), so translating back to $z$, we see that

**Theorem 14.3.7.** *Let $\mathcal{O}$ be a second-order differential operator of the form (14.3.23). Suppose $a_0(z), a_1(z)$ are analytic outside some $\overline{\mathbb{D}_r(0)}$. Then $\infty$ is a regular singular point if and only if near infinity,*

$$|a_1(z)| = O(z^{-1}), \qquad |a_0(z)| = O(z^{-2}) \tag{14.3.37}$$

*If*

$$a_1(z) = c_1z^{-1} + O(z^{-2}), \qquad a_0(z) = c_0z^{-2} + O(z^{-3}) \tag{14.3.38}$$

*then the indicial equation at infinity (in terms of $z^{-\gamma}$) is*

$$\gamma(\gamma-1) + (2+c_1)\gamma - c_0 = 0 \tag{14.3.39}$$

**Example 14.3.8** (Hypergeometric Equation). The equation (14.0.5) has

$$a_1(z) = -\frac{c - (a+b+1)z}{z(1-z)}, \qquad a_0(z) = \frac{ab}{z(1-z)} \tag{14.3.40}$$

$a_1$ has first-order poles at $z = 0$ and $z = 1$ and is $O(z^{-1})$ at infinity, while $a_0$ has only first-order poles at $z = 0$ and $z = 1$ and is $O(z^{-2})$ at infinity. Thus, $0, 1, \infty$ are all regular singular points. An equation on $\widehat{\mathbb{C}}$, all of whose singular points are regular, is called *Fuchsian*. That $a_0$ has only first-, not second-order poles, at $z = 0$ and $z = 1$ means one index at each point is 0, that is, there are solutions analytic at each point (but normally different solutions). We'll say a lot more about this in the next section. □

**Example 14.3.9** (Bessel's Equation). The equation (14.0.8) has

$$a_1(z) = -\frac{1}{z}, \qquad a_0(z) = -\left(1 - \frac{\alpha^2}{z^2}\right) \tag{14.3.41}$$

Since $a_1$ has a first-order pole at 0 and $a_0$ a second-order pole, $z = 0$ is a regular singular point. Since $a_0(z)$ is $O(1)$ and not $O(z^{-2})$ at infinity, $z = \infty$ is an irregular singular point, and the equation is not Fuchsian. The indicial equation at 0 is

$$\lambda(\lambda - 1) + \lambda - \alpha^2 = 0 \tag{14.3.42}$$

solved by $\lambda = \pm\alpha$. If $2\alpha \notin \mathbb{Z}$, the circuit matrix has distinct eigenvalues, so there are solutions of the form $z^{\pm\alpha}g(z)$ with $g$ analytic and $g(0) \neq 0$. In Section 14.5, we'll see that it remains true if $\alpha + \frac{1}{2} \in \mathbb{Z}$ but if $\alpha \in \mathbb{Z}$, there are logarithmic terms in the solutions which is $O(z^{-|\alpha|})$ at $z = 0$. □

**Example 14.3.10** (Airy's Equation). In (14.0.10), all points in $\mathbb{C}$ are regular. Since $a_0(z)$ is not $O(z^{-2})$ at infinity, we see $\infty$ is an irregular singular point. □

**Notes and Historical Remarks.** The theory of this section is associated with Lazarus Fuchs (1833–1902) and Ferdinand Georg Frobenius (1849–1917), but relied heavily on the earlier work on the hypergeometric equation by Euler, Gauss, Jacobi, Kummer, and Riemann discussed in the Notes to the next section.

Fuchs [**143, 144, 145**] in 1865–68 had essentially complete results based on the study of the monodromy possibilities. Frobenius [**139, 140**] in 1873, when he was only 24, simplified Fuchs' proof by systematic use of generalized power series, $\sum_{n=0}^\infty a_n z^{\lambda+n}$, and direct substitution. Because the "Frobenius method" was more palatable to many, his role is often overemphasized to the point that he is sometimes given credit for work of Fuchs, such as the indicial equation; Frobenius was always scrupulous in giving Fuchs the credit he deserves.

The two new ideas in Frobenius are direct construction of solutions near a regular singular point by estimates of the series and the notion of irreducibility of a differential equation, which he defined as having no solutions that solve a lower-order equation—near individual points, there are always such solutions if $n \geq 2$ as we have seen, but one looks at global solutions on $\widehat{\mathbb{C}} \setminus \{\text{singular points}\}$.

Fuchs was a student of Kummer at Berlin and was influenced by Weierstrass. He began his famous work as a Privatdozent at Berlin, but then moved to other universities until returning to Berlin as Kummer's successor in 1884. Fuchs' students include Landau and Schur, students he shared with Frobenius. Besides the work on ODEs, he is known for Fuchsian groups (see Section 8.3 of Part 2A) and Fuchsian functions, a name invented by Poincaré in reference to their use in some of Fuchs' work on ODEs.

Like Fuchs, Frobenius was a student at Berlin, but of Weierstrass rather than Kummer. Also like Fuchs, he eventually returned to Berlin. He did so in 1893 to fill the vacancy caused by Kronecker's death. Frobenius was the dominant figure in Berlin's math department for twenty-five years, where he saw with chagrin the rise of Göttingen under Klein and later Hilbert with a style of mathematics he disliked. He had bad relations with Klein and Lie. Even more than his work on ODEs, Frobenius' reputation relies on his work in number theory and especially on representations of finite groups. The Notes to Section 6.8 of Part 4 has a capsule biography of Frobenius.

For a discussion of the history of ODEs in the complex domain, see Gray [**160**]. For a more extensive textbook discussion, see Hille [**190**] or Teschl [**383**]. Euler–Cauchy differential equations go back to Euler [**128**] in 1769.

An interesting insight on the solutions of the Euler–Cauchy equation can be seen by the change of variables $z = e^u$ which turns (14.3.24) into

$$g^{(n)}(u) = \sum_{k=0}^{n-1} b_k g^{(k)}(u) \tag{14.3.43}$$

if $g(u) = f(\log z)$. This equation (also first solved in generality by Euler) has solutions of the form $e^{\lambda_j u}$, where $(\lambda_j^n - \sum b_k \lambda_j^k) = 0$. If there are multiple roots, $u^\ell e^{\lambda_j u}$ also occurs. Since $e^{\lambda_j u} = z^{\lambda_j}$ and $u^\ell e^{\lambda_j u} = z^{\lambda_j}(\log z)^\ell$, we see the two equations and their solutions are essentially equivalent.

Fuchsian ODEs are also the essence of Hilbert's twenty-first problem which says: "To show that there always exists a linear differential equation of the Fuchsian class, with given singular points and monodromic group." That is, given finitely many points, $\zeta_1, \dots, \zeta_n \in \widehat{\mathbb{C}}$ and a finite-dimensional representation of $\pi_1(\widehat{\mathbb{C}} \setminus \{\zeta_1, \dots, \zeta_n\})$, find a Fuchsian system with that monodromy group. Note that we've phrased it as a system, that is, matrix-valued

ODEs all of whose solutions have at most power blow-up at the singular points. After work of Riemann, this is sometimes called the Riemann–Hilbert problem.

The history is complicated (see Yandell [**414**]). This was widely believed to have been solved by the Slovenian mathematician, Josip Plemelj (1873–1967), in 1908. In fact, Plemelj only solved a slightly different problem: A system is called Fuchsian if $\|A(z)\| \leq C|z-z_0|^{-1}$ near any singular point, and regular if all solutions have at most polynomial growth in any sector near a singular point. Plemelj proved there was a regular ODE with the prescribed monodromy and remarked that the system he found was Fuchsian except at perhaps one point. In 1989, the Russian mathematician, Andrei Bolibruch (1950–2003), found a counterexample to the precise Hilbert problem. The conjecture is true if the representation of $\pi_1$ is known to be irreducible.

## Problems

1. (a) Verify (14.3.22).

   (b) Prove that (14.3.23) holds.

2. Let $f$ obey the differential equation

$$z^2 f'' - (4z - z^2)f' + (4 - z)f = 0 \tag{14.3.44}$$

   (a) Find and solve the indicial equation.

   (b) Let $g = z^{-1}f$. Find the differential equation for $g$ and then the solutions of (14.3.44).

   (c) Why is $z = 0$ a singular point even though all solutions are analytic? Such singularities are called *accidental singularities.*

3. Consider Legendre's equation ($\alpha$ and $\beta$ are complex parameters)

$$(1 - z^2)\frac{d^2 f}{dz^2} + 2z\frac{df}{dz} + \left[\alpha(\alpha+1) - \frac{\beta^2}{1 - z^2}\right] f = 0 \tag{14.3.45}$$

   Find the singular points in $\widehat{\mathbb{C}}$, determine if they are regular. And if they are regular, determine the indices at the point.

   **Note.** Some authors use "Legendre equation" for the case $\beta = 0$ and call the general $\beta$ case the "associated Legendre equation."

4. Find the singular points in $\widehat{\mathbb{C}}$ and classify them for

   (a) $z^3(z^2 - 1)f'' - 6z(z + 1)f' + (z^2 + 7)f = 0$

   (b) $z^2 f'' + (e^{-z} - 1)f' + 7f = 0$

   (c) $z(z^2 - 1)f'' - 2zf' + f = 0$

## 14.4. Hypergeometric Functions

> Gelfand started our "negotiations" with a frontal attack. "Well, you are doing some homological algebra but we already have Beilinson for that. If you are going to work with me, you have to start from scratch. In medieval times painter's pupils worked for years just preparing paints for the master. Do you know what a hypergeometric function is? No? Very well, you can work with me on hypergeometric functions." After a few days Gelfand changed tactics. He asked me to open the celebrated handbook of Bateman and Erdélyi and point out the formulas I liked. He reacted to my choices quite positively: "Well, you have some taste. Why were you so interested in that abstract nonsense?"
>
> —*Vladimir Retakh* [**326**][2]

In this section, we'll study the simplest Fuchsian ODE. We'll begin by showing invariance under FLTs and then the triviality of the degree-1 and of the two-point case. We'll then turn to the simplest nontrivial example: the second-order case with three regular singular points, which we can suppose are $0, 1, \infty$.

We'll next prove a theorem of Papperitz that for any choice of indices, $\alpha_1, \alpha_2, \beta_1, \beta_2, \gamma_1, \gamma_2$, at $0, 1, \infty$, respectively, there is a unique degree-2 Fuchsian ODE with $0, 1, \infty$ as singular points and those indices respectively at $0, 1, \infty$ if and only if

$$\alpha_1 + \alpha_2 + \beta_1 + \beta_2 + \gamma_1 + \gamma_2 = 1 \tag{14.4.1}$$

We'll then turn to a viewpoint of Riemann that relates solutions with different $\alpha, \beta, \gamma$ and reduce to the case $\alpha_1 = \beta_1 = 0$. So long as a particular combination is not a negative integer, we'll be able to write the solution regular at $z = 0$ as an explicit power series which we'll identify as ${}_2F_1$, the basic hypergeometric series.

Then we'll define general ${}_pF_q$ and then turn to properties of ${}_2F_1$, and next to ${}_1F_1$, the conformal hypergeometric function which has an irregular singular point at infinity. Finally, we'll see hypergeometric functions include many other functions, including the complete elliptic integrals, $K(k)$ and $E(k)$.

**Theorem 14.4.1.** *Let $\Omega_1, \Omega_2$ be two regions in $\widehat{\mathbb{C}}$ and $F\colon \Omega_1 \to \Omega_2$ an analytic bijection. Suppose $z_1, z_2$ are isolated points in $\widehat{\mathbb{C}} \setminus \Omega_1$ and $\widehat{\mathbb{C}} \setminus \Omega_2$, respectively, and that $F$ extends to an analytic bijection of $\Omega_1 \cup \{z_1\}$ to $\Omega_2 \cup \{z_2\}$. Let $V$ be a monodromic family on $\Omega_2$. Then $\widetilde{V} \equiv \{f \circ F \mid f \in V\}$ is a monodromic family on $\Omega_1$. $z_2$ is a regular singular point of $V$ if and only if $z_1$ is a regular singular point of $\widetilde{V}$, and in that case, the indices are the same.*

[2]describing his initial work with Israel Gelfand

**Proof.** Immediate, given that near $z_1$, $F(z) - z_2 = C(z - z_1) + O((z - z_1)^2)$ with $C \neq 0$. □

In particular, we can use an FLT to reduce a Fuchsian ODE with two or three singular points to another one with any particular singular points we wish. We begin with the degree-1 ODE, that is, $\dim(V) = 1$.

**Proposition 14.4.2.** *Let $\mathcal{O}$ be a first-order Fuchsian ODE with singular points $z_1, \dots, z_{n-1}, z_n = \infty$ and indices $\lambda_1, \dots, \lambda_n$. Then*

$$\sum_{j=1}^{n} \lambda_j = 0 \tag{14.4.2}$$

*The function $f$ solving $\mathcal{O}f = 0$ has the form*

$$f(z) = c \prod_{j=1}^{n-1} (z - z_j)^{\lambda_j} \tag{14.4.3}$$

*and $\mathcal{O}$ has the form (for a constant $k$)*

$$\mathcal{O}g = k\biggl(g' - \sum_{j=1}^{n-1} \frac{\lambda_j}{z - z_j} g\biggr) \tag{14.4.4}$$

*Moreover, if $\{\lambda_j\}_{j=1}^{n}$ obey (14.4.2), there is such a Fuchsian ODE.*

**Proof.** By hypothesis,

$$h(z) = \frac{f(z)}{\prod_{j=1}^{n-1} (z - z_j)^{\lambda_j}} \tag{14.4.5}$$

$h$ is analytic at each $z_j$ and nonvanishing there, and so analytic and nonvanishing on all of $\mathbb{C}$. Since $\infty$ is a regular singular point, $|h(z)| \leq D|z|^N$ near infinity. This implies first that $h$ is a polynomial and then a constant since it has no zeros. So $f$ has the form (14.4.3). Such an $f$ is $O(z^{-\lambda_n})$ if and only if (14.4.2) holds. □

If $\deg(\mathcal{O}) = 2$ and there are only two singular points, we can suppose they are 0 and $\infty$, where we have

**Proposition 14.4.3.** *If $\mathcal{O}$ is a second-order Fuchsian ODE whose only two singular points are 0 and $\infty$, then the indices $\lambda_1, \lambda_2$ at 0 and $\mu_1, \mu_2$ at $\infty$ obey $\lambda_j = -\mu_j$ and the solutions are*

$$f(z) = cz^{\lambda_1} + dz^{\lambda_2} \tag{14.4.6}$$

*if $\lambda_1 \neq \lambda_2$ and*

$$f(z) = cz^{\lambda_1}(1 + dL(z)) \tag{14.4.7}$$

*if* $\lambda_1 = \lambda_2$. *The ODE has the form (for a constant* $k$*)*

$$\mathcal{O}g = k\biggl(g'' + \biggl(\frac{1-\lambda_1-\lambda_2}{z}\biggr)g' + \frac{\lambda_1\lambda_2}{z^2}\,g\biggr) \tag{14.4.8}$$

**Proof.** The ODE has the form

$$\mathcal{O}g = k(g'' + P(z)g' + Q(z)g) \tag{14.4.9}$$

where $P, Q$ are rational with poles only at 0 and behavior $P(z) = O(1/z)$, $Q(z) = O(1/z^2)$ at infinity (by Theorem 14.3.7). Since $P$ (respectively, $Q$) only have first-order (respectively, second-order) poles at 0, we see that

$$\mathcal{O}g = k\biggl(g'' + \frac{A}{z}\,g' + \frac{B}{z^2}\,g\biggr) \tag{14.4.10}$$

This is an Euler differential equation with solutions $z^{\lambda_1}, z^{\lambda_2}$ if $\lambda_1, \lambda_2$ solve $\lambda(\lambda-1) + A\lambda + B = 0$, with a log solution if $\lambda_1 = \lambda_2$. □

Thus, the case $\deg(\mathcal{O}) = 2$ with three critical points is the simplest nontrivial case. We may as well take the critical points as $0, 1, \infty$ (see Problem 1). Here is the main classification theorem:

**Theorem 14.4.4** (Papperitz's Theorem)**.** *There exists a second-order Fuchsian ODE with singular points,* $0, 1, \infty$*, and indices,* $\alpha_1, \alpha_2$ *at* $0$*,* $\beta_1, \beta_2$ *at* $1$*, and* $\gamma_1, \gamma_2$ *at* $\infty$*, if and only if* (14.4.1) *holds. It is unique (up to a constant) and has the form*

$$\begin{aligned} f'' &+ \biggl[\frac{1-\alpha_1-\alpha_2}{z} + \frac{1-\beta_1-\beta_2}{z-1}\biggr]f' \\ &+ \biggl[\frac{\alpha_1\alpha_2}{z^2} + \frac{\beta_1\beta_2}{(z-1)^2} + \frac{\gamma_1\gamma_2-\alpha_1\alpha_2-\beta_1\beta_2}{z(z-1)}\biggr]f = 0 \end{aligned} \tag{14.4.11}$$

**Remark.** (14.4.11) is sometimes called the *Papperitz–Riemann equation.*

**Proof.** If the ODE is $f'' - a_1(z)f' - a_0(z)f = 0$, then $a_1$ is $O(z^{-1})$ at $\infty$ with simple poles allowed only at 0 and 1. It follows (Problem 2) that $a_1(z) = -\frac{A}{z} - \frac{B}{z-1}$. Similarly, $a_0$ is $O(z^{-2})$ at $\infty$ and at worst double poles at 0 and 1. It thus follows (Problem 2(c)) that $a_0(z) = -\frac{C}{z^2} - \frac{D}{z(z-1)} - \frac{E}{(z-1)^2}$, so the ODE is

$$f'' + \biggl(\frac{A}{z} + \frac{B}{z-1}\biggr)f' + \biggl(\frac{C}{z^2} + \frac{D}{z(z-1)} + \frac{E}{(z-1)^2}\biggr)f = 0 \tag{14.4.12}$$

The indicial equations are

$$\alpha(\alpha-1) + A\alpha + C = 0 \tag{14.4.13}$$

$$\beta(\beta-1) + B\beta + E = 0 \tag{14.4.14}$$

$$\gamma(\gamma+1) - (A+B)\gamma + (C+D+E) = 0 \tag{14.4.15}$$

Thus,

$$A = 1 - \alpha_1 - \alpha_2\,, \qquad C = \alpha_1\alpha_2 \tag{14.4.16}$$

$$B = 1 - \beta_1 - \beta_2\,, \qquad E = \beta_1\beta_2 \tag{14.4.17}$$

$$A + B = \gamma_1 + \gamma_2 + 1\,, \qquad C + D + E = \gamma_1\gamma_2 \tag{14.4.18}$$

For this to have a solution, it is necessary and sufficient that (14.4.1) holds, and then (14.4.12) becomes (14.4.4). □

We now shift to Riemann's point of view. Riemann realized that in worrying about transformation properties, one should look at the totality of solutions of an ODE. So the *Riemann $\mathcal{P}$-function* which is defined with the symbol

$$\mathcal{P}\begin{Bmatrix} \zeta_1 & \zeta_2 & \zeta_3 & \\ \alpha_1 & \beta_1 & \gamma_1 & z \\ \alpha_2 & \beta_2 & \gamma_2 & \end{Bmatrix} \tag{14.4.19}$$

where $z, \zeta_j \in \widehat{\mathbb{C}}$, $\alpha_j, \beta_j, \gamma_j \in \mathbb{C}$, and $\zeta_1 \neq \zeta_2 \neq \zeta_3 \neq \zeta_1$ represents the two-dimensional family of germs at $z$ associated to the Fuchsian system, with singularities at $\zeta_j$ and indices $\alpha, \beta, \gamma$. Here are the basic relations, which are all obvious:

**Theorem 14.4.5.** (a) *$\mathcal{P}$ is invariant under permutations of the first three columns and interchanging of $\alpha_1$ and $\alpha_2$ (or $\beta_1$ and $\beta_2$ or $\gamma_1$ and $\gamma_2$).*

(b) *For $Q$ any FLT, $\mathcal{P}$ is invariant under $\zeta_j \to Q(\zeta_j)$, $z \to Q(z)$ for all $j$.*

(c)
$$(z - \zeta_1)^\mu \mathcal{P}\begin{Bmatrix} \zeta_1 & \zeta_2 & \infty & \\ \alpha_1 & \beta_1 & \gamma_1 & z \\ \alpha_2 & \beta_2 & \gamma_2 & \end{Bmatrix} = \mathcal{P}\begin{pmatrix} \zeta_1 & \zeta_2 & \infty & \\ \alpha_1 + \mu & \beta_1 & \gamma_1 - \mu & z \\ \alpha_2 + \mu & \beta_2 & \gamma_2 - \mu & \end{pmatrix} \tag{14.4.20}$$

In particular,

$$\begin{aligned} &\mathcal{P}\begin{Bmatrix} 0 & 1 & \infty & \\ \alpha_1 & \beta_1 & \gamma_1 & z \\ \alpha_2 & \beta_2 & \gamma_2 & \end{Bmatrix} \\ &= z^{\alpha_1}(1-z)^{\beta_1}\mathcal{P}\begin{Bmatrix} 0 & 1 & \infty & \\ 0 & 0 & \gamma_1 + \alpha_1 + \beta_1 & z \\ \alpha_2 - \alpha_1 & \beta_2 - \beta_1 & \gamma_2 + \alpha_1 + \beta_1 & \end{Bmatrix} \end{aligned} \tag{14.4.21}$$

This means that we can express all solutions in terms of the case $\alpha_1 = 0$, $\beta_1 = 0$, in which case there remain three free parameters which for now we take as

$$a = \gamma_1\,, \qquad b = \gamma_2 \tag{14.4.22}$$

and $\alpha_2$. $\beta_2$ is then determined by

$$\alpha_2 + \beta_2 + a + b = 1 \tag{14.4.23}$$

which is (14.4.1) when $\alpha_1 = \beta_1 = 0$. Then (14.4.11) becomes

$$f'' + \left[\frac{1-\alpha_2}{z} + \frac{1-\beta_2}{z-1}\right] f' + \frac{ab}{z(z-1)}\, f = 0 \tag{14.4.24}$$

Multiply by $z(1-z)$ and note that

$$(1-\alpha_2)(1-z) + (1-\beta_2)z = -c + (a+b+1)z \tag{14.4.25}$$

where

$$c \equiv 1 - \alpha_2 \tag{14.4.26}$$

Since $2 - \alpha_2 - \beta_2 = a + b + 1$ by (14.4.23), we have thus proven:

**Proposition 14.4.6.** *If $\alpha_1 = \beta_1 = 0$ and $a, b, c$ are given by* (14.4.22) *and* (14.4.26)*, then the Papperitz–Riemann equation becomes the hypergeometric equation in Gauss form* (*which we saw as* (14.0.4))

$$z(1-z)f'' + [c - (1+a+b)z]f' - abf = 0 \tag{14.4.27}$$

For later purposes, we note that

$$\alpha_1 = 0, \quad \alpha_2 = 1-c, \quad \beta_1 = 0, \quad \beta_2 = c-a-b, \quad \gamma_1 = a, \quad \gamma_2 = b \tag{14.4.28}$$

Since $\alpha_1 = 0$, this equation has a unique solution which is regular at $z = 0$ and has value $f(z = 0) = 1$—at least so long as $\alpha_2 \neq 1, 2, 3, \dots$ (when there are potential log terms), that is, $c \neq 0, -1, -2$. We write this solution as $F(a, b, c; z)$ or as ${}_2F_1\left({a, b \atop c}; z\right)$. The reason for ${}_2F_1$ and the name *hypergeometric function* applied to the function will be discussed shortly.

To compute the power series, we need to apply $\frac{d}{dz}$ and $z$ to power series. For *formal power series*, we define $\{g\}_n$ by requiring

$$g(z) = \sum_{n=0}^{\infty} \frac{\{g\}_n}{n!}\, z^n \tag{14.4.29}$$

so

$$\{g\}_n = g^{(n)}(0) \tag{14.4.30}$$

if the series converges. In that case, this implies that

$$\{g^{(k)}\}_n = \{g\}_{n+k} \tag{14.4.31}$$

Since $z^\ell(z^n/n!) = (n+1)\dots(n+\ell)\frac{z^{n+\ell}}{(n+\ell)!}$, we obtain

$$\{z^\ell g\}_n = n(n-1)\dots(n-\ell+1)\{g\}_{n-\ell} \tag{14.4.32}$$

where $n - \ell < 0$ is no problem since then one factor of $n, n-1, \dots, (n-\ell+1)$ vanishes.

Thus,

$$\begin{aligned}\{z(1-z)f''\}_n &= n\{f''\}_{n-1} - n(n-1)\{f''\}_{n-2} \\ &= n\{f\}_{n+1} - n(n-1)\{f\}_n \end{aligned} \tag{14.4.33}$$

$$\{f'\}_n = \{f\}_{n+1}, \qquad \{zf'\}_n = n\{f\}_n$$

so (14.4.27) becomes

$$\begin{aligned}(n+c)\{f\}_{n+1} &= [n(n-1) + (1+a+b)n + ab]\{f\}_n \\ &= (n+a)(n+b)\{f\}_n \end{aligned} \tag{14.4.34}$$

so the solution with $\{f\}_0 = 1$ has, by induction,

$$\{f\}_n = \frac{[a(a+1)\dots(a+n-1)]\,[b(b+1)\dots(b+n-1)]}{c(c+1)\dots(c+n-1)} \tag{14.4.35}$$

We have therefore proven

**Theorem 14.4.7.** *Let $c \neq 0, -1, \dots$. The unique solution of (14.4.27) regular at $z = 0$ with $f(0) = 1$ is given by (14.0.6), that is,*

$${}_2F_1\left(\begin{matrix} a\,,\, b \\ c \end{matrix}; z\right) = \sum_{n=0}^{\infty} \frac{(a)_n (b)_n}{(c)_n n!}\, z^n \tag{14.4.36}$$

*converging for all $z \in \mathbb{D}$.*

To understand the name, recall a *geometric series* is a formal sum $\sum_{n=0}^{\infty} t_n$ where

$$\frac{t_{n+1}}{t_n} = a \tag{14.4.37}$$

a constant. A *hypergeometric series* is one where $t_{n+1}/t_n = R(n)$, a rational function of $n$. Writing $R(n)$ as a ratio of polynomials and factoring them, we see that

$$R(n) = \frac{(\alpha_1 + n)\dots(\alpha_p + n)}{(\beta_1 + n)\dots(\beta_q + n)n}\, z \tag{14.4.38}$$

where we can include the extra $1/n$ in $R(n)$, if need be, by adding $\alpha_p = 0$ (and so cancel the $1/n$). If $t_0 = 1$, we see the sum is

$${}_pF_q\left(\begin{matrix} \alpha_1, \dots, \alpha_p \\ \beta_1, \dots, \beta_q \end{matrix}; z\right) = \sum_{n=0}^{\infty} \frac{\prod_{j=1}^{p} (\alpha_j)_n}{\prod_{k=1}^{q} (\beta_k)_n} \frac{1}{n!}\, z^n \tag{14.4.39}$$

A simple use of the Cauchy radius formula (Problem 4) shows the series has radius of convergence $\infty$ if $p < q+1$, 1 if $p = q+1$, and 0 if $p > q+1$. ${}_2F_1$ occurs most often and ${}_1F_1$ next most often, which we'll discuss later in this section. ${}_2F_1$ is called the *hypergeometric function* and ${}_pF_q$ is called a *generalized hypergeometric function.*

By (14.4.21), we see:

**Theorem 14.4.8.** *If $\alpha_2 - \alpha_1 \notin \mathbb{Z}$, a basis of solutions for* $\mathcal{P}\left\{\begin{matrix} 0 & 1 & \infty & \\ \alpha_1 & \beta_1 & \gamma_1 & z \\ \alpha_2 & \beta_2 & \gamma_2 & \end{matrix}\right\}$ *is*

$$z_1^{\alpha_1}(1-z)^{\beta_1}{}_2F_1\left(\begin{matrix}\gamma_1+\alpha_1+\beta_1\,,\ \gamma_2+\alpha_1+\beta_1\\ 1-\alpha_2+\alpha_1\end{matrix};z\right) \qquad \textit{and}$$

$$z_1^{\alpha_2}(1-z)^{\beta_1}{}_2F_1\left(\begin{matrix}\gamma_1+\alpha_2+\beta_1\,,\ \gamma_2+\alpha_2+\beta_1\\ 1-\alpha_1+\alpha_2\end{matrix};z\right) \tag{14.4.40}$$

For simplicity of notation, we consider what happens if $\alpha_1 - \alpha_2 \in \mathbb{Z}$ in the case of (14.4.27), where if $c \notin \mathbb{Z}$, the solutions are $F(a,b,c;z)$ and $z^{1-c}F(a+1-c,b+1-c,2-c;z)$. We also note that if $c-a-b \notin \mathbb{Z}$, by Theorem 14.4.5(b) with $Q(z) = 1-z$, the solutions are also spanned (with one of these regular at $z=1$) $F(a,b,a+b+1-c;1-z)$ and $(1-z)^{c-a-b}F(c-a,c-b,c+1-a-b;1-z)$.

As we've seen, if $c \in \{0,-1,-2,\dots\}$, $F(a,b,c;z)$ blows up, that is, it has a pole at these values of $c$. It is natural to divide by $\Gamma(c)$ to cancel these poles, so we define

$$u_1(a,b,c;z) = \frac{\Gamma(a)\Gamma(b)}{\Gamma(c)}\,F(a,b,c;z) \tag{14.4.41}$$

$$u_2(a,b,c;z) = z^{1-c}u_1(a+1-c,b+1-c,2-c;z) \tag{14.4.42}$$

It is easy to see that if $c_0 \in \{0,1,2,\dots\}$, then $u_1(a,b,c_0;z) = u_2(a,b,c_0;z)$. Thus,

$$\lim_{c\to c_0}\frac{u_1-u_2}{c-c_0} = \frac{\partial}{\partial c}(u_1-u_2)\bigg|_{c=c_0}$$

is a second solution. Calculations (see the reference in the Notes) show that it has the form $\log(z)u_1(a,b,c;z) + g(z)$, where $g$ is meromorphic at $z=0$ (regular if $c_0 = 0,1$ and pole of order $c_0 - 1$ otherwise) with explicit power or Laurent series.

This completes the discussion of ${}_2F_1$ as solutions of the ODE. We next want to discuss some of the many properties and relations (with some proofs left to the Notes and Problems), introduce ${}_1F_1$, and finally, write down some special cases of ${}_2F_1$ and ${}_1F_1$. We always suppose $c \notin \{0,-1,-2,\dots\}$ without explicitly stating it. We begin by summarizing what we'll say about ${}_2F_1$:

(1) *Euler integral formula.* This says that if $\operatorname{Re} c > \operatorname{Re} b > 0$, then

$${}_2F_1\left(\begin{matrix}a\,,\ b\\ c\end{matrix};z\right) = \frac{\Gamma(c)}{\Gamma(b)\Gamma(c-b)}\int_0^1 t^{b-1}(1-t)^{c-b-1}(1-zt)^{-a}\,dt \tag{14.4.43}$$

for $z \in \mathbb{C} \setminus [1, \infty)$ with the branch of $(1 - zt)^{-a}$ is taken with value 1 at $z = 0$. The proof is simple: for $|z| < 1$, one expands $(1 - zt)^{-a}$ by the binomial theorem, gets a sum of beta integrals, and uses $\Gamma(n + d) = (d)_n \Gamma(d)$ (by the functional equation for $\Gamma$) for $d = b, c$. Then one analytically continues both sides. The details are left to the Problems (see Problem 5). One use of (14.4.43) is to allow analytic continuation of ${}_2F_1$ in $z$ if it is defined by the convergent power series when $|z| < 1$.

(2) *Gauss' contiguous relations.* Gauss wrote down fifteen relations giving values of ${}_2F_1\left({a,\,b \atop c}; z\right)$ if $z$ is fixed and $a, b$ and/or $c$ are changed by $\pm 1$. We'll discuss four below (see Theorem 14.4.9) involving derivatives with respect to $z$. By exploiting the ODE, some involve only values and give various three-term recursions (see Problem 6(c)). These equations can be regarded as analogs of the functional equation for $\Gamma$. In particular, if one defines ${}_2F_1$ in the region $\operatorname{Re} c > \operatorname{Re} b > 0$ by (14.4.43), these formulae can then allow the extension to all $b, c$.

(3) *Pfaff and Euler relations, Kummer's 24 functions and quadratic relations.* Using invariance of the Riemann $\mathcal{P}$-function under $\beta_1 \leftrightarrow \beta_2$ and under $1 \leftrightarrow \infty$, one gets relations of Euler and Pfaff between apparently different ${}_2F_1$'s that are equal (see Theorem 14.4.10 below). Combining this with permutations of $0, 1, \infty$, Kummer found 24 single ${}_2F_1$ expressions that solve the same hypergeometric ODE; see the Notes. Finally, there are special relations relating some ${}_2F_1$'s at different values of $z$; see the Notes.

(4) *Jacobi's yypergeometric formula.* If $b$ is a negative integer, ${}_2F_1\left({a,\,b \atop c}; z\right)$ is a polynomial in $z$, for which Jacobi found an elegant formula

$$
{}_2F_1\left({a\,,\ -n \atop c}\,; z\right) = \frac{z^{1-c}(1-z)^{c+n-a}}{(c)_n} \frac{d^n}{dz^n} \left[z^{c+n-1}(1-z)^{a-c}\right] \tag{14.4.44}
$$

which we'll prove below (Theorem 14.4.11). This will lead to Rodrigues' formula for Legendre and Jacobi polynomials (see Example 14.4.17).

**Theorem 14.4.9** (Gauss' Contiguous Relations). *We have that*

$$
\text{(a)}\ \frac{d}{dz} F(a, b, c; z) = \frac{ab}{c} F(a+1, b+1, c+1; z) \tag{14.4.45}
$$

$$
\text{(b)}\ z \frac{d}{dz} F(a, b, c; z) = a(F(a+1, b, c; z) - F(a, b, c; z)) \tag{14.4.46}
$$

$$
= b(F(a, b+1, c; z) - F(a, b, c; z)) \tag{14.4.47}
$$

$$
\text{(c)}\ z \frac{d}{dz} F(a, b, c; z) = (c-1)(F(a, b, c-1; z) - F(a, b, c; z)) \tag{14.4.48}
$$

**Proof.** (a) By (14.4.34),

$$\{F'\}_n = \frac{(a)_{n+1}(b)_{n+1}}{(c)_{n+1}} = \frac{ab}{c}\,\frac{(a)_n(b)_n}{(c)_n}$$

proving (14.4.45).

(b) By (14.4.34),

$$\left\{z\,\frac{d}{dz}\,F\right\}_n = n\,\frac{(a)_n(b)_n}{(c)_n} = a\left[\frac{[(a+1)_n-(a)_n](b)_n}{(c)_n}\right]$$

since $a(a+1)_n - a(a)_n = (a+n-a)(a)_n$. This proves (14.4.46), and (14.4.47) is similar.

(c) This is similar to (b) if we note that

$$\frac{1}{(c-1)_n} - \frac{1}{(c)_n} = \left[\frac{1}{c-1} - \frac{1}{c+n-1}\right]\frac{1}{(c)_{n+1}} = \frac{n}{c-1}\,\frac{1}{(c)_{n-1}} \qquad \square$$

**Theorem 14.4.10.** *We have*

(a) (Euler transformation)

$$F(a,b,c;z) = (1-z)^{c-a-b}F(c-a,c-b,c;z)$$

(b) (Pfaff transformation)

$$F(a,b,c;z) = (1-z)^{-a}F\left(a,c-b;c;\frac{z}{z-1}\right) \tag{14.4.49}$$

$$F(a,b,c;z) = (1-z)^{-b}F\left(c-a,b,c;\frac{z}{z-1}\right) \tag{14.4.50}$$

**Proof.** (a) By (14.4.20), the invariance of $\mathcal{P}$ under $\beta_1 \leftrightarrow \beta_2$ or $\gamma_1 \leftrightarrow \gamma_2$, (and the fact that since the Riemann $\mathcal{P}$-function is a vector space, it is invariant under changing $(z-1)^\mu$ to $(1-z)^\mu$),

$$(1-z)^{a+b-c}\mathcal{P}\begin{Bmatrix} 0 & 1 & \infty & \\ 0 & 0 & a & z \\ 1-c & c-a-b & b & \end{Bmatrix}$$

$$= \mathcal{P}\begin{Bmatrix} 0 & 1 & \infty & \\ 0 & 0 & c-a & z \\ 1-c & a+b-c & c-b & \end{Bmatrix}$$

Using the fact that $c-(c-a)-(c-b) = a+b-c$, we see that $(1-z)^{a+b-c}F(a,b,c;z) = F(c-a,c-b,c;z)$ as the elements of $\mathcal{P}$ regular at $z=0$ with value 1 there.

(b) From invariance of $\mathcal{P}$ under FLT (Theorem 14.4.5(b)), we have (since $z \to \frac{z}{z-1}$, take $0, 1, \infty$ to $0, \infty, 1$)

$$\mathcal{P}\left\{\begin{matrix} 0 & 1 & \infty & \\ 0 & 0 & a & \frac{z}{z-1} \\ 1-c & b-a & c-b & \end{matrix}\right\}$$

$$= \mathcal{P}\left\{\begin{matrix} 0 & \infty & 1 & \\ 0 & 0 & a & z \\ 1-c & b-a & c-b & \end{matrix}\right\}$$

$$= (1-z)^a \mathcal{P}\left\{\begin{matrix} 0 & 1 & \infty & \\ 0 & 0 & a & z \\ 1-c & c-a-b & b & \end{matrix}\right\}$$

by invariance under interchange of columns and (14.4.20). Identifying the solutions regular at $z = 0$ with value 1 there leads to (14.4.49). We get (14.4.50) by $a \leftrightarrow b$ symmetry. $\square$

**Remark.** The more common proofs use Euler's integral formula (14.4.43).

As the final general property, we turn to Jacobi's formula (14.4.44). The key observation is that (Problem 7(a))

$$\left[z(1-z)\frac{d}{dz} + c(1-z) - (a+b+1-c)z\right]g = Q^{-1}\frac{d}{dz}\,z(1-z)Qg \tag{14.4.51}$$

where

$$Q(z) = z^{c-1}(1-z)^{a+b-c} \tag{14.4.52}$$

Thus, if $f = f(a,b,c;z)$, then the hypergeometric equation (14.4.29) says that (Problem 7(b)),

$$\frac{d}{dz}\left(z(1-z)Q(z)f'\right) = abQf \tag{14.4.53}$$

Changing indices and using the fact that, by (14.4.45), $f^{(k)}$ is a multiple of $F(a+k, b+k, c+k; z)$, we get (Problem 7(c))

$$\frac{d}{dz}\left(z^k(1-z)^kQf^{(k)}\right) = (a+k-1)(b+k-1)z^{k-1}(1-z)^{k-1}Qf^{(k-1)} \tag{14.4.54}$$

which implies (Problem 7(d))

$$\frac{d^k}{dz^k}\left(z^k(1-z)^kQ(z)f^{(k)}\right) = (a)_k(b)_kQf \tag{14.4.55}$$

Since $f^{(k)}(z) = \frac{(a)_k(b)_k}{(c)_k} f(a+k, b+k, c+k; z)$ by (14.4.45), we have the first part of

**Theorem 14.4.11** (Jacobi's Hypergeometric Formula). (a) *Let $Q(z)$ be given by* (14.4.52). *Then*

$$\frac{d^k}{dz^k}\left[z^k(1-z)^k Q(z) f(a+k, b+k, c+k; z)\right] = (c)_k Q(z) f(a, b, c; z) \quad (14.4.56)$$

(b) *If $n$ is any nonnegative integer, $f(a, -n, c; z)$ is a polynomial of degree $n$ given by*

$$f(a, -n, c; z) = \frac{z^{1-c}(1-z)^{c+n-a}}{(c)_n}\left(\frac{d}{dz}\right)^n z^{c+n-1}(1-z)^{a-c} \quad (14.4.57)$$

**Remark.** The $f$ in (14.4.57) is sometimes called a *hypergeometric polynomial.* We'll focus soon instead on these polynomials after a change of variables when they become Jacobi polynomials.

**Proof.** (a) is proven above. Since, for $n \in \{0, 1, 2, \dots\}$, $(-n)_k = 0$ for $k \geq n+1$, we see that $f(a, -n, c; z)$ is a polynomial of degree $n$. In particular, $f(a, 0, c; z) \equiv 1$, so in (14.4.55), $f^{(n)}(z) = (a)_n(-n)_n/(c)_n$, which yields (14.4.57). $\square$

We turn next to the *confluent hypergeometric function*, ${}_1F_1\left(\begin{smallmatrix} a \\ b \end{smallmatrix}; z\right)$, which we'll also denote $M(a, b; z)$. It is useful to understand ${}_1F_1$ as a limit of ${}_2F_1$'s:

**Theorem 14.4.12.** (a) *We have that*

$$M(a, c; z) = \lim_{b\to\infty} {}_2F_1\left(\begin{matrix} a\ b \\ c \end{matrix}; \frac{z}{b}\right) \quad (14.4.58)$$

*where the limit is uniform for $z$ in each $\mathbb{D}_R(0)$.*

(b) *$M(a, c; z)$ solves*

$$zf''(z) + (c-z)f'(z) - af(z) = 0 \quad (14.4.59)$$

(c) *For the differential equation* (14.4.59), *0 is a regular singular point and $\infty$ an irregular singular point.*

(d) $$M(a, c; z) = \sum_{n=0}^{\infty} \frac{(a)_n}{(c)_n n!} z^n \quad (14.4.60)$$

**Remark.** $F(a, b, c; \frac{z}{b})$ has $0, \infty$, and $b$ as regular singular points. Thus, as $b \to \infty$, two regular singularities coalesce. The name "confluent" comes from the confluence of the two singularities.

**Proof.** (a) For any fixed $k$, $(b)_k/b^k = \prod_{j=0}^{k-1}(1 - j/b) \to 1$. Moreover, by the estimate in Problem 8, the right side is uniformly bounded as $b \to \infty$ on any ball $|z| \leq R$. Thus, we have the claimed uniform convergence.

(b) Let $g_m(z)$ be the right side of (14.4.58) for $b = m \in \mathbb{Z}_+$, then (14.4.27) becomes

$$(z - m^{-1}z^2)g_m''(z) + [(c-z) - m^{-1}(1+a)z]g_m'(z) - ag_m(z) = 0$$

Taking $m \to \infty$ using the fact that, by the uniform convergence and analyticity, derivatives converge, leads to (14.4.59).

(c) In canonical form, (14.4.59) has $a_1(z) = 1 - c/z$, $a_0(z) = -a/z$. Since $a_1$ is $O(1/z)$ near 0 and $a_0(z) = o(1/z^2)$, 0 is a regular singular point. Since $a_1(z) \neq O(1/z)$ at $\infty$, we see $\infty$ is irregular.

(d) is immediate from the proof of (a). □

One can also compute the limit of the Euler integral (14.4.45) (Problem 9): If $\operatorname{Re} b > \operatorname{Re} a > 0$,

$$ {}_1F_1\left(\begin{matrix} a \\ b \end{matrix}; z\right) = \frac{\Gamma(b)}{\Gamma(a)\Gamma(a-b)} \int_0^1 t^{a-1}(1-t)^{b-a-1}e^{zt}\, dt \tag{14.4.61}$$

As a final topic, we want to show that a large number of useful functions are special cases of hypergeometric or conformal hypergeometric functions, or can be simply expressed in terms of them. This occurs both because many power series are hypergeometric and because many special functions obey ODEs with two regular singular points, or a regular singular point and an irregular singularity at infinity. We'll also see that all the so-called classical orthogonal polynomials are expressible in terms of hypergeometric or conformal hypergeometric polynomials.

**Example 14.4.13** (Hypergeometric Examples). Here are some functions singular at $z = 1$. Since $\frac{(2)_n}{(1)_n} = n+1$, we have

$$\log(1-z) = -z\sum_{n=0}^{\infty} \frac{z^n}{n+1} = -zF(1,1,2;z) \tag{14.4.62}$$

Since

$$\binom{-a}{n} = \frac{(-a)(-a-1)\dots(-a-(n-1))}{n!} = (-1)^n \frac{(a)_n}{n!} \tag{14.4.63}$$

we see that for any $b$,

$$(1-z)^{-a} = F(a,b,b;z) \tag{14.4.64}$$

that is, $(1-z)^{-a}$ is really a ${}_1F_0$, which we can write as a ${}_2F_1$.

Using $\arcsin(z) = \int_0^z (1-t^2)^{-1/2}\, dt$ and expanding the integrand with the binomial theorem, one gets (Problem 12)

$$\arcsin(z) = zF(\tfrac{1}{2}, \tfrac{1}{2}, \tfrac{3}{2}; z^2) \tag{14.4.65}$$

Similarly, $\arctan(z) = zF(\frac{1}{2}, 1, \frac{3}{2}; -z^2)$. □

**Example 14.4.14** (Legendre Functions). As noted in Problem 3 of Section 14.3, (14.3.45) has regular singularities at $1, -1, \infty$. As seen from the calculations of the indices in that problem, the solution regular at $z = 1$ (with the conventional normalization) is

$$P_\alpha^\beta(z) = \frac{1}{\Gamma(1-\beta)} \left[\frac{1+z}{1-z}\right]^{\beta/2} F\left(-\alpha, \alpha+1, 1-\beta; \frac{1-z}{2}\right) \tag{14.4.66}$$

called a *Legendre function.* The traditional second solution, called the *Legendre function of the second kind,* is given by a difference of ${}_2F_1$'s; see the references in the Notes. □

**Example 14.4.15** (Complete Elliptic Integrals). Recall (see (10.5.90) and (10.5.105) of Part 2A) that the complete elliptic integrals of the first and second kind are given by

$$K(k) = \int_0^1 \frac{dt}{\sqrt{(1-k^2t^2)(1-t^2)}} \tag{14.4.67}$$

and

$$E(k) = \int_0^1 \sqrt{\frac{1-k^2t^2}{1-t^2}}\, dt \tag{14.4.68}$$

By expanding the square roots involving $k$ in a binomial series and computing the integrals as beta functions, one sees that (Problem 13)

$$K(k) = \frac{\pi}{2} F(\tfrac{1}{2}, \tfrac{1}{2}, 1; k^2), \qquad E(k) = \frac{\pi}{2} F(-\tfrac{1}{2}, \tfrac{1}{2}, 1; k^2) \tag{14.4.69}$$

□

**Example 14.4.16** (Confluent Hypergeometric Examples). The *error function* is defined by

$$\operatorname{erf}(z) = \frac{2}{\sqrt{\pi}} \int_0^z e^{-t^2}\, dt \tag{14.4.70}$$

and is important in statistics for $z > 0$. By expanding the power series, one sees that (Problem 14)

$$\operatorname{erf}(z) = \frac{2z}{\sqrt{\pi}} M(\tfrac{1}{2}, \tfrac{3}{2}; -z^2) \tag{14.4.71}$$

Since the Fresnel functions of (5.7.58) of Part 2A are related to error functions (at complex values) by power series expansions, we have

$$C(z) + iS(z) = zM(\tfrac{1}{2}, \tfrac{3}{2}; iz^2) \tag{14.4.72}$$

*Parabolic cylinder functions* solve *Weber's equation*

$$\frac{d^2 f}{dz^2} - (\tfrac{1}{4} z^2 + a) f = 0 \tag{14.4.73}$$

This is regular with entire solutions. The ones even and odd under $z \to -z$ are given by (Problem 15)

$$y_1(a;z) = e^{-z^2/4} M\left(\frac{1}{2}\,a + \frac{1}{4}, \frac{1}{2}; \frac{z^2}{2}\right) \tag{14.4.74}$$

$$y_2(a;z) = e^{-z^2/4} M\left(\frac{1}{2}\,a + \frac{3}{4}, \frac{3}{2}; \frac{z^2}{2}\right) \tag{14.4.75}$$

Recall that in Section 6.4 of Part 1, we used the harmonic oscillator basis, $\varphi_n(x)$, which solves for $n = 0, 1, 2, \dots,$

$$-\frac{1}{2}\,\frac{d^2\varphi_n}{dx^2} + \frac{1}{2}\,x^2\varphi_n = \left(n + \frac{1}{2}\right)\varphi_n \tag{14.4.76}$$

with

$$\varphi_n(-x) = (-1)^n \varphi_n(x) \tag{14.4.77}$$

The *Hermite polynomials* can be defined by

$$\varphi_n(x) = \frac{1}{\sqrt{2^n n!}}\,\frac{1}{\sqrt[4]{\pi}}\,e^{-x^2/2} H_n(x) \tag{14.4.78}$$

Noting that (14.4.73) and (14.4.76) differ by a scaling, one finds that (Problem 16)

$$H_{2n}(x) = (-1)^n\,\frac{(2n)!}{n!}\,M\left(-n, \frac{1}{2}; x^2\right) \tag{14.4.79}$$

$$H_{2n+1}(x) = (-1)\frac{(2n+1)!}{n!}\,(2x) M\left(-n, \frac{3}{2}; x^2\right) \tag{14.4.80}$$

Hermite polynomials are an orthogonal basis in $L^2(\mathbb{R}, \pi^{-1/2} e^{-x^2}\,dx)$ with

$$\int_{-\infty}^{\infty} H_m(x) H_n(x) \pi^{-1/2} e^{-x^2}\,dx = n!\,2^n \delta_{nm} \tag{14.4.81}$$

Using $e^{x^2/2} \frac{d}{dx} e^{-x^2/2} = \frac{d}{dx} - x$, (14.4.76) becomes

$$-H_n''(x) + 2xH_n'(x) = 2nH_n(x) \tag{14.4.82}$$

Moreover,

$$H_n(x) = 2^n x^n + O(x^{n-1}) \tag{14.4.83}$$

and the generating function is

$$F(x,t) \equiv \sum_{n=0}^{\infty} H_n(x) \frac{t^n}{n!} = \exp(2xt - t^2) \tag{14.4.84}$$

(as can be seen by noting both sides obey the partial differential equation $(\frac{\partial^2}{\partial x^2} + 2x\frac{\partial}{\partial x} - 2t\frac{\partial}{\partial t})F = 0$, are both polynomials of degree $n$ in $x$ times $t^n$, and obey (14.4.83)). This, in turn. implies the recursion relation

$$H_{n+1}(x) = 2xH_n(x) - 2nH_{n-1}(x) \tag{14.4.85}$$

Similarly, the *generalized Laguerre polynomials*, $L_n^{(\alpha)}(x)$ ($\alpha = 0$ are called *Laguerre polynomials*), solve Laguerre's differential equation

$$zf''(z) + (\alpha - 1 - x)f'(x) + nf(x) = 0 \tag{14.4.86}$$

and are given by

$$L_n^{(\alpha)}(z) = \binom{n+\alpha}{n} M(-n, \alpha+1; z) \tag{14.4.87}$$

and obey (for $\alpha > -1$)

$$\int_0^\infty x^\alpha e^{-x} L_n^{(\alpha)}(x) L_m^{(\alpha)}(x)\, dx = \frac{\Gamma(n+\alpha+1)}{n!}\,\delta_{nm} \tag{14.4.88}$$

With its conventional normalization, $L_n^{(\alpha)}$ is a polynomial of degree $n$, but the leading coefficients may not be positive. Indeed,

$$L_n^{(\alpha)}(x) = \frac{(-1)^n}{n!}\, x^n + O(x^{n-1}) \tag{14.4.89}$$

and has the generating function

$$\sum_{n=0}^\infty L_n^{(\alpha)}(x) t^n = (1-t)^{-\alpha-1} \exp\biggl(-\frac{xt}{1-t}\biggr) \tag{14.4.90}$$

which leads to the three-term recursion relation

$$(n+1)L_{n+1}^{(\alpha)}(x) + (x - \alpha - 2n - 1)L_n^{(\alpha)}(x) + (n+\alpha)L_{n-1}^\alpha(x) = 0 \tag{14.4.91}$$

*Whittaker functions* are solutions of

$$\frac{d^2w}{dz^2} + \biggl(-\frac{1}{4} + \frac{\alpha}{z} + \frac{\frac{1}{4} - \beta^2}{z^2}\biggr) w = 0 \tag{14.4.92}$$

related to the solution of a Schrödinger equation in a Coulomb field. The one regular at $z = 0$ is given by

$$M_{\alpha,\beta}(z) = e^{-z/2} z^{\beta+1/2} M(\beta - \alpha + \tfrac{1}{2}, 1 + 2\beta; z) \tag{14.4.93}$$

The *incomplete gamma function* $\gamma(z, w) = \int_0^w t^{z-1} e^{-t}\, dt$ is given by

$$\gamma(z, w) = z^{-1} w^z e^{-w} M(1, z+1, w) \tag{14.4.94}$$

Finally, we note that *Bessel functions*, $J_\alpha(x)$, given by (14.0.9), can be expressed as a generalized hypergeometric function

$$J_\alpha(x) = \frac{(\frac{x}{2})^\alpha}{\Gamma(\alpha+1)}\, {}_0F_1\biggl(\alpha + 1; -\frac{x^2}{4}\biggr) \tag{14.4.95}$$

□

**Example 14.4.17** (Jacobi Polynomials). As a final example, we consider Jacobi polynomials, which include Legendre, Chebyshev, and Gegenbauer polynomials as special cases. They are defined for each $\alpha, \beta \in (-\frac{1}{2}, \infty)$ via

$$P_n^{(\alpha,\beta)}(z) = \frac{(\alpha+1)_n}{n!}\, {}_2F_1\left(\begin{matrix} n+\alpha+\beta+1, -n \\ \alpha+1 \end{matrix}; \frac{1-z}{2}\right) \tag{14.4.96}$$

Thus, except for an affine change of variables, these are the polynomials of (14.4.57). By that formula, $P_n^{(\alpha,\beta)}$ is given by *Rodrigues' formula:*

$$P_n^{(\alpha,\beta)}(z) = \frac{(-1)^n}{2^n n!}(1-z)^{-\alpha}(1+z)^{-\beta}\frac{d^n}{dz^n}\left[(1-z)^{n+\alpha}(1+z)^{n+\beta}\right] \tag{14.4.97}$$

which implies (Problem 17)

$$\begin{aligned} &\int_{-1}^{1} P_n^{(\alpha,\beta)}(x)P_m^{(\alpha,\beta)}(x)(1-x)^\alpha(1+x)^\beta\,dx \\ &\qquad = \frac{2^{\alpha+\beta+1}\Gamma(n+\alpha+1)\Gamma(n+\beta+1)}{(2n+\alpha+\beta+1)\Gamma(n+\alpha+\beta+1)n!}\,\delta_{mn} \end{aligned} \tag{14.4.98}$$

$P_n^{(\alpha,\beta)}(x)$ has leading coefficient

$$\frac{(\alpha+\beta+n+1)_n}{2^n n!}\,x^n + O(x^{n-1}) \tag{14.4.99}$$

obeys the differential equation ($u = P^{(\alpha,\beta)}(x)$)

$$(1-x^2)u'' + [\beta - \alpha - (\alpha+\beta+2)x]u' + n(n+\alpha+\beta+1)u = 0 \tag{14.4.100}$$

They have the generating functions

$$F^{(\alpha,\beta)}(x,t) \equiv \sum_{n=0}^{\infty} P_n^{(\alpha,\beta)}(x)t^n = 2^{\alpha+\beta}R^{-1}(1-t+R)^{-\alpha}(1+t+R)^{-\beta} \tag{14.4.101}$$

where

$$R = (1-2xt+t^2)^{1/2} \tag{14.4.102}$$

and the three-term recursion relation

$$\begin{aligned} 2n(n+\alpha+\beta)(2n+\alpha+\beta-2)P_n^{(\alpha,\beta)}(x) &= (2n+\alpha+\beta-1) \\ &\quad [(2n+\alpha+\beta)(2n+\alpha+\beta-2)x + \alpha^2 - \beta^2]P_{n-1}^{(\alpha,\beta)}(x) \\ &\quad - 2(n+\alpha-1)(n+\beta-1)(2n+\alpha+\beta)P_{n-2}^{(\alpha+\beta)}(x) \end{aligned} \tag{14.4.103}$$

Special cases of interest are:

(1) *Chebyshev polynomials of the first kind* (see Problem 8 of Section 3.1 of Part 2A). $\alpha = \beta = -\frac{1}{2}$

$$T_n(x) = \frac{2^{2n}(n!)^2}{(2n)!}\,P_n^{(-\frac{1}{2},-\frac{1}{2})}(x) \tag{14.4.104}$$

with the normalization picked so that

$$T_n(\cos\theta) = \cos(n\theta) \tag{14.4.105}$$

(2) *Chebyshev polynomials of the second kind* (see Problem 8 of Section 3.1 of Part 2A). $\alpha = \beta = \frac{1}{2}$

$$U_n(x) = \frac{(2)_n}{(\frac{3}{2})_n} P_n^{(\frac{1}{2},\frac{1}{2})}(x) \tag{14.4.106}$$

The normalization is such that

$$U_n(\cos\theta) = \frac{\sin((n+1)\theta)}{\sin(\theta)} \tag{14.4.107}$$

(3) *Legendre polynomials.* $\alpha = \beta = 0$, so the measure in (14.4.98) is $dx$. The standard normalization is

$$P_n(x) = P_n^{(1,1)}(x) \tag{14.4.108}$$

(4) *Gegenbauer or ultraspherical polynomials.* $\alpha = \beta = a - \frac{1}{2}$, so this includes the above examples for $a \neq 0$ (i.e., $\alpha \neq -\frac{1}{2}$). The normalization is

$$C_n^{(a)}(x) = \frac{(2a)_n}{(a+\frac{1}{2})_n} P^{(a-\frac{1}{2},a-\frac{1}{2})}(x) \tag{14.4.109}$$

The normalization is picked so that one has the generating function convergent for all $|t| < 1$,

$$(1 - 2xt + t^2)^{-a} = \sum_{n=0}^{\infty} C_n^{(a)}(x) t^n \tag{14.4.110}$$

This is valid for $a \neq 0$. The generating function for Chebyshev polynomials of the first kind, $T_n(x)$, is given by

$$\log(1 - 2x + t^2)^{-1/2} = \sum_{n=1}^{\infty} T_n(x) \frac{t^n}{n} \tag{14.4.111}$$

consistent with (14.4.110) and the relation (see Problem 20)

$$\frac{2}{n} T_n(x) = \lim_{\alpha\to 0} \frac{C_n^{(\alpha)}(x)}{\alpha} \tag{14.4.112}$$

(14.4.110) and (14.4.111) are proven in Problem 18 of this section and Problem 9 of Section 3.1 of Part 2A.

Gegenbauer polynomials enter in the study of spherical harmonics; see Section 3.5 of Part 3.

(5) *Chebyshev polynomials of the third and fourth kind.* These obey

$$V_n(\cos\theta) = \frac{\sin[(n+\frac{1}{2})\theta]}{\sin[\frac{\theta}{2}]}, \qquad W_n(\cos\theta) = \frac{\cos[(n+\frac{1}{2})\theta]}{\cos[\frac{\theta}{2}]}$$

and are multiples of $P^{(\frac{1}{2},-\frac{1}{2})}$ and $P^{(-\frac{1}{2},\frac{1}{2})}$, respectively.

□

**Notes and Historical Remarks.** There are a number of specialized books on hypergeometric functions and their applications to number theory, combinatorics, and physics—see [**24, 66, 136, 147, 343, 356, 357**]. There are also several standard compendiums on special functions ([**4, 118, 156, 305, 411**]) that have lots of formulae on hypergeometric functions. The chapters of [**15**] on the subject are a wonderful introduction. In particular, [**15**] provides details on issues we only hint at, such as log solutions when $\alpha_1 - \alpha_2$ is an integer, quadratic relations, and special values.

The term "hypergeometric" goes back to Wallis [**400**] in 1656. In 1767, Euler [**128**] included a detailed look at the function defined by his integral, (14.4.43), including a form of the ODE and the power series.

Johann Friedrich Pfaff (1765–1825), who was one of Gauss' teachers, found his transformation in 1797 [**315**] and presented it in his calculus text. The subject became a central one after Gauss' 1813 treatise [**149**] on the subject. In 1836, Kummer found his famous 24 solutions [**236**] of Gauss' differential equation. In modern language, he found a group of 24 elements (isomorphic to $S_4$) which acts on the set of solutions and which takes $F(a, b, c; z)$ into single $F$'s rather than sums.

All the early works focused on individual solutions, their series, and continuation. Riemann's 1857 work, which looked at the totality of solutions and introduced the notion of monodromy, had a profound influence. The explicit form of the ODEs obeyed by Riemann's solutions was found by E. Papperitz in 1889 [**311**].

Carl Friedrich Gauss (1777–1855) was undoubtedly the greatest mathematician between Euler and Riemann, and to many, of all time. As a young prodigy, he came to the attention of the Duke of Brunswick who supported his attendance at what is now the Technical University of Braunschweig, then at Göttingen, and finally at Helmstadt where he submitted his dissertation (his first proof of the fundamental theorem of algebra) to Pfaff. In 1801, he published his masterpiece, *Disquisitiones Arithmeticae* [**148**] on number theory, and also in that year gained great fame for the following: the asteroid Ceres was discovered early in that year, but there were only a few observations before the planetoid went behind the sun. With then current techniques, the orbit could not be accurately predicted. Using what we now call the method of least squares, Gauss determined the orbit well enough for the asteroid to be found. On the basis of this fame, Gauss, after the death of his patron, was able to get an appointment as director of the observatory at Göttingen. He spent his career there and, in addition to his "pure mathematics," developed techniques in magnetism, geodesy, and potential theory. Indeed, his work on Gaussian curvature and Gauss' law (on div and integrals) had roots in this applied work.

With *Disquisitiones Arithmeticae*, Gauss made number theory a central subject of mathematics. His other contributions include the notion of curvature, the work on hypergeometric functions mentioned in this section, and, as we have seen, the unpublished work on elliptic functions, non-Euclidean geometry, the Poisson summation formula, the prime number conjecture, and the Cauchy integral theorem.

Ernst Eduard Kummer (1810–93) entered the University of Halle to be a theologian but shifted to mathematics. He taught at a gymnasium in Liegnitz for ten years starting in 1832 where his students included Kronecker. While there, he wrote the paper on hypergeometric functions we mentioned in this section. This brought him to the attention of Dirichlet and Jacobi who helped him get a position in Breslau in 1842. When Dirichlet moved from Berlin to Göttingen in 1855, he recommended Kummer as his successor in Berlin. With the addition of Weierstrass and Kronecker to the faculty, Berlin became a center. Kummer's most important work was in number theory, leading to the notion of ideal in algebraic number rings. His students included (some joint with Weierstrass) Georg Cantor, Elwin Christoffel, Paul de Bois-Raymond, Georg Frobenius, Lazarus Fuchs, Justus Grassmann, Wilhelm Killing, Hans van Mangoldt, Franz Mertens, Carle Runge, Arthur Schoenflies, and Hermann Schwarz, who was his son-in-law.

We mention briefly four pieces of hypergeometric technology we have not discussed. First, quadratic relations, of which we mention:

$$F(2a, 2b, a+b+\tfrac{1}{2}; z) = F(a, b, a+b+\tfrac{1}{2}; 4z(1-z))$$

$$F(a, b, 2b; z) = (1-z)^{-a/2} F\left(\frac{1}{2}\,a, b - \frac{1}{2}\,a, b + \frac{1}{2}; \frac{z^2}{4z-4}\right)$$

These were found initially by Kummer [**236**] and later by Goursat [**155**]; these are discussed in texts like [**15**] and [**147**]. Even earlier, Landen [**244, 245**], who spent most of his life as a land agent (he is the land agent of Watson [**406**]; the marquis is Fagnano), found a quadratic relation for elliptic integrals that, given (14.4.69), is a special case of quadratic relations for hypergeometric functions. See Newman [**297**] for a simple proof of Landen's relation in Gauss's form.

The second theme is that sometimes ${}_2F_1$ can be evaluated at special values of $z$, especially $z = \pm 1$. The most famous example is Gauss' result that if $\mathrm{Re}(c-a-b) < 0$, then

$$\lim_{x \uparrow 1} \frac{F(a, b, c; x)}{(1-x)^{c-a-b}} = \frac{\Gamma(c)\Gamma(a+b-c)}{\Gamma(a)\Gamma(b)}$$

The third is the work of Schwarz (and then Klein) on when a hypergeometric function is algebraic (equivalently, the global monodromy group is finite). This is discussed in the Notes to Section 7.3 of Part 2A.

The fourth is that Ramanujan, largely in his unpublished notebooks (see Berndt [**55**]), developed numerous connections of elliptic and theta functions to ${}_2F_1$, so much so that Berndt refers to his theory of elliptic functions distinct from Abel, Jacobi, and Weierstrass.

Detailed derivations of generating functions, recursion relations, and differential equations for Hermite, Laguerre, and Jacobi polynomials can be found in Andrews, Askey, and Roy [**15**, Ch. 6] or in Szegő's book [**377**]. Chapter 4 of Part 4 discusses some aspects of the general theory of orthogonal polynomials.

Legendre introduced his polynomials [**261**], the first orthogonal polynomials, in 1782 in connection with multipole expansions. If $|y| = \rho$, $|x| = r > \rho$, and $x \cdot y/|x||y| = \cos\theta$, then

$$\frac{1}{|x-y|} = \sum_{\ell=0}^{\infty} \frac{\rho^\ell}{r^{\ell+1}} P_\ell(\cos\theta) \tag{14.4.113}$$

as follows from (14.4.110) with $\alpha = \frac{1}{2}$. Ivory [**212**] seems to be the first to note their orthogonality in $L^2([-1,1],dx)$. For higher-dimensional analogs of the multipole expansion, see Section 3.5, especially Example 3.5.12 of Part 3.

(14.4.97) is known as Rodrigues' formula, after its initial appearance for the case $\alpha = \beta = 1$, that is, for Legendre polynomials, in the 1816 thesis [**333**] of Olinde Rodrigues (1795–1851), a student of Laplace. It lay unnoticed until its rediscovery by J. Ivory [**213**] and Jacobi [**216**]. Jacobi suggested that he and Ivory publish their result in French because it was unknown there (!) and they did in 1837. For many years, it was known as the Ivory–Jacobi formula until Hermite rediscovered Rodrigues' paper and Heine [**187**] campaigned for a change of name. Ironically, in 1810, Laplace [**252**] in his work in celestial mechanics had what we would now call Rodrigues' formula for Hermite polynomials. As a sidelight, we note that Rodrigues, who spent most of his life as a banker, discussed something close to quaternions three years before Hamilton (see Altmann [**12**] and Altmann–Ortiz [**13**]). As an added historical insult, E. Cartan and E. T. Bell referred to Rodrigues and his collaborator Olinde.

For a discussion of solutions of Legendre's equation, including the second solution, see, for example, [**4**].

**Problems**

1. By using a fractional linear transformation, find the general form of the Papperitz–Riemann equation for general singular points $\zeta_1, \zeta_2, \zeta_3$ all finite.

2. Let $a_1, a_0$ be the coefficients of an analytic second-order ODE.

(a) If $0, 1$ are the only singular points in $\mathbb{C}$, prove that

$$a_1(z) = -\frac{A}{z} - \frac{B}{z-1} + g_1(z)$$

$$a_0(z) = -\frac{C}{z^2} - \frac{E}{(z-1)^2} - \frac{F}{z} - \frac{G}{z-1} + g_0(z)$$

where $g_0, g_1$ are entire functions.

(b) If $a_1(z) = O(1/z)$ at infinity, prove $g_1 \equiv 0$.

(c) If $a_0(z) = O(1/z^2)$ at infinity, prove that $g_0(z) \equiv 0$ and $G = -F$, so $a_0(z)$ has the form in (14.4.12).

3. This problem will provide insight into (14.4.1). This argument is from Riemann's original paper on monodromy!

   (a) Let $C_0, C_1$ be the circuit matrices for an ODE singular only at $0, 1$, given by small circles around 0 and 1. Let $C_2$ be the circuit matrix for $|z| = 2$. Prove that $C_2 = C_0 C_1$.

   (b) If the ODE is second-order and regular at $0, 1, \infty$, prove $\det(C_0) = \exp(2\pi i(\alpha_1 + \alpha_2))$, $\det(C_1) = \exp(2\pi i(\beta_1 + \beta_2))$, $\det(C_2) = \exp(-2\pi i(\gamma_1 + \gamma_2))$.

   (c) Prove that $\alpha_1 + \alpha_2 + \beta_1 + \beta_2 + \gamma_1 + \gamma_2$ is in $\mathbb{Z}$.

4. For any $\alpha, \beta \in \mathbb{C} \setminus \{0, 1, \dots\}$, prove that $\lim_{n\to\infty} \sqrt[n]{(\alpha)_n/(\beta)_n} = 1$ and $\lim_{n\to\infty} \sqrt[n]{1/(\alpha)_n} = 0$. Verify that ${}_pF_q$ has a radius of convergence in $z$ of $\infty$ if $p < q+1$, 1 if $p = q+1$, and 0 if $p > q+1$.

5. (a) Show that the power series about $w = 0$ for $(1-w)^{-a}$ is

$$(1-w)^{-a} = \sum_{n=0}^{\infty} \frac{(a)_n}{n!}\, w^n \tag{14.4.114}$$

   (b) Verify that the power series about $z = 0$ for both sides of (14.4.43) agree.

   (c) Prove (14.4.43) for all $z \in \mathbb{C} \setminus [1, \infty)$.

6. Prove that

$$\text{(a)}\quad z\,\frac{dF}{dz}(a,b,c;z) = (1-z)^{-1}[(c-a)F(a-1,b,c;z) - (a-c+bz)F(a,b,c;z)] \tag{14.4.115}$$

$$\text{(b)}\quad z\,\frac{dF}{dz}(a,b,c;z) = zc^{-1}(1-z)^{-1}[(c-a)(c-b)F(a,b,c+1;z) - c(a+b-c)F(a,b,c;z)] \tag{14.4.116}$$

(c) $$a(1-z)F(a+1,b,c;z)+[c-2a-(b-a)z] \\ F(a,b,c;z)-(c-a)F(a-1,b,c;z)=0 \quad (14.4.117)$$

7. This problem fills in the details of the proof of Jacobi's hypergeometric formula.
   (a) Verify (14.4.51).
   (b) Verify (14.4.53).
   (c) Verify (14.4.55).
   (d) Verify (14.4.56).
8. (a) Let $b \in [1,\infty)$. For $n \leq 2[b]$, prove that we have $\frac{(b)_n}{b^n n!} \leq \frac{3^n}{n!}$.
   (b) Let $b \in [1,\infty)$. For $n \geq 2[b]$, prove that we have $\frac{(b)_n}{b^n n!} \leq \frac{2^n}{b^{n/2}}$. (*Hint:* For all $j$, prove that $b+j \leq b(j+1)$; and for $\frac{n}{2} \leq j \leq n$, that $b+j \leq 2j$.)
   (c) Prove that
$$\sup_{\substack{|z|\leq R \\ b\geq (4R)^2}} \sum_{n=1}^\infty \left| \frac{(a)_n (b)_n}{(c)_n n!} \left(\frac{z}{b}\right)^n \right| < \infty$$
9. (a) Prove (14.4.61) as a limit of (14.4.43).
   (b) Prove (14.4.61) by expanding $e^{zt}$ in a power series.
10. Prove the following for $M(a,b;z) = {}_1F_1\left(\begin{smallmatrix} a \\ b \end{smallmatrix}; z\right)$:
   (a) $\dfrac{dM}{dz}(a,b;z) = \dfrac{a}{b} M(a+1,b+1;z)$
   (b) $z\dfrac{dM}{dz}(a,b;z) = a(M(a+1,b;z) - M(a,b;z))$
   (c) $z\dfrac{dM}{dz}(a,b;z) = (b-1)(M(a,b-1;z) - M(a,b;z)$
   (d) $z\dfrac{dM}{dz}(a,b;z) = (b-a)M(a-1,b;z) + (a-b+z)M(a,b;z)$
   (e) $z\dfrac{dM}{dz}(a,b;z) = \dfrac{z}{b}(a-b)M(a,b+1;z) + zM(a,b;z)$
11. (a) Prove that $e^{-z}(e^z g)' = g' + g$.
   (b) Using (a), prove *Kummer's transformation*, $M(a,b;z) = e^z M(b-a,b;-z)$.
12. Verify (14.4.65) by the method mentioned in the text.
13. By expanding $(1-k^2t^2)^{\pm 1/2}$ in a binomial series, verify (14.4.69).

14. (a) Prove that the error function, (14.4.70), is given by the convergent power series

$$\operatorname{erf}(z) = \frac{2}{\sqrt{\pi}} \sum_{n=0}^{\infty} \frac{(-1)^n}{n!} \frac{1}{2n+1} z^{2n+1}$$

(b) Verify (14.4.71).

15. Check that the functions on the right side of (14.4.74)/(14.4.75) solve Weber's ODE, (14.4.73).

16. By comparing (14.4.73)/(14.4.76) and using (14.4.77)/(14.4.78), verify (14.4.79)/(14.4.80).

17. Use (14.4.97) and integration by parts to obtain (14.4.98).

18. (a) Prove that

$$\sum_{n=0}^{\infty} \cos(n\theta) \frac{t^n}{n!} = \tfrac{1}{2} \left[\exp(e^{i\theta} t) + \exp(e^{-i\theta} t)\right]$$

(b) Prove that

$$\sum_{n=0}^{\infty} T_n(x) \frac{t^n}{n!} = \tfrac{1}{2} \left[\exp\bigl(t\bigl[x + \sqrt{x^2-1}\,\bigr]\bigr) + \exp\bigl(t\bigl[x - \sqrt{x^2-1}\,\bigr]\bigr)\right] \quad (14.4.118)$$

(c) Prove that

$$\sum_{n=0}^{\infty} \cos(n\theta) t^n = \tfrac{1}{2} \left[\frac{1}{1-te^{i\theta}} + \frac{1}{1-te^{-i\theta}}\right]$$

(d) Prove that

$$\sum_{n=0}^{\infty} T_n(x) t^n = \frac{1-xt}{1-2xt+t^2}$$

**Remark.** See also Problem 9 of Section 3.1 of Part 2A.

19. (a) By writing

$$(1-2xt+t^2)^{-\alpha} = (1+t^2)^{-\alpha} \left(1 - \frac{2xt}{1+t^2}\right)^{-\alpha}$$

and expanding the second factor via the binomial theorem and then powers $(1+t^2)^{-n-\alpha}$ by the binomial theorem, prove that

$$(1-2xt+t^2)^{-\alpha} = \sum_{n,k} \frac{(\alpha)_{k+n}}{k!\,n!} (2xt)^k (-1)^n t^{2n} \quad (14.4.119)$$

(b) Prove that

$$(1-2xt+t^2)^{-\alpha} = \sum_{n=0}^{\infty} \frac{(\alpha)_n}{n!}\, t^n (2x)^n {}_2F_1\left(\begin{matrix} -\frac{n}{2}, \frac{1-n}{2} \\ 1-n-\alpha \end{matrix}; \frac{1}{x^2}\right) \tag{14.4.120}$$

(c) Using (14.4.96) and (14.4.109), verify that

$$C_n^{(\alpha)}(x) = \frac{(\alpha)_n}{n!} (2x)^n {}_2F_1\left(\begin{matrix} -\frac{n}{2}, \frac{1-n}{2} \\ 1-n-\alpha \end{matrix}; \frac{1}{x^2}\right)$$

and so conclude (14.4.110).

20. Using (14.4.104) and (14.4.109), verify (14.4.112) for $n \geq 1$ and then prove (14.4.111) from (14.4.110).

## 14.5. Bessel and Airy Functions

In this section, we'll study the solutions of the Bessel equation, (14.0.8), and Airy equation, (14.0.10). In many ways, this will be primarily a taxonomy of solutions—as with trigonometric functions, there are a large number of different-named solutions of Bessel's equation, for example, and we introduce and name them and state some of their basic properties. We'll also introduce spherical Bessel functions and explain how regular and spherical Bessel functions arise in the expansion of plane waves in cylindrical and spherical coordinates and how eigenvalues of the Laplacian in a disk (respectively, ball) (with Dirichlet boundary conditions) are exactly the zeros of suitable Bessel (respectively, spherical Bessel) functions.

We start with Bessel's equation, (14.0.8). As we saw in Example 14.3.9, the indices are $\pm\alpha$; hence, so long as $2\alpha \notin \mathbb{Z}$, we are guaranteed a solution of the form

$$f(z) = \sum_{n=0}^{\infty} c_n z^{\alpha+n} \tag{14.5.1}$$

(and since we consider $\alpha \in \mathbb{C}$, we can replace $\alpha$ by $-\alpha$ to get the other solution). The differential equation

$$z^2 f'' + z f' + (z^2 - \alpha^2) f = 0 \tag{14.5.2}$$

yields

$$(\alpha+n)(\alpha+n-1)c_n + (\alpha+n)c_n - \alpha^2 c_n + c_{n-2} = 0 \tag{14.5.3}$$

or

$$c_n = -\frac{c_{n-2}}{n(n+2\alpha)} \tag{14.5.4}$$

Thus, since $c_{-1} = 0$, we see $c_1 = c_3 = \cdots = c_{2n+1} \cdots = 0$ and

$$c_{2n} = \frac{(-1)^n c_0}{2^{2n}\Gamma(n+1)(n+\alpha)(n+\alpha-1)\dots(\alpha+1)} \tag{14.5.5}$$

Taking $c_0 = 1/\big[2^\alpha\Gamma(\alpha+1)\big]$, we see that we can define the *Bessel function* of order $\alpha$ (sometimes called the *Bessel function of the first kind*)

$$J_\alpha(z) = \sum_{n=0}^{\infty} \frac{(-1)^n}{\Gamma(n+1)\Gamma(n+\alpha+1)} \left(\frac{z}{2}\right)^{\alpha+2n} \tag{14.5.6}$$

**Theorem 14.5.1.** *$z^{-\alpha}J_\alpha(z)$ is an entire function of $z$ and $\alpha$, and $J_\alpha(z)$ solves* (14.5.1). *If $\alpha \notin \mathbb{Z}$, $J_\alpha$ and $J_{-\alpha}$ are independent and are the solutions which are $O(z^{\pm\alpha})$ at $z=0$.*

**Remarks.** 1. By allowing $c_0$ to have zeros, we have extended $J_\alpha$ to $\{\alpha \mid 2\alpha \in \mathbb{Z}\}$.

2. As we'll see, $\{\alpha \mid \alpha + \frac{1}{2} \in \mathbb{Z}\}$ still have $J_\alpha$ and $J_{-\alpha}$ independent, even though $\alpha-(-\alpha) \in \mathbb{Z}$ and there might have been logs according to the general theory. As we'll also see, for $\alpha = n \in \mathbb{Z}$,

$$J_{-n}(z) = (-1)^n J_n(z) \tag{14.5.7}$$

**Proof.** For any $\alpha \in \mathbb{C}$, by Stirling's formula, $\Gamma(\alpha+n+1) \to \infty$ as $n \to \infty$ uniformly for $\alpha$ in compact subsets of $\mathbb{C}$, so because of the $1/n!$, the series in (14.5.6) (multiplied by $z^{-\alpha}$) converges uniformly in $\alpha$ and $z$ on compacts, proving analyticity. Running the calculation in (14.5.5) backwards shows $J_\alpha$ solves (14.5.1) for $\alpha \notin \mathbb{Z}$ and then, by continuity, for $\alpha \in \mathbb{Z}$ also.

If $\alpha \notin \mathbb{Z}$, the first term in (14.5.4) is nonzero, so $J_{\pm\alpha}(z)$ is $O(z^{\pm\alpha})$ at zero, proving independence. This is true even if $\alpha + \frac{1}{2} \in \mathbb{Z}$.

If $\alpha = \pm m$, we have

$$\Gamma(n-m+1)\Gamma(n+1)\big|_{n=k+m} = \Gamma(k+1)\Gamma(k+m+1)$$

is infinity if $k = -m, \dots, -1$, and we get (14.5.7). □

The following relates $J_{\alpha\pm n}$ to $J_\alpha$:

**Theorem 14.5.2.** *We have that*

(a) $J'_\alpha(z) = \frac{1}{2}(J_{\alpha-1}(z) - J_{\alpha+1}(z))$ (14.5.8)

(b) $\dfrac{2\alpha}{z}\, J_\alpha(z) = J_{\alpha+1}(z) + J_{\alpha-1}(z)$ (14.5.9)

(c) $J_{\alpha\pm 1}(z) = \dfrac{\alpha}{z}\, J_\alpha(z) \mp J'_\alpha(z)$ (14.5.10)

(d) $J_{\alpha\pm n}(z) = z^{\pm\alpha} z^n (\mp 1)^n \left(\dfrac{1}{z}\dfrac{d}{dz}\right)^n (z^{\mp\alpha} J_n(z))$ (14.5.11)

(e) $J_{1/2}(z) = \sqrt{\dfrac{2}{\pi z}}\, \sin(z), \quad J_{-1/2}(z) = \sqrt{\dfrac{2}{\pi z}}\, \cos(z)$ (14.5.12)

(f) *For* $n = 0, 1, 2, \dots$, $J_{\pm(n+1/2)}(z)$ *is a sum of* $z^{-1/2}\cos(z)$ *times a polynomial of degree at most* $n$ *in* $z^{-1}$ *and* $z^{-1/2}\sin(z)$ *times a polynomial of degree at most* $n$ *in* $z^{-1}$.

**Proof.** We start with (d) for $n = 1$, namely,

$$J_{\alpha\pm 1}(z) = \mp z^{\pm\alpha}\,\frac{d}{dz}\,(z^{\mp\alpha}J_\alpha(z)) \tag{14.5.13}$$

which follows from looking at the expansions and checking term by term (Problem 1(a)).

This can be rewritten as (14.5.10), which implies (14.5.8)/(14.5.9). A simple induction in $n$ proves (14.5.11) (Problem 1(b)). The second formula in (14.5.12) follows from $2^{2n}\Gamma(n+1)\Gamma(n+\frac{1}{2}) = \sqrt{\pi}\,\Gamma(2n+1)$ and the first from $2^{2n+1}\Gamma(n+1)\Gamma(n+\frac{3}{2}) = \sqrt{\pi}\,\Gamma(2n+2)$. These $\Gamma$ relations are special cases of (9.6.26) of Part 2A. (f) follows from (e) and (d). □

We now turn towards the second solution when $\alpha \in \mathbb{Z}$. We begin with

**Proposition 14.5.3.** (a) (14.0.8) *can be rewritten as*

$$\frac{d}{dz}\left(z\,\frac{d}{dz}\right)f + \left(z - \frac{\alpha^2}{z}\right)f = 0 \tag{14.5.14}$$

(b) *For any two solutions,* $f$ *and* $g$*, of* (14.5.14)*, we have*

$$\widetilde{W}(f,g)(z) = z[f(z)g'(z) - g(z)f'(z)] \tag{14.5.15}$$

*is constant.*

(c) $$\widetilde{W}(J_\alpha, J_{-\alpha}) = -\frac{2}{\pi}\,\sin(\alpha\pi) \tag{14.5.16}$$

**Proof.** (14.5.14) is immediate from $z(zf')' = z^2f'' + zf'$. (14.5.14) implies

$$f(zg')' - g(zf')' = 0$$

and the left is $\frac{d}{dz}\widetilde{W}(f,g)(z)$. Finally, (14.5.16) follows (Problem 2) by taking $z \downarrow 0$ in the power series and using the Euler reflection formula, (9.6.25) of Part 2A. □

This suggests the definition

$$Y_\alpha(z) = \frac{J_\alpha(z)\cos(\pi\alpha) - J_{-\alpha}(z)}{\sin(\pi\alpha)} \tag{14.5.17}$$

called the *Neumann function of order* $\alpha$ or the *Weber function of order* $\alpha$ or the *Bessel function of the second kind.* By l'Hôpital's rule, it has a removable singularity as $\alpha \to n \in \mathbb{Z}$:

$$Y_n(z) = \frac{1}{\pi}\,\frac{\partial J_\alpha}{\partial\alpha}\bigg|_{\alpha=n}(z) - (-1)^n\,\frac{\partial J_{-\alpha}}{\partial\alpha}\bigg|_{\alpha=n} \tag{14.5.18}$$

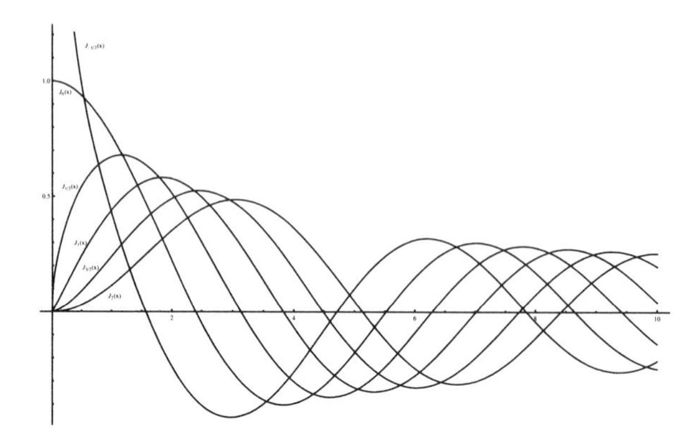

**Figure 14.5.1.** Bessel $J_\alpha$ for $a = -\frac{1}{2}, 0, \frac{1}{2}, 1, \frac{3}{2}, 2$.

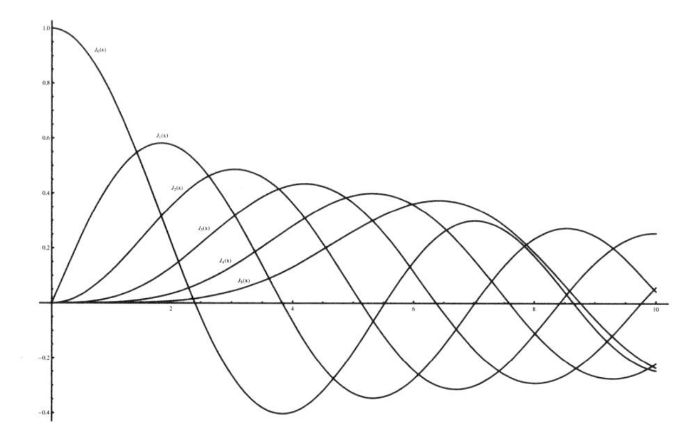

**Figure 14.5.2.** Bessel $J_n$ for $n = 0, 1, 2, 3, 4, 5$.

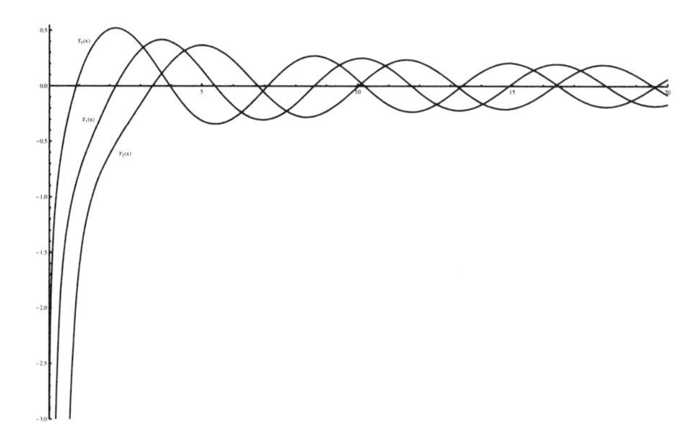

**Figure 14.5.3.** Neumann $Y_n$, $n = 0, 1, 2$.

$J$ and $Y$ are graphed in Figures 14.5.1, 14.5.2, and 14.5.3.

**Theorem 14.5.4.** (a) *$Y_\alpha(z)$ is analytic for $(\alpha, z) \in \mathbb{C} \times \mathbb{C} \setminus (-\infty, 0]$.*

(b) *$Y_\alpha$ solves (14.0.8) and is independent of $J_\alpha$.*

(c) $$\widetilde{W}(J_\alpha, Y_\alpha) = \frac{2}{\pi} \tag{14.5.19}$$

(d) *For $\alpha \equiv n = 0, 1, 2, \dots,$*

$$Y_n(z) = \frac{2}{\pi} J_n(z) \log\left(\frac{z}{2}\right) + z^{-n} g_n(z) \tag{14.5.20}$$

*for an entire function, $g_n$, of $z$ with $g_n(0) \neq 0$.*

**Remark.** (14.5.20) has the log behavior allowed by the general theory if $2\alpha \in \mathbb{Z}$.

**Proof.** (a) is immediate from the definition.

(b) $Y_\alpha$ solves (14.0.8) for $\alpha \notin \mathbb{Z}$ by (14.5.17), and then by continuity for $\alpha \in \mathbb{Z}$. Independence follows from (c).

(c) is immediate from $\widetilde{W}(J_\alpha, J_\alpha) = 0$ and (14.5.17) for $\alpha \notin \mathbb{Z}$ and, in general, by continuity.

(d) follows from (14.5.18), the power series, and $\frac{\partial}{\partial\alpha}(\frac{z}{2})^\alpha = (\frac{z}{2})^\alpha \log(\frac{z}{2})$. $\square$

Next, we want to state two integral representations for Bessel functions and explain how they lead to asymptotics for the Bessel functions. The Hankel contour, $H(c, R)$, is shown in Figure 14.7.1. If this contour is $H$, then one has, for any $\alpha$ and for $\operatorname{Re} z > 0$,

$$J_\alpha(z) = \frac{1}{2\pi i} \oint_H w^{-\alpha-1} \exp(\tfrac{1}{2}\, z[w - w^{-1}])\, dw \tag{14.5.21}$$

called the *Schläfli integral.* Another representation is only valid for $\operatorname{Re} \alpha > -\frac{1}{2}$,

$$J_\alpha(z) = \frac{1}{\sqrt{\pi}\, \Gamma(\alpha + \frac{1}{2})} \left(\frac{z}{2}\right)^\alpha \int_{-1}^{1} e^{izt} (1 - t^2)^{\alpha - 1/2}\, dt \tag{14.5.22}$$

From either of these, one gets the asymptotic formula valid as $x \to \infty$ along $(0, \infty)$:

$$J_\alpha(x) \sim \sqrt{\frac{2}{\pi x}} \cos\left(x - \frac{\alpha\pi}{2} - \frac{\pi}{4}\right) + o(x^{-1/2}) \tag{14.5.23}$$

We'll prove this in Example 15.2.11 from (14.5.39) below, but we note that we'll prove it in a sector and also that, while (14.5.22) only proves this if $\operatorname{Re} \alpha > -\frac{1}{2}$, (14.5.9) shows that if (14.5.23) is known for $\alpha$ and $\alpha + 1$, it holds for $\alpha - 1$ so we can go from that region to the entire $\alpha$ plane. This asymptotic formula and (14.5.17) imply

$$Y_\alpha(x) \sim \sqrt{\frac{2}{\pi x}} \sin\left(x - \frac{\alpha\pi}{2} - \frac{\pi}{4}\right) + o(x^{-1/2}) \tag{14.5.24}$$

This suggests we define solutions

$$H_\alpha^{(1)}(z) = J_\alpha(z) + iY_\alpha(z), \qquad H_\alpha^{(2)}(z) = J_\alpha(z) - iY_\alpha(z) \tag{14.5.25}$$

called *Hankel functions* or *Bessel functions of the third and fourth kind*, which have plane wave asymptotics

$$\begin{aligned} H_\alpha^{(1)}(x) &= \sqrt{\frac{2}{\pi x}}\, e^{i(x-\frac{\alpha\pi}{2}-\frac{\pi}{4})} + o(x^{-1/2}) \\ H_\alpha^{(2)}(x) &= \sqrt{\frac{2}{\pi x}}\, e^{-i(x-\frac{\alpha\pi}{2}-\frac{\pi}{4})} + o(x^{-1/2}) \end{aligned} \tag{14.5.26}$$

In fact (see the Notes to Section 15.2 and (15.2.57)/(15.2.58)), these functions have asymptotic series as $x \to \infty$ along $(0,\infty)$.

By (14.5.17), we have

$$i\sin(\pi\alpha)H_\alpha(z) = J_{-\alpha}(z) - J_\alpha(z)e^{-i\alpha\pi} \tag{14.5.27}$$

When $\alpha \notin \mathbb{Z}$, the integrand in (14.5.22) has a branch cut along $(-\infty, 0)$. But when $\alpha \in \mathbb{Z}$, we only have a singularity at $w = 0$ and we see $J_n(z)$ are Laurent coefficients, that is,

$$\exp(\tfrac{1}{2}\, z(w - w^{-1})) = \sum_{n=-\infty}^{\infty} w^n J_n(z) \tag{14.5.28}$$

the definition of $J_n(z)$ in Problem 2 of Section 3.7 of Part 2A. This leads to some simple addition formulae (see Problem 3).

This in turn leads to (see Problem 2(d) of Section 3.7 of Part 2A)

$$e^{i\vec{k}\cdot\vec{x}} = J_0(kx) + \sum_{n=1}^{\infty} 2i^n J_n(kx)\cos(n\theta) \tag{14.5.29}$$

where $k = |\vec{k}|$, $x = |\vec{x}|$, and $\vec{k}\cdot\vec{x} = kx\cos(\theta)$, or equivalently, if $\vec{k} = (k\cos(\theta_k), k\sin(\theta_k))$, $\vec{x} = (x\cos(\theta_x), x\sin(\theta_x))$, we have

$$e^{i\vec{k}\cdot x} = J_0(kx) + \sum_{n=-\infty}^{\infty} i^{|n|} J_n(kx)e^{in\theta_k}e^{-in\theta_x} \tag{14.5.30}$$

To give the analog in three dimensions, one defines *spherical Bessel functions* by

$$j_n(x) = \sqrt{\frac{\pi}{2x}}\, J_{n+1/2}(x), \qquad n = 0, 1, 2, \dots \tag{14.5.31}$$

$$= (-x)^n \left(\frac{1}{x}\frac{d}{dx}\right)^n \frac{\sin(x)}{x} \tag{14.5.32}$$

by (14.5.11)/(14.5.12). Then (see Example 3.5.20 of Part 3), if $\vec{k}, \vec{x}$ are in $\mathbb{R}^3$ with

$$|\vec{k}| = k, \quad |\vec{x}| = x, \quad \vec{k}\cdot\vec{x} = kx\cos(\theta)$$
$$\vec{k} = (k\sin(\theta_k)\cos(\varphi_k), k\sin(\theta_k)\sin(\varphi_k), k\cos(\theta_k))$$

and similarly for $x$ ($(k, \theta_k, \varphi_k)$ are *spherical coordinates*),

$$e^{i\vec{k}\cdot\vec{x}} = \sum_\ell (i)^\ell(2\ell+1)j_\ell(kr)P_\ell(\cos(\theta)) \tag{14.5.33}$$
$$= \frac{1}{4\pi}\sum_{\ell,m}(i)^\ell j_\ell(kr)\,\overline{Y_{\ell m}(\theta_k,\varphi_k)}\,Y_{\ell m}(\theta_x,\varphi_x) \tag{14.5.34}$$

formulae of considerable use in quantum scattering theory.

Bessel and spherical Bessel functions arise in problems with circular symmetry (respectively, spherical symmetry) in $\mathbb{R}^2$ (respectively, $\mathbb{R}^3$) also because of separation of variables. Suppose that one wants to solve $\Delta u = k^2 u$ in $\mathbb{R}^2$ for $|x| < 1$ with $u(\vec{x}) = 0$ if $|\vec{x}| = 1$. Because of the natural rotation symmetry, it is natural to try $u(\vec{x}) = f(x)e^{in\theta}$. Then (see the references in the Notes), $\Delta u = k^2 u$ is solved by $f(x) = J_n(kx)$ and the $f(x)|_{x=1} = 0$ boundary condition is given by $J_n(k) = 0$, that is, the eigenvalues of $\Delta$ are given by squares of the zeros of $J_n$. Moreover, it is known (see the Notes) that all solutions of $J_n(z) = 0$ are on $(0,\infty)$. In $\mathbb{R}^3$, the eigenfunctions are given by $j_\ell(kr)Y_{\ell m}(\theta,\varphi)$, where $j_\ell$ is a spherical Bessel function and $Y_{\ell m}$ a spherical harmonic.

We summarize with

**Theorem 14.5.5.** *The eigenvalues of* $-\Delta$ *in* $\{\vec{x} \in \mathbb{R}^\nu \mid |\vec{x}| < 1\}$ *with* $u(\vec{x}) = 0$ *if* $|\vec{x}| = 1$ *boundary conditions are given by*

(a) *The squares of zeros of* $J_n(k)$, $n = 0, 1, 2, \dots$ *if* $\nu = 2$ *(with multiplicity* $1$ *if* $n = 0$ *and* $2$ *if* $n = 1, 2, \dots$*).*

(b) *The squares of the zeros of* $j_\ell(k)$, $\ell = 0, 1, \dots$ *if* $\nu = 3$ *with multiplicity* $\ell + 1$.

Finally in the discussion of Bessel functions, we note that Bessel functions along the imaginary axis arise so often that one defines the *Bessel functions of imaginary argument* or *modified Bessel functions* by

$$I_\alpha(z) = e^{-\alpha\pi i/2}J_\alpha(ze^{i\pi/2}) \tag{14.5.35}$$

$$K_\alpha(z) = \frac{\pi}{2\sin(\alpha\pi)}\,[I_{-\alpha}(z) - I_\alpha(z)] \tag{14.5.36}$$

$K_\alpha(x)$, called the *MacDonald function* or *modified Bessel function of the second kind* ($I_\alpha$ is "of the first kind"), is related to $H_\alpha^{(1)}$ by

$$K_\alpha(z) = \tfrac{1}{2}\pi i e^{i\alpha\pi/2} H_\alpha^{(1)}(ze^{\pi i/2}) \tag{14.5.37}$$

Notice that $K_{-\alpha}(z) = K_\alpha(z)$ by (14.5.36).

The phase in (14.5.35) is picked so $I_\alpha(x)$ is real on $[0,\infty)$. $K_\alpha$ and $I_\alpha$ both solve the differential equation

$$z^2 f'' + zf' - (z^2 + \alpha^2)f = 0 \tag{14.5.38}$$

For example,

$$I_{1/2}(z) = \sqrt{\frac{2}{\pi z}}\,\sinh(z), \quad I_{-1/2}(z) = \sqrt{\frac{2}{\pi z}}\,\cosh(z), \quad K_{1/2}(z) = \sqrt{\frac{\pi}{2z}}\,e^{-z}$$

More generally, $K_\alpha(z)$ is the solution of (14.5.38) that has $K_\alpha(x) \to 0$, $K'_\alpha(x) \to 0$ as $x \to \infty$. Indeed, we'll prove (see the argument after Theorem 14.7.3) that for $|\arg(z)| < \frac{1}{2}\pi$, one has

$$K_\alpha(z) = \int_0^\infty e^{-z\cosh t}\cosh(\alpha t)\,dt \tag{14.5.39}$$

which implies, by the monotone convergence theorem, that for $x$ real as $x \to \infty$, $|K_\alpha(x)| + |K'_\alpha(x)| \to 0$. It implies much more; see Example 15.2.12. In a sense, $J$ is like cos and $I$ like cosh, $H^{(1)}(x)$ is $e^{ix}$, and $K(x)$ is like $e^{-x}$.

Finally, we turn to Airy's equation

$$y'' - zy = 0 \tag{14.5.40}$$

This has no singular points, so its solutions are entire functions of $z$. $y(z) = \sum_{n=0}^\infty a_n z^n$ obeys (14.5.40) if and only if for $n = 0, 1, 2, \dots$,

$$n(n-1)a_n = a_{n-3} \tag{14.5.41}$$

This implies $a_2 = 0$ since $a_{-1} = 0$, and then $a_{3k+2} = 0$ for $k = 0, 1, 2, \dots$. We get

**Theorem 14.5.6.** *The solutions of* (14.5.40) *are given by*

$$\begin{aligned} y(z) &= y(0)\left(1 + \frac{z^3}{3\cdot 2} + \frac{z^6}{6\cdot 5\cdot 3\cdot 2} + \frac{z^9}{9\cdot 8\cdot 6\cdot 5\cdot 3\cdot 2} + \cdots\right) \\ &\quad + y'(0)\left(z + \frac{z^4}{4\cdot 3} + \frac{z^7}{7\cdot 6\cdot 4\cdot 3} + \frac{z^{10}}{10\cdot 9\cdot 7\cdot 6\cdot 4\cdot 3} + \cdots\right) \end{aligned} \tag{14.5.42}$$

Two special solutions are $Ai(z)$ and $Bi(z)$. We'll give definitions for $z$ real as real integrals and, more generally in Section 14.7, as contour integrals for $z \in \mathbb{C}$. $Ai$ is singled out by going to zero as $x \to \infty$ along the positive real axis.

Recall (see (14.0.11)) that the Airy function is defined for $x$ real by

$$Ai(x) = \frac{1}{\pi} \int_0^\infty \cos(xt + \tfrac{1}{3}\,t^3)\,dt \tag{14.5.43}$$

The same integration by parts argument that shows the improper integral exists shows that (Problem 4(a))

$$Ai(x) = \lim_{\varepsilon \downarrow 0} \frac{1}{\pi} \int_0^\infty e^{-\varepsilon t} \cos(xt + \tfrac{1}{3}\,t^3)\,dt \tag{14.5.44}$$

and that (Problem 4(b)) $Ai$ is $C^2$ with

$$Ai''(x) = \lim_{\varepsilon \downarrow 0} -\frac{1}{\pi} \int_0^\infty e^{-\varepsilon t} t^2 \cos(xt + \tfrac{1}{3}\,t^3)\,dt \tag{14.5.45}$$

so that

$$\begin{aligned} Ai''(x) - xAi(x) &= \lim_{\varepsilon \downarrow 0} -\frac{1}{\pi} \int_0^\infty e^{-\varepsilon t} (x + t^2) \cos(xt + \tfrac{1}{3}\,t^3)\,dt \\ &= \lim_{\varepsilon \downarrow 0} -\frac{1}{\pi} \int_0^\infty e^{-\varepsilon t} \frac{d}{dt} \sin(xt + \tfrac{1}{3}\,t^3)\,dt \\ &= 0 \end{aligned}$$

using an integration by parts and $\sin(xt + \frac{1}{3}t^3)\big|_{t=0} = 0$. Thus, $Ai$ solves (14.5.40) and so is an entire function (the contour integral method of Section 14.7 provides an easier proof that $Ai$ obeys the differential equation).

In Section 15.4 (see Example 15.4.3), we'll prove the integral representation, valid for $|\arg(z)| < \pi$,

$$Ai(z) = \frac{\exp(-\frac{2}{3} z^{3/2})}{2\pi} \int_0^\infty \exp(-z^{1/2} t) \cos(\tfrac{1}{3}\,t^{3/2}) t^{-1/2}\,dt \tag{14.5.46}$$

Another solution is given by

$$Bi(x) = \frac{1}{\pi} \int_0^\infty \sin(xt + \tfrac{1}{3}\,t^3)\,dx + \frac{1}{\pi} \int_0^\infty \exp(xt - \tfrac{1}{3}\,t^3)\,dt \tag{14.5.47}$$

As above, one can show (Problem 4(c)) that $Bi$ solves (14.5.40). The first term alone does not because now there is a boundary term in the integration by parts, but the other integral also has a boundary term and they cancel. $Ai$ and $Bi$ are graphed in Figure 14.5.4.

One can obtain the power series expansions of the entire functions, $Ai$ and $Bi$ from (14.5.42) and

$$Ai(0) = \left[3^{2/3}\,\Gamma\Big(\frac{2}{3}\Big)\right]^{-1}, \quad Ai'(0) = -\left[3^{1/3}\,\Gamma\Big(\frac{1}{3}\Big)\right]^{-1} \tag{14.5.48}$$

$$Bi(0) = \left[3^{1/6}\,\Gamma\Big(\frac{2}{3}\Big)\right]^{-1}, \quad Bi' = -\left[3^{-1/6}\,\Gamma\Big(\frac{1}{3}\Big)\right]^{-1} \tag{14.5.49}$$

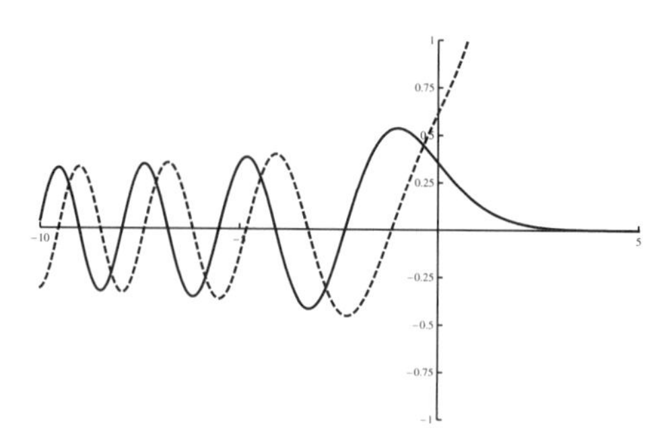


**Figure 14.5.4.** $Ai$ (solid) and $Bi$ (dotted).

An easy argument (Problem 5) shows $\lim_{x\to\infty} Ai(x) = 0$, while $\lim_{x\to\infty} Bi(x) = \infty$, so they are independent solutions. Indeed, detailed asymptotics (see Example 15.2.12) shows that as $x \to \infty$,

$$Ai(x) = \frac{\exp(-\frac{2}{3}x^{3/2})}{2\sqrt{\pi}\, x^{1/4}}\,(1 + O(x^{-3/2})) \tag{14.5.50}$$

$$Ai'(x) = -\frac{x^{1/4}\exp(-\frac{2}{3}x^{3/2})}{2\sqrt{\pi}}\,(1 + O(x^{-3/2})) \tag{14.5.51}$$

$$Bi(x) = \frac{\exp(\frac{2}{3}x^{3/2})}{\sqrt{\pi}\, x^{1/4}}\,(1 + O(x^{-3/2})) \tag{14.5.52}$$

$$Bi'(x) = \frac{x^{1/4}\exp(\frac{2}{3}x^{3/2})}{\sqrt{\pi}}\,(1 + O(x^{-3/2})) \tag{14.5.53}$$

and as $x \to -\infty$,

$$Ai(x) = \frac{1}{\sqrt{\pi}|x|^{1/4}} \cos\biggl(\frac{2}{3}\,|x|^{3/2} - \frac{\pi}{4}\biggr)(1 + O(x^{-3/2})) \tag{14.5.54}$$

$$Ai'(x) = -\frac{|x|^{1/4}}{\sqrt{\pi}} \sin\biggl(\frac{2}{3}\,|x|^{3/2} - \frac{\pi}{4}\biggr)(1 + O(x^{-3/2})) \tag{14.5.55}$$

$$Bi(x) = -\frac{1}{\sqrt{\pi}\,|x|^{1/4}} \sin\biggl(\frac{2}{3}\,|x|^{3/2} - \frac{\pi}{4}\biggr)(1 + O(x^{-3/2})) \tag{14.5.56}$$

$$Bi'(x) = -\frac{|x|^{1/4}}{\sqrt{\pi}} \cos\biggl(\frac{2}{3}\,|x|^{3/2} - \frac{\pi}{4}\biggr)(1 + O(x^{-3/2})) \tag{14.5.57}$$

The Wronskian, $W(f,g)(x) = f(x)g'(x) - f'(x)g(x)$, is constant for two solutions of (14.5.40), and in this case, using the asymptotics at either $x = \infty$ or $x = -\infty$, shows that

$$W(Ai, Bi) = \frac{1}{\pi} \tag{14.5.58}$$

## Notes and Historical Remarks.

> From a letter to a young lady, London, March 19, 1857: "I have been doing what I guess you won't let me do when we are married, sitting up till 3 o'clock in the morning fighting hard against a mathematical difficulty. Some years ago I attacked an Integral of Airy's .... But there was one difficulty about it .... I took it up again a few days ago, and after two or three days' fight, the last of which I sat up till 3, I at last mastered it ...."
>
> *Sir George G. Stokes* [**372**]

Special cases of Bessel functions appeared in eighteenth-century work of the Bernoullis, Euler, and Lagrange on physical models. The first systematic study was in 1816 by Bessel [**56**], who used them in astronomical studies. The generating function (3.7.14) of Part 2A was found in 1857 by Schlömilch [**341**]. There is a classic book of Watson [**405**] covering all aspects of their study. Chapter 10 of the NIST digital library [**308**] is on the subject.

Friedrich Wilhelm Bessel (1784–1846) was a German astronomer. Without a university education, he first came to the notice of professionals with his calculations of the orbit of Halley's comet when he was twenty, calculations based on two- hundred-year-old data. By 1809, he was established enough for the King of Prussia to appoint him director of the Königsberg Observatory where he spent the rest of his career. Besides his special functions, he is best known for the first measurements of stellar distances by parallax.

For expansions like (14.5.33), see various books on the theoretical physics of scattering, for example, [**152, 282, 299**]. For a discussion of zeros of Bessel functions, including the absence of nonreal zeros for $J_n$, see [**381**, Ch. 9].

Airy functions are named after George Biddell Airy (1801–92), a British astronomer and physicist. Early in his career, Airy was Lucasian Professor of Mathematics at Cambridge, the chair held by Newton, Dirac, and Hawking, but the income from the chair (£100/yr) was insufficient for his prospective father-in-law to allow him to marry, so Airy shifted to astronomy. He became the Plumian Professor of Astronomy which paid five times the Lucasian chair! Not long after, he became Astronomer Royal, established the Greenwich observatory as a scientific center and as the zero point for longitude.

In his work in optics in 1838 [**10**], he introduced (up to a linear change in the $x$ variable) the function in the integral (14.5.43). The symbol $Ai$ and the modern normalization is due to Jeffreys [**220**].

There are so many applications to classical and quantum physics; there is an entire book on the subject [**388**] and a whole chapter in the NIST Digital Library [**307**].

### Problems

1. (a) Verify (14.5.13) by checking the power series coefficients.

   (b) Provide the inductive step in the proof of (14.5.11).

2. Verify (14.5.16).

3. Use (14.5.24) to prove

$$J_n(z+\zeta) = \sum_{m=-\infty}^{\infty} J_m(z) J_{n-m}(\zeta) \tag{14.5.59}$$

   (proving absolute and uniform (in compact subsets of $z$ and $\zeta$) convergence of the sum).

4. (a) Prove (14.5.44) by using the same integration by parts that was used in Problem 1 of Section 14.0.

   (b) Prove (14.5.45).

   (c) Show $Bi(x)$ solves (14.5.40).

5. (a) By integrating by parts using

$$e^{i(tx+\frac{1}{3}t^3)} = \frac{1}{i(x+t^2)} \frac{d}{dt} e^{i(tx+\frac{1}{3}t^3)}$$

   prove $Ai(x) = O(x^{-N})$ for all $N$ as $x \to +\infty$.

   (b) Prove the first integral in $Bi$ goes to zero as $x \to \infty$, but the second goes to infinity.

   (c) Why doesn't this prove $Ai(x) = O(|x|^{-N})$ as $x \to -\infty$?

## 14.6. Nonlinear ODEs: Some Remarks

Thus far, we have focused on linear ODEs in the complex domain with analytic coefficients. In this section, we want to say something about the nonlinear case. It is so brief that we can't even call it introduction. Rather we want to focus on one phenomenon that doesn't happen in the linear case: movable singularities. We begin, however, with a common feature with the linear case, local solubility:

**Theorem 14.6.1.** *Let $F(w,z)$, defined with an $m$-component complex vector, $w$, and complex number, $z$, be analytic in some $\mathbb{D}_{\delta_1}(w_0) \times \mathbb{D}_{\delta_2}(z_0)$ with values in $\mathbb{C}^m$ for some $m$. Then there is $\delta_3 < \delta_2$ and $f(z)$ analytic in $\mathbb{D}_{\delta_3}(z_0)$ so that $f[\mathbb{D}_{\delta_3}(z_0)] \subset \mathbb{D}_{\delta_1}(w_0)$ and $f$ obeys*

$$f'(z) = F(f(z), z) \tag{14.6.1}$$

$$f(z_0) = w_0 \tag{14.6.2}$$

*Moreover, $f$ is the unique solution of (14.6.1)/(14.6.2) in the sense that if $g$ also obeys these relations in some $\mathbb{D}_{\delta_4}(z_0)$, then $g = f$ on $\mathbb{D}_{\min(\delta_3,\delta_4)}(z_0)$.*

One proves this by rewriting (14.6.1)/(14.6.2) as an integral equation

$$f(z) = w_0 + \int_0^1 (z - z_0)F(f((1-t)z_0 + tz), (1-t)z_0 - tz)\, dt \qquad (14.6.3)$$

and uses a contraction mapping theorem. The argument is in Problem 1. We note that a scalar $n$th-order equation $f^{(n)}(z) = G(f(z), \dots, f^{(n-1)}(z), z)$ can be rewritten in the form (14.6.2) (see Problem 2).

In the linear case, we could analytically continue a solution along any simple curve on which all the coefficients are analytic. Thus, the singularities are *fixed singularities*, that is, their location is the same under small perturbations of the initial conditions. This need not be true in the nonlinear case, as can be seen in the simplest examples:

**Example 14.6.2** (Riccati Equation). Consider the ODE

$$f'(z) = -f(z)^2 \qquad (14.6.4)$$

whose coefficients are constant and so defined on all of $\mathbb{C}$. For $a \in \mathbb{C}$, $a \neq 0$ and note that

$$f_a(z) = -(a - z)^{-1} \qquad (14.6.5)$$

solves (14.6.4), with the initial condition $f_a(0) = -a^{-1}$. $f_a(z)$ has a pole at $z = a$. A small change of initial conditions makes a small but nonzero change in the localization of the pole. This is a *movable singularity*. □

There are even third-order scalar equations with movable natural boundaries! (see the Notes). There are nonlinear equations associated to elliptic functions, for example, (10.4.45) of Part 2A, that have natural movable poles. Painlevé classified a class of second-order equations with only movable poles and found those solutions which can be written in terms of known functions, like elliptic functions, and the solutions of eight equations which define new transcendent functions, now called the Painlevé functions. They arise in problems of completely integrable systems and in random matrix theory. We'll provide some pointers to them in the Notes.

**Notes and Historical Remarks.** About 1910, Chazy [**79, 80, 81**] showed that various third-order ODEs, including

$$\frac{d^3 f}{dz^3} - 2f\,\frac{d^2 f}{dz^2} - \left(\frac{df}{dz}\right)^2$$

could have natural boundary; see, for example, Ablowitz–Clarkson [**3**].

For more on the Painlevé functions, see [**137, 167, 215**].

**Problems**

1. Use the contraction mapping on (14.6.3) to prove Theorem 14.6.1.

   **Remark.** You should use the proof of the contraction mapping theorem to get analytic approximations that converge uniformly.

2. Prove that an $n$th-order equation can be rewritten in the form (14.6.2) by considering $g(z) = \big(f(z), \dots, f^{(n-1)}(z)\big)$ and writing the $n$th-order equation as $g'(z) = G\big(g(z), z\big)$.

## 14.7. Integral Representation

In this section, we'll see that some of the special functions of this chapter have representations as contour or as ordinary integrals. These are of special interest because they often allow an analysis of asymptotics as some variable or parameters in the functions go to infinity. Indeed, this section provides a link to the next chapter, which will discuss the asymptotic analysis of contour and ordinary integrals as parameters go to infinity.

We begin with the gamma function of Section 9.6 of Part 2A. Of course, we already have an integral representation, Euler's famous formula, (9.6.1) of Part 2A, but it only converges if $\operatorname{Re} z > 0$. The problem if $\operatorname{Re} z \le 0$ is an infinity at $t = 0$ in the integral. We already know how to avoid singularities in contour integrals by moving the contour, but that seems difficult for an endpoint. However, if we make the line of integration into a branch cut, we can change to a contour around the cut and so avoid the singularity. It is traditional (although a minority of authors take the negative of our choice) to place the cut on $(-\infty, 0]$ rather than $[0, \infty)$, but that's the basic strategy.

Thus, we define the *Hankel contour* (aka *Hankel loop*), $H(c, R)$, for two real parameters $0 < c < R$, as seen in Figure 14.7.1, as follows. Let

$$\alpha_\pm(c, R) = -\sqrt{R^2 - c^2} \pm ic$$

$H$ goes along $\operatorname{Im} w = -c$ from $-\infty - ic$ to $\alpha_-$, counterclockwise along the circle $|w| = R$ from $\alpha_-$ to $\alpha_+$, and then along the line $\operatorname{Im} w = c$ from $\alpha_+$ to $-\infty + ic$.

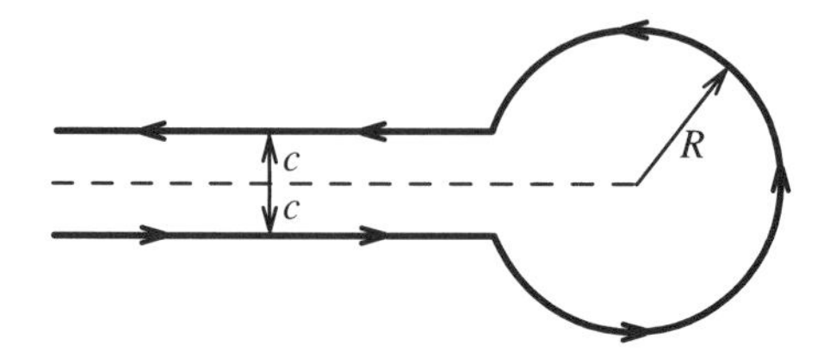


**Figure 14.7.1.** The Hankel contour, $H(c, R)$.

**Theorem 14.7.1.** *For any $0 < c < R$, we have that*

(a) *For $z \in \mathbb{C} \setminus \{0, -1, -2, \dots\}$,*

$$\Gamma(z) = \frac{1}{2i \sin \pi z} \int_{H(c,R)} w^{z-1} e^w \, dw \tag{14.7.1}$$

(b) *For $z \in \mathbb{C}$,*

$$\frac{1}{\Gamma(z)} = \frac{1}{2\pi i} \int_{H(c,R)} w^{-z} e^w \, dw \tag{14.7.2}$$

**Remarks.** 1. These are called *Hankel's formulae.*

2. For $z_0 \in \{1, 2, \dots\}$, (14.7.1) needs to be interpreted as a limit $z \to z_0$ (see Problem 1); otherwise it is $0/0$.

3. We interpret $w^z$ (with $w^{-z} = 1/w^z$) as $e^{z \log w}$, where the branch of $\log w$ is taken in $\mathbb{C} \setminus (-\infty, 0]$ with $|\operatorname{Im} \log w| < \pi$.

4. This theorem also has a proof using Wielandt's theorem (see Problem 2).

**Proof.** We note first that

$$|w^z e^w| \le e^{\pi|y|} |w|^x e^{\operatorname{Re} w}, \qquad z = x + iy \in \mathbb{C} \setminus (-\infty, 0] \tag{14.7.3}$$

since $|w^z| = \exp(\operatorname{Re}(z \log w)) \le \exp(x \log|w| + |y|\pi)$ on account of $|\operatorname{Im}(\log w)| \le \pi$. Thus, the integrals in (14.7.1)/(14.7.2) converge for all $z \in \mathbb{C}$ and, by Theorem 3.1.6 of Part 2A, define entire functions. By the CIT, it is easy to see the integrals are independent of $c$ and $R$.

Let $Q(z)$ stand for the integral in (14.7.1) and let $\operatorname{Re} z > 0$. Then one can first take $c \downarrow 0$ and then $R \downarrow 0$ (since $\operatorname{Re} z > 0$, the potential singularity at $z = 0$ is integrable). Since for $t \in (0, \infty)$,

$$(-t - i0)^{z-1} - (-t + i0)^{z-1} = |t|^{z-1}(e^{-\pi i(z-1)} - e^{\pi i(z-1)}) \tag{14.7.4}$$

$$= |t|^{z-1}(-2i) \sin \pi(z-1) \tag{14.7.5}$$

$$= (2i)|t|^{z-1} \sin \pi z \tag{14.7.6}$$

we see that for $\operatorname{Re} z > 0$,

$$Q(z) = [(2i) \sin \pi z] \Gamma(z) \tag{14.7.7}$$

on account of (9.6.1) of Part 2A. This proves (14.7.1) for $\operatorname{Re} z > 0$ and then for $z \in \mathbb{C} \setminus \{0, -1, -2, \dots\}$ by analyticity.

By (14.7.1),

$$\text{RHS of (14.7.2)} = \frac{\sin \pi z}{\pi} \Gamma(1 - z) \tag{14.7.8}$$

so (14.7.2) follows from the Euler reflection formula, (9.6.25) of Part 2A. □

This in turn implies

**Theorem 14.7.2.** *For any $\alpha, z \in \mathbb{C}$ with $\operatorname{Re} z > 0$, Schläfli's integral (14.5.21) holds, that is,*

$$J_\alpha(z) = \frac{1}{2\pi i} \oint_H w^{-\alpha-1} \exp(\tfrac{1}{2}\, z[w - w^{-1}])\, dw \tag{14.7.9}$$

**Proof.** Both sides of (14.5.21) are entire and analytic in $\operatorname{Re} z > 0$, so it suffices to prove the result for $z \in (0, \infty)$. In (14.5.6), use (14.7.2) (with $z$ of that formula equal to $n + \alpha + 1$). We get

$$J_\alpha(z) = \sum_{n=0}^{\infty} \frac{(-1)^n}{n!} \int_H \left(\frac{z}{2}\right)^{\alpha+2n} w^{-\alpha-n-1} e^w\, dw \tag{14.7.10}$$

Since $z \in (0, \infty)$, we can change variables from $w$ to $\frac{1}{2}wz$, replacing $H(c, R)$ by $H(cz^{-1}, Rz^{-1})$, and then use the $(c, R)$-independence to return to $H(c, R)$.

A simple estimate (Problem 3) allows one to exchange the sum and integral. Then using $\sum_{n=0}^{\infty} \frac{(-1)^n}{n!}(zw^{-1})^n = \exp(-zw^{-1})$, we obtain (14.5.21). $\square$

Taking $R = 1$ and $c \downarrow 0$, we get a sum of two integrals. For the piece from $-\infty$ to $-1$, we let $w = -e^x$, $0 < x < \infty$, and get

$$\frac{\sin(\pi\alpha)}{\pi} \int_{-\infty}^{0} e^{-\alpha x - z \sinh x}\, dx \tag{14.7.11}$$

For the circle, we add the two semicircles with $w = e^{\pm i\theta}$, $0 < \theta < \pi$, and get $\frac{1}{\pi} \int_0^\pi \cos(\alpha\theta - z \sin\theta)\, d\theta$. The result is a formula, also of Schläfli,

$$J_\alpha(z) = \frac{1}{\pi} \int_0^\pi \cos(\alpha\theta - z \sin\theta)\, d\theta - \frac{\sin(\pi\alpha)}{\pi} \int_0^\infty e^{-\alpha x - z \sinh x}\, dx \tag{14.7.12}$$

which generalizes (3.7.17) of Part 2A from $\alpha \in \mathbb{Z}$.

There is also a useful general change of variables in (14.7.9). Let $w = e^t$. Then $H$ goes into a $t$ contour, $S_0$, shown in Figure 14.7.2, and we have

$$J_\alpha(z) = \frac{1}{2\pi i} \int_{S_0} e^{-\alpha t} e^{z \sinh t}\, dt \tag{14.7.13}$$

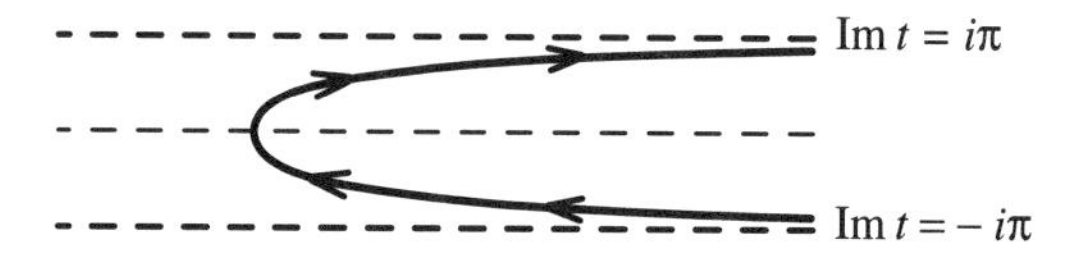


**Figure 14.7.2.** The Sommerfeld contour, $S_0$.

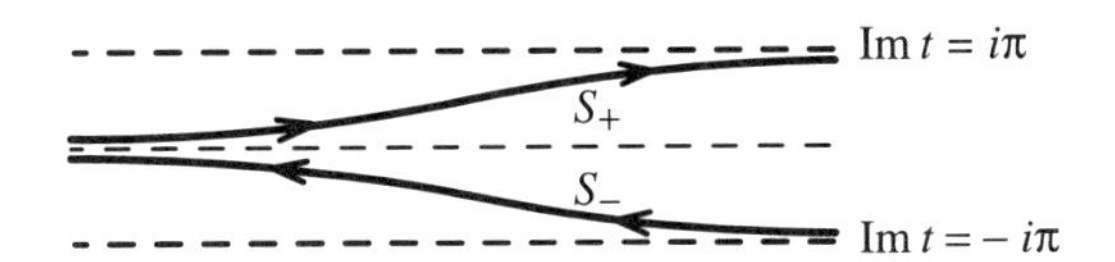


**Figure 14.7.3.** The Sommerfeld contours, $S_\pm$.

Here $S_0$ goes from $\infty - i\pi$ to $\log R \in \mathbb{R}$ and out to $\infty + i\pi$. Since $\sinh t$ for $t = x \pm i\pi$ is $-\sinh x$, the integrals converge at the ends. Since $\sinh(-x) = -\sinh x$, we can also pull the contour to the one $S$ shown in Figure 14.7.3, called the *Sommerfeld contour*. It has two pieces, $S_+$ and $S_-$. $S_+$ runs from $-\infty + i0$ to $\infty + i\pi$ and $S_-$ from $\infty - i\pi$ to $-\infty - i0$. Remarkably, the integrals over $S_+$ and $S_-$ separately give $H_\alpha^{(1)}$ and $H_\alpha^{(2)}$ up to a factor of 2 (since $J_\alpha = \frac{1}{2}(H_\alpha^{(1)} + H_\alpha^{(2)})$)!

**Theorem 14.7.3** (Sommerfeld's Formula). *For* $\operatorname{Re} z > 0$,

$$H_\alpha^{(1)}(z) = \frac{1}{\pi i} \int_{S_+} e^{-\alpha t} e^{z \sinh t}\, dt \tag{14.7.14}$$

*and* $H_\alpha^{(2)}$ *is similar, with* $S_+$ *replaced by* $S_-$.

**Remark.** The integral is called *Sommerfeld's integral.*

**Proof.** Let $C_\alpha^+(z)$ be the right side of (14.7.14) and $C_\alpha^-(z)$ the same integral with $S_+$ replaced by $S_-$.

The change of variables for $C_{-\alpha}^\pm$, $t = -w \pm i\pi$, takes $S_\pm$ to $S_\pm$ since $w = -t \pm i\pi$ means $-\infty \to \infty \pm i\pi$ and $\infty \pm i\pi \to -\infty$. The direction of the contour is reversed but $dw = -dt$. $\sinh(-t \pm i\pi) = \sinh t$, with $e^{-(-\alpha)t} = e^{-\alpha w} e^{\pm i\pi\alpha}$, we conclude

$$C_{-\alpha}^\pm(z) = e^{\pm i\pi\alpha} C_\alpha(z) \tag{14.7.15}$$

By (14.7.13),

$$J_\alpha(z) = C_\alpha^+(z) + C_\alpha^-(z) \tag{14.7.16}$$

and so, by (14.7.15),

$$J_{-\alpha}(z) = e^{i\pi\alpha} C_\alpha^+(z) + e^{-i\pi\alpha} C_\alpha^-(z) \tag{14.7.17}$$

It follows that

$$-J_\alpha(z) e^{-i\alpha\pi} + J_{-\alpha}(z) = 2i \sin(\alpha\pi) C_\alpha^+(z) \tag{14.7.18}$$

By (14.5.27), we conclude for $\alpha \notin \mathbb{Z}$ that $2C_\alpha^+(z) = H_\alpha^{(1)}(z)$. By analyticity, this also holds for $\alpha \in \mathbb{Z}$. $\square$

(14.7.14) holds for $S_+$ if $-\frac{\pi}{2} < \arg(z) < \frac{\pi}{2}$. It is easy to see that we can change $\arg(z)$ if we also move $S_+$ to keep the end in the region of decay. In particular, if $0 < \arg(z) < \pi$, we have (14.7.14) modified if $S_+$ is replaced by $\widetilde{S}_+$ running for $-\infty + i\frac{\pi}{2}$ to $\infty + i\frac{\pi}{2}$, that is, for $0 < \arg(z) < \pi$,

$$H_\alpha^{(1)}(z) = \frac{1}{\pi i} \int_{-\infty}^{\infty} e^{z \sinh(t+i\pi/2)} e^{-\alpha t} e^{-i\pi\alpha/2} \tag{14.7.19}$$

Given (14.5.37), $\sinh(t+i\pi/2) = i\cosh(t)$ and $\frac{1}{2}(e^{-\alpha t} + e^{\alpha t}) = \cosh(\alpha t)$, we find that (14.5.39) holds, that is,

$$K_\alpha(z) = \int_0^\infty e^{-z\cosh t} \cosh(\alpha t)\, dt \tag{14.7.20}$$

for all $\alpha$ when $\operatorname{Re} z > 0$.

Related to this is a representation for $I_n(z)$ for $n \in \mathbb{Z}$:

**Theorem 14.7.4.** *For $n \in \mathbb{Z}$, we have*

$$I_n(z) = \pi^{-1} \int_0^\pi e^{z\cos\theta} \cos(n\theta)\, d\theta \tag{14.7.21}$$

**Proof.** We start with (3.7.17) of Part 2A, which we rewrite (using $z\sin(-\theta) - n(-\theta) = -(z\sin\theta - n\theta)$)

$$J_n(z) = \frac{1}{2\pi} \int_{-\pi}^{\pi} e^{i(z\sin\theta - n\theta)}\, d\theta \tag{14.7.22}$$

so, by (14.5.35),

$$I_n(z) = \frac{1}{2\pi} \int_{-\pi}^{\pi} e^{-z\sin\theta - in(\theta+\frac{\pi}{2})}\, d\theta \tag{14.7.23}$$

$$= \frac{1}{2\pi} \int_{-\pi}^{\pi} e^{z\cos\theta} e^{-in\theta}\, d\theta \tag{14.7.24}$$

since $\sin(\theta - \frac{\pi}{2}) = -\cos(\theta)$, which implies (14.7.21). $\square$

**Theorem 14.7.5.** *Let $\operatorname{Re}\alpha > -\frac{1}{2}$ and $z \in \mathbb{C}$. Then* (14.5.22) *holds.*

**Proof.** Since $\operatorname{Re}\alpha > -\frac{1}{2}$, the integral converges, and thus, by Theorem 3.1.6 of Part 2A, the integral defines an entire function of $z$, and we need only expand $e^{izt} = \sum_{n=0}^\infty \frac{(iz)^n}{n!} t^n$ and can interchange the sum and integral (by identifying power series). If $n$ is odd, the integrand is odd under $t \to -t$, so the integral vanishes.

Replacing $t^2$ by $s$, the coefficient of $z^{2n}$ in the integral is

$$\frac{(-1)^n}{(2n)!} \int_0^1 (1-s)^{\alpha-\frac{1}{2}} s^{n-\frac{1}{2}}\, ds = \frac{(-1)^n}{2n!} \frac{\Gamma(\alpha+\frac{1}{2})\Gamma(n+\frac{1}{2})}{\Gamma(n+\alpha+1)} \tag{14.7.25}$$

by the $\beta$-function integral, (9.6.43) of Part 2A. By the Legendre duplication formula, (9.6.26) of Part 2A, we have

$$\frac{\Gamma(n+\frac{1}{2})}{\sqrt{\pi}\,(2n)!} = \frac{1}{2^{2n}\Gamma(n+1)} \tag{14.7.26}$$

so

$$\text{RHS of (14.5.22)} = \sum_{n=0}^{\infty} \frac{(-1)^n}{\Gamma(n+1)\Gamma(n+\alpha+1)} \left(\frac{z}{2}\right)^{\alpha+2n} \tag{14.7.27}$$

which is $J_\alpha(z)$. □

The final topic for this section is integral formulae for Airy functions, which are solutions of the ODE

$$y''(z) - zy(z) = 0 \tag{14.7.28}$$

We claim that formulae for such $y$'s should be given by integrals of the form

$$y(z) = \frac{1}{2\pi i} \int_\Gamma e^{-zt+\frac{1}{3}t^3}\, dt \tag{14.7.29}$$

We'll just pull this out of the hat, but it is natural to think of Fourier or Laplace transforms, and so, some integral of the form $e^{-zt}f(t)$, and the calculations below would lead to $f(z) = e^{t^3/3}$. Our initial calculations will be formal and imagine there are no boundary terms in integration by parts.

Formally, we take second derivatives to see

$$y'' = \frac{1}{2\pi i} \int_\Gamma t^2 (e^{-zt+\frac{1}{3}t^3})\, dt \tag{14.7.30}$$

On the other hand,

$$zy = \frac{1}{2\pi i} \int_\Gamma -\frac{d}{dt}(e^{-zt}) e^{\frac{1}{3}t^3}\, dt \tag{14.7.31}$$

$$= \frac{1}{2\pi i} \int_\Gamma e^{-zt} \frac{d}{dt} e^{\frac{1}{3}t^3}\, dt \tag{14.7.32}$$

$$= \frac{1}{2\pi i} \int_\Gamma t^2 (e^{-zt+\frac{1}{3}t^3})\, dt \tag{14.7.33}$$

so, formally, integrals of the form (14.7.29) solve (14.7.28).

To understand when there is no boundary term, we note that $e^{\frac{1}{3}t^3}$ goes to zero as $|t| \to \infty$ precisely in the shaded region in Figure 14.7.4, that is,

$$\arg(z) \in \left(\frac{\pi}{6}, \frac{\pi}{2}\right) \cup \left(\frac{5\pi}{6}, \frac{7\pi}{6}\right) \cup \left(-\frac{\pi}{2}, -\frac{\pi}{6}\right) = Q \tag{14.7.34}$$

To justify the integration by parts, it suffices that the contour have asymptotic phase inside $Q$. If the ends are in the same of the three components of $Q$, one can close the contour inside a region where the integrand decays rapidly and see that the integral is zero. Thus, we need contours like $\Gamma_1, \Gamma_2, \Gamma_3$ shown in Figure 14.7.4. By the Cauchy integral formula, the integrals are independent of the precise details for $\Gamma_j$ (so long as we only

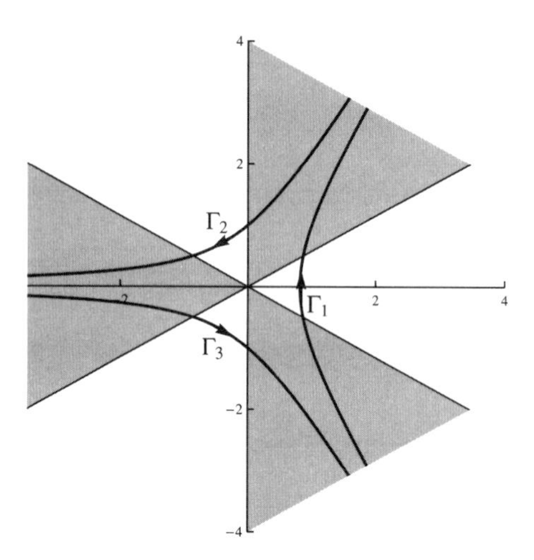


**Figure 14.7.4.** Contours for $Ai$ and $Bi$.

deform through the shaded regions), but for definiteness, we can take the asymptotic phases to be $\frac{\pi}{6}$, $\pi$, and $-\frac{\pi}{6}$. For example, for definiteness, take $\Gamma_2$ to be the hyperbola, $z = x + iy$, $y(y - x\sqrt{3}) = 1$. $\Gamma_1$ and $\Gamma_3$ are obtained by rotations by $\pm 120°$. Thus, we define

$$y_j(t) = \frac{1}{2\pi i} \int_{\Gamma_j} \exp(-zt + \tfrac{1}{3}\, t^3)\, dt \tag{14.7.35}$$

**Theorem 14.7.6.** *The functions, $y_j$, of* (14.7.35) *solve Airy's equation* (14.7.28) *and obey*

$$y_1 + y_2 + y_3 = 0 \tag{14.7.36}$$

*Any two are linearly independent; indeed,*

$$Ai(z) = y_1(z) \tag{14.7.37}$$

$$Bi(z) = \tfrac{1}{2}\, [y_3(z) - y_2(z)] \tag{14.7.38}$$

**Proof.** The formal argument that $y_j$ solves (14.7.38) is justified by the rapid falloff of the integrand near infinity. (14.7.36) holds because the combined contour can be closed near infinity. Linear independence follows from (14.7.37).

By using the same argument that shows the integral (14.0.11) converges, one can show (Problem 4) that the contours asymptotic to $\arg(z) = \pm\frac{\pi}{3}$ can be pushed for $z$ real to the imaginary axis leading to $y_1$ being given by (14.0.11) and $\frac{1}{2}(y_3 - y_2)$ by the integral (14.5.47) for $Bi(z)$. $\square$

One obtains (14.5.46) by suitably deforming the contour, $\Gamma_1$. Since the proper contour to deform to has to do with the saddle-point method, we defer the proof to Section 15.4 (see Example 15.4.3).

**Notes and Historical Remarks.** The Hankel contour was found by Hankel [**172**] in 1864. A closely related integral representation was found by Laplace [**253**]. The Hankel contour can also be used to analytically continue $\zeta$ and prove its functional equation (see Problems 5 and 6).

The use of the Hankel contour for $J_\alpha$ and (14.7.12) is due to Schläfli [**339**]. The change from $H$ to $S$ and, in particular, Theorem 14.7.3 is due to the physicist A. Sommerfeld [**363**]. In addition to his own contributions to mathematical physics, Arnold Sommerfeld (1868–1951) was notable for his impact on students. His students included Bethe, Brillouin, Debye, Heisenberg, Heitler, Landé, Meissner, Pauli, and Peierls, and his postdocs included Condon, von Laue, Morse, Pauling, and Rabi.

### Problems

1. By using (14.7.7), verify that (14.7.1) holds at $z = 1, 2, \dots$ as a limit.

2. (a) Let $F(z)$ be the right side of (14.7.1). Prove that $F(z+1) = zF(z)$ by an integration by parts.

   (b) Prove that $F(z)$ is bounded on $\{z \mid 1 \le \operatorname{Re} z \le 2, |\operatorname{Im} z| \ge 1\}$ and then on $\{z \mid 1 \le \operatorname{Re} z \le 2\}$.

   (c) Prove that $\lim_{x \uparrow 1} F(x) = 1$ and then use Wielandt's theorem (Proposition 9.6.3 of Part 2A) to prove that $F(z) = \Gamma(z)$.

3. Justify the interchange of sum and integral in (14.7.10).

4. If (14.0.11) is defined by an integration by parts, show by first deforming $\Gamma_1$ to $\arg(z) = \pm\frac{\pi}{3}$ and then $\arg(z) = \pm\varphi$, $\varphi \in [\frac{\pi}{3}, \frac{\pi}{2})$, that one can prove $y_1(z) = Ai(z)$. Do the same for (14.7.28).

5. Let
$$G(z, w) = \frac{w^{(z-1)} e^w}{1 - e^w} \tag{14.7.39}$$

   (a) For $\operatorname{Re} z > 1$, prove that
$$2i\zeta(z)\Gamma(z)\sin(\pi z) = \int_{H(c,R)} G(z, w)\, dw \tag{14.7.40}$$
   for $0 < c < R < 2\pi$, where $H$ is the Hankel contour. Why is $R < 2\pi$ needed? (*Remark*: You'll need (9.6.121) of Part 2A.)

   (b) Use (14.7.40) to prove that $\zeta(z)$ has a meromorphic continuation to $\mathbb{C}$ with a pole only at $z = 1$ where the residue is 1.

   (c) Recall the Bernoulli numbers are given by (3.1.48) of Part 2A. Prove (13.3.18) and, in particular, that $\zeta(-2n) = 0$.

6. For $R > \frac{1}{2}$, $R \neq 2\pi n$ ($n = 1, 2, \dots$), define

$$Q_R(z) = \int_{H(c=\frac{1}{2},R)} G(z,w)\,dw \tag{14.7.41}$$

so $Q_\pi(z) = $ LHS of (14.7.40).

(a) For $n = 1, 2, \dots$, prove that

$$Q_\pi(z) - Q_{(2n+1)\pi}(z) = (4\pi i)\cos\biggl(\frac{\pi}{2}(z-1)\biggr)\sum_{k=1}^{n}(2\pi k)^{z-1} \tag{14.7.42}$$

(*Hint*: Residue calculus.)

(b) For $\operatorname{Re} z < 0$, prove that $\lim_{n\to\infty} Q_{(2n+1)\pi}(z) = 0$. (*Hint*: Show that $\sup_n \sup_{|w|=(2n+1)\pi} |e^w/(1-e^w)| < \infty$.)

(c) For all $z \in \mathbb{C}$, prove that

$$\zeta(1-z) = 2\zeta(z)\Gamma(z)\biggl[\cos\biggl(\frac{\pi}{2}z\biggr)\biggr](2\pi)^{-z} \tag{14.7.43}$$

(d) Show that (14.7.43) is equivalent to (13.3.19). (*Hint*: Use the Euler reflection formula.)

7. (a) Prove the following for the Hurwitz zeta function (13.2.19) when $\operatorname{Re} s > 1$:

$$\zeta(s;x)\Gamma(s) = \int_0^\infty \frac{t^{s-1}e^{-xt}}{e^t - 1} \tag{14.7.44}$$

which is an analog of (9.6.121) of Part 2A.

(b) By following Problem 5, show that $\zeta(s;x)$ has a meromorphic continuation to all of $\mathbb{C}$.

(c) By following Problem 6, prove that for all $z \in \mathbb{C}$,

$$\zeta(z;x) = \frac{\Gamma(1-z)}{(2\pi)^{1-z}}\,\{e^{-i\pi(1-z)/2}\beta(x,1-z) + e^{i\pi(1-z)/2}\beta(-x,1-z)\} \tag{14.7.45}$$

where $\beta$ is given by (13.2.55). Thus, deduce (13.2.54).

(d) Prove the functional equation for $\zeta(s;x)$ when $x$ is rational, that is, for $1 \le p \le q$ integers,

$$\zeta\biggl(1-s;\frac{p}{q}\biggr) = \frac{2\Gamma(s)}{(2\pi q)^s}\sum_{r=1}^{q}\cos\biggl(\frac{\pi s}{2} - \frac{2\pi pr}{q}\biggr)\zeta\biggl(s;\frac{r}{q}\biggr) \tag{14.7.46}$$

*Chapter 15*

# Asymptotic Methods

> Asymptotic analysis, that branch of mathematics devoted to the study of the behavior of functions within chosen limits, was once thought of more as a specialized art than a necessary discipline.
>
> —*From the back cover of Bleistein and Handelsman* [**59**]

**Big Notions and Theorems:** Asymptotic Series, Any Series (Even Divergent) Is an Asymptotic Series, Laplace's Method, Gaussian Approximation, Watson's Lemma, Analytic Watson Lemma, Complete Stirling Approximation via Binet's Formula, Stokes' Phenomenon, Stokes Lines, Asymptotics of Special Functions, Stationary Phase, Wave Packet Asymptotics, van der Corput's Lemma, Decay of Fourier Transform of Measures on Curved Hypersurfaces, Saddle-Point Method, Steepest Descent, Partitions, Bell Numbers, Hardy–Ramanujan–Uspensky Asymptotics, Hardy–Littlewood Circle Method, WKB Approximation, Riccati Equation, Variation of Parameters, ODE Asymptotics for $L^1$ and for Polynomial $V$

Our goal in this chapter is to introduce some of the techniques used in the study of two related problems: asymptotics of solutions of ODEs and asymptotics of integrals of the form

$$Q(s) = \int f(x) e^{-sg(x)}\, d\mu(x) \tag{15.0.1}$$

as $s \to \infty$. While many of the ideas are not specifically connected to complex analysis, the subject belongs here because one method (the saddle-point method) depends on the ability to move the contour in a contour integral and because some of the results on ODEs involve analytic ODEs in the full complex plane. Indeed, many of the examples of integral asymptotics will use representations of special functions by integrals, the subject of Section 14.7.

We have made this a bonus chapter because it is not standard material in basic graduate courses in analysis, but the topic really is a central one in mathematics, especially areas allied to applications (as can be seen that much of this material is a part of the basic graduate curriculum in applied mathematics).

Many asymptotic results involve not just leading order but systematic series that are not convergent but only asymptotic. Section 15.1 discusses such series—its most interesting theorem is one that says any series, even those with zero radius of convergence, can occur even for functions analytic in $\mathbb{C} \setminus (-\infty, 0]$.

Section 15.2 turns to the simplest case of (15.0.1), where $g$ and $f$ are real, with $g$ bounded from above and $f$ strictly positive. If both are continuous and $d\mu$ is finite, it is easy to see that as $s \to \infty$, we have that $(Q(s))^{1/s} \to e^{\max g(x)}$. The corrections will be studied when the underlying space is $[a, b]$ or $[0, \infty)$ or $\mathbb{R}$, via two methods: first, a Gaussian approximation, and second, using something called Watson's lemma.

The paradigm, although it does not quite fit the above, is

$$n! = \int_0^\infty x^n e^{-x}\, dx \tag{15.0.2}$$

The region is all of $[0, \infty)$, and more important, with $f(x) = e^{-x}$, $g(x)$ is $\log x$ which is not bounded above. If we think of $x^n e^{-x}$ as $e^{n \log x - x}$, the function $n \log x - x$ is bounded above with a maximum at $x = n$. This suggests we change $x$ to $ny$ and write

$$n! = n^{n+1} \int_0^\infty \exp(n[\log y - y])\, dx \tag{15.0.3}$$

Since $y = 1$ is the maximum point for $\log y - y$, this suggests the leading behavior is $n! \sim n^{n+1} e^{-n}$ and the higher-order analysis gives the leading $n! \sim \sqrt{2\pi}\, n^{n+\frac{1}{2}} e^{-n}$ of Stirling's formula (9.7.1) of Part 2A.

In Section 15.3, we turn to the next simplest case of $Q(s)$ where $g(x)$ is pure imaginary. For $s$ large, there are faster and faster oscillations which cancel at nearby points if $g'(x_0) \neq 0$. Indeed, we'll show the integral near points with $g'(x_0) \neq 0$ is $O(s^{-N})$ for all $N$. Thus, the points that matter are those with $g'(x_0) = 0$, called points of stationary phase.

If $g$ is fully complex, there is a tradeoff between oscillations and growth of the real part, which is difficult to analyze directly. But if $g$ is analytic, one can shift the contour of integration, the subject of Section 15.4. The idea will push the contour to a point, $z_0$, with $g'(z_0) = 0$. If $g(z_0) \neq 0$ there, then $\vec{\nabla}|g(z_0)| = \text{Re}(\overline{g(z_0)}\, \vec{\nabla} g(z_0))/|g(z_0)| = 0$, so $z_0$ is a critical point of

$|g(z_0)|$. By the maximum principle, it cannot be a minimum or maximum, so it must be a saddle point, which gives one name to this method. It will turn out that if $f''(z_0) \neq 0$ and the contour is taken in the direction of the decrease of $|f(z_0)|$, the calculation will be easier (because the phase will be constant and we will essentially have Laplace's method)—hence, the other name—the method of steepest descent.

We will illustrate the methods of Sections 15.2–15.4 by applying them to integral representations of special functions and so get their asymptotics. Section 15.5 will then turn to another way of getting asymptotics for ODEs, the WKB approximation.

Besides asymptotics of the special functions of Chapter 14, there are several other applications, among them the high points of this chapter. Section 15.2 obtains Stirling's formula to all orders from Binet's formula; Section 15.3 includes estimates on free wave packet spreading in quantum mechanics and also decay estimates for measures supported on uniformly curved hypersurfaces. Section 15.4 proves asymptotics for $p(n)$, the number of partitions of $n$ indistinguishable objects. Section 15.5 finds the asymptotics at infinity for solutions of $-u''(x) + P(x)u(x) = 0$ when $P$ is a polynomial.

**Notes and Historical Remarks** For books on asymptotics of integrals, see [**59, 91, 95, 117, 413**].

## 15.1. Asymptotic Series

This is a preliminary section to set the language for an asymptotic statement like

$$Ai(z) \sim \frac{\exp(-\frac{2}{3}z^{3/2})}{2\sqrt{\pi}\, z^{1/4}} \sum_{k=0}^{\infty} a_k z^{-3k/2}, \qquad |\arg z| < \frac{\pi}{3} - \varepsilon \tag{15.1.1}$$

Recall that if $f$ and $g$ are two functions on a region $\Omega$ and $z_0 \in \overline{\Omega}$, we write

$$f = O(g) \qquad \text{at } z_0 \tag{15.1.2}$$

to mean that $f(z)/g(z)$ is bounded for $z$ near $z_0$ and $z \in \Omega$ and

$$f = o(g) \qquad \text{at } z_0 \tag{15.1.3}$$

if

$$\lim_{\substack{z \to z_0 \\ z \in \Omega}} |f(z)/g(z)| = 0 \tag{15.1.4}$$

The $z_0$ and region $\Omega$ are often implicit. The functions are typically analytic in $\Omega$ but need not be for the notation to make sense. Sometimes $\Omega$ is an interval in $\mathbb{R}$ rather than an open subset of $\mathbb{C}$.

A sequence of functions $\{h_n\}_{n=0}^\infty$ is called *decreasing at* $z_0$ if and only if for each $n = 0, 1, 2, \dots$, $h_{n+1} = o(h_n)$. We write

$$f(z) \sim g(z) \sum_{n=0}^\infty a_n h_n(z) \tag{15.1.5}$$

at $z_0$ on $\Omega$ if and only if for each fixed $N$,

$$\frac{f(z)}{g(z)} - \sum_{n=0}^N a_n h_n(z) = O(h_{N+1}(z)) \tag{15.1.6}$$

(15.1.5) is called an *asymptotic series* for $f$. If $g = 1$, $h_n(z) = (z - z_0)^n$ (or if $z_0 = \infty$, $h_n(z) = z^{-n}$), then (15.1.5) is called an *asymptotic power series*. Of course, we can absorb $a_n$ into $h_n$, but typically the $h_n$ are simple functions like $z^n$.

We emphasize that (15.1.5) says nothing about $f(z)$ for $z$ fixed and, in particular, the infinite sum may diverge for all $z$ (see the example immediately below). The sum may converge but not to $f$, for example, if $f(z) = (1-z)^{-1} + e^{-1/z}$ and $\Omega = \{z \mid |\arg(z)| < \frac{\pi}{3}, |z| \neq 0\}$, then $f(z) \sim \sum_{n=0}^\infty z^n$, but the convergent sum is not $f$. What (15.1.5) does say is that the difference for the truncated sum is very small for $z$ very close to $z_0$. In many cases, the finite sums for $z$ fixed get close to $f(z)$ but only until some large value of $N$, and then they diverge wildly. Getting good numerical results from an asymptotic series is an art. Sometimes, as discussed in the Notes, a summation procedure will get $f$ from the series.

**Example 15.1.1.** For $z \in \mathbb{C} \setminus (-\infty, 0]$, define

$$f(z) = \int_0^\infty \frac{e^{-t}}{1+zt}\, dt \tag{15.1.7}$$

We claim if any $\varepsilon > 0$, at 0, in the sector $\{z \mid |\arg(z)| < \pi - \varepsilon\}$,

$$f(z) \sim \sum_{n=0}^\infty (-1)^n n!\, z^n \tag{15.1.8}$$

This is because the geometric series with remainder

$$(1+zt)^{-1} = \sum_{n=0}^N (-zt)^n + (-zt)^{N+1}(1+zt)^{-1}$$

so

$$\left| f(z) - \sum_{n=0}^N (-1)^n n!\, z^n \right| \le |z|^{N+1} \int_0^\infty \frac{t^{N+1}}{|1+zt|}\, dt \tag{15.1.9}$$

and the integral is bounded in $\{z \mid |\arg(z)| < \pi - \varepsilon, |z| < 1\}$. The power series in (15.1.8) has zero radius of convergence, so this illustrates the difference between asymptotic and convergent series. However, the function $f$ can be recovered by various summability methods, for example, Borel and Padé; see the Notes. $\square$

While we generally intend (15.1.5) as (15.1.6), one sometimes writes (15.1.5) in a situation where $g$ has zeros; for example, as $x \to \infty$ in $\mathbb{R}$,

$$Ai(-x) \sim \frac{1}{\sqrt{\pi}\, x^{1/4}} \cos\left(\frac{2}{3}\, x^{3/2} - \frac{1}{4}\, \pi\right) \sum_{n=0}^{\infty} \tilde{a}_n x^{-3n/2} \tag{15.1.10}$$

This doesn't mean (15.1.6), which would force the zeros of $f$ to be exactly those of $g$. Rather, one typically has this in case $g = g_1 g_2$ with $g_1$ nonvanishing and $g_2(z) = O(1)$ and $\limsup_{z\to z_0} |g_2(z)| > 0$ and intends (15.1.5) to mean

$$\frac{f(x)}{g_1(x)} - g_2(x) \sum_{n=0}^{N} a_n h_n(x) = O(h_{N+1}(x)) \tag{15.1.11}$$

Sometimes $g_1 = 1$.

Before leaving this general subject, we want to note that any power series—no matter how divergent—is an asymptotic series for an analytic function in a large sector (but not in a punctured disk because of the Riemann removable singularities theorem, Theorem 3.8.1 of Part 2A).

**Theorem 15.1.2.** *Let $\{a_n\}_{n=0}^{\infty}$ be an arbitrary sequence of complex numbers. Then there exists $f(z)$ analytic in $\Omega = \mathbb{C} \setminus (-\infty, 0]$ so that in $\Omega$ at $0$,*

$$f(z) \sim \sum_{n=0}^{\infty} a_n z^n \tag{15.1.12}$$

**Proof.** Let $\widetilde{\Omega} = \{z \mid |\arg(z)| < \frac{\pi}{3}\}$. We'll first prove the result with $\Omega$ replaced by $\widetilde{\Omega}$. We claim that we can pick $b_n$ and $c_n > 0$ inductively, with $b_0 = a_0$, $c_0 = 1$, so that for $n \geq 1$,

$$\sup_{z\in\widetilde{\Omega}} |z b_n e^{-c_n z}| \leq 2^{-n}, \qquad \sup_{z\in\widetilde{\Omega}} |z^n b_n e^{-c_n z}| \leq 2^{-n} \tag{15.1.13}$$

$$\sum_{m=0}^{n} b_m z^m e^{-c_m z} = \sum_{m=0}^{n} a_m z^m + O(z^{n+1}) \tag{15.1.14}$$

For given $\{b_m, c_m\}_{m=0}^{n-1}$, first pick $b_n$ so (15.1.14) holds (if it holds for one $c_n$, it holds for all). Since $|ze^{-c_n z}| + |z^n e^{-c_n z}| \to 0$ uniformly in $\widetilde{\Omega}$ as $c_n \to \infty$, we can then pick $c_n$ so that (15.1.13) holds.

Now let

$$f(z) = \sum_{n=0}^{\infty} b_n z^n e^{-c_n z} \tag{15.1.15}$$

which converges uniformly on $\widetilde{\Omega}$ by (15.1.13) and so defines an analytic function in $\widetilde{\Omega}$. Moreover, for any $N$, by (15.1.14), in $\widetilde{\Omega}$,

$$\begin{aligned} f(z) - \sum_{n=0}^{N} a_n z^n = O(z^{N+1}) + b_{N+1} z^{N+1} e^{-c_{N+1} z} \\ + z^{N+1} \sum_{j=1}^{\infty} z^j b_{N+j} e^{-c_{N+j} z} \end{aligned} \tag{15.1.16}$$

The second term is $O(z^{N+1})$ and, by (15.1.13) and $|z^j| \le \max(|z|, |z|^{N+j})$, we see the third term is $O(z^{N+1})$. Thus, (15.1.12) holds.

To get the result for $\Omega$, define $\tilde{a}_n$ by

$$\tilde{a}_n = \begin{cases} a_{n/3} & \text{if } n \equiv 0 \bmod 3 \\ 0 & \text{if } n = 1, 2 \bmod 3 \end{cases} \tag{15.1.17}$$

and $g(z)$ analytic in $\widetilde{\Omega}$ with $g(z) \sim \sum_{n=0}^{\infty} \tilde{a}_n z^n$ there. If $f(z) = g(z^{1/3})$, then $f$ is analytic in $\Omega$ and obeys (15.1.12) at 0 in $\Omega$. □

**Remark.** By using $z^{1/m}$ for $m > 3$, we can even get $f$ analytic in a multi-sheeted surface with the proper asymptotic series there.

**Notes and Historical Remarks.** Special cases of asymptotic series appeared in the nineteenth century. The general formal definition goes back to 1886 work of Poincaré [**317**] (who gave the name "asymptotic series") and Stieltjes [**367**]. They are sometimes called *Poincaré series*. Example 15.1.1 is due to Euler [**126**].

For books on asymptotic expansions, see [**91, 103, 117, 402**].

Some basic integrals of the form $\int_x^\infty g(y)\,dy$ have asymptotic series in $x^{-1}$ that can be computed by integration by parts. Problem 2 does the complementary error function and Problem 4 the Fresnel integrals $C(x)$ and $S(x)$ of (5.7.58) of Part 2A.

$$\begin{aligned} \text{Erfc}(x) &\equiv \frac{2}{\sqrt{\pi}} \int_x^\infty e^{-y^2}\,dy \\ &\sim \frac{e^{-x^2}}{\pi x} \left( 1 - \frac{1}{2x^2} + \frac{1 \cdot 3}{(2x^2)^2} - \frac{1 \cdot 3 \cdot 5}{(2x^2)^3} + \frac{1 \cdot 3 \cdot 5 \cdot 7}{(2x^2)^4} \cdots \right) \end{aligned} \tag{15.1.18}$$

Laplace [**253**], who found this series, noted the terms give alternate upper and lower bounds for the true value, allowing us to illustrate the divergent asymptotic series. The following table, computed in Mathematica, shows $F(x) = (\pi x) e^{x^2} \text{Erfc}(x)$ for $x = 1, 2, 3, 5, 10$ and compares with the first $(n+1)$ terms of the asymptotic series in (15.1.18). The entry with $*$ (respectively, #) is the best lower (respectively, upper) bound.

| | $x = 1$ | $x = 2$ | $x = 3$ | $x = 5$ | $x = 10$ |
|---|---|---|---|---|---|
| $F(x)$ | 0.75787 | 0.90535 | 0.95181 | 0.98109430731 | 0.9950731878244697473807 |
| $n = 1$ | 0.5 * | 0.8750 | 0.94444 | 0.98000000000 | 0.9950000000000000000000 |
| $n = 2$ | 1.25 # | 0.9218 | 0.95370 | 0.98120000000 | 0.9950750000000000000000 |
| $n = 3$ | −0.625 | 0.8925 * | 0.95113 | 0.98108000000 | 0.9950731250000000000000 |
| $n = 4$ | 5.937 | 0.9182 # | 0.95213 | 0.98109680000 | 0.9950731906250000000000 |
| $n = 5$ | −23.593 | 0.8894 | 0.95163 | 0.98109377600 | 0.9950731876718750000000 |
| $n = 6$ | 138.8281 | 0.9290 | 0.95194 # | 0.98109444128 | 0.9950731878342968750000 |
| $n = 7$ | −916.914 | 0.8646 | 0.95172 * | 0.98109426831 * | 0.9950731878237394531250 |
| $S$ values | 8, 9, 10 | 8, 9, 10 | 8, 9, 10 | 18, 19, 20 | 108, 109, 1000 |
| $S1$ | 7001.15 | 0.9854 | 0.95556 | 0.98109430734 | 0.9950731878244697473807 * |
| $S2$ | −60302 | 0.7287 | 0.94398 | 0.98109430730 | 0.9950731878244697473807 # |
| $S3$ | 579081 | 1.3384 | 0.96967 | 0.98109430733 # | $6.53 \times 10^{565}$ |

We show the values for $n = 1, 2, 3, \dots, 10$ for $x = 1, 2, 3$ but $n = 8, 9, 10$ are replaced by $n = 18, 19, 20$ for $x = 5$ and $n = 108, 109$ and $1000$ for $x = 10$. You can see the alternating bounds and that the values don't get too close for $n = 1, 2$, but these agree to four places for $x = 3$, ten places for $x = 5$, and over twenty for $x = 10$. Indeed, if you only look at $n$ up to 109, you would be sure the series was convergent, but as the value of $n = 1000$ shows, it is not!

Just because a series is divergent doesn't mean that it can't be coaxed to give an exact answer! We saw this in our study of Fourier series where a marginally divergent series could be Cesàro summable (see Section 3.5, especially Theorem 3.5.6, of Part 1). But many of the divergent series in this section go termwise to infinity in absolute value faster than any polynomial, and so cannot be Cesàro summable (see Problem 4). There are, however, other summability methods that might work, of which we mention two: Padé and Borel. We refer the reader to some of the many books in this area [**25, 26, 61, 63, 92, 175, 346, 376**] for more on these subjects and on other summability methods. In particular, for the results mentioned below, Baker's books [**25, 26**] discuss Stieltjes' theorem on Padé approximants and Hardy [**175**] proves Watson's theorem.

Given a formal series, $\sum_{n=0}^{\infty} a_n z^n$, the Padé approximants $f^{[k,\ell]}(z)$ are the unique rational function $P(z)/Q(z)$ with $\deg(P) = k$, $\deg(Q) = \ell$, and $f^{[k,\ell]}(z) - \sum_{n=0}^{k+\ell} a_n z^n = O(z^{k+\ell+1})$. $f^{[k,0]}$ are the Taylor approximations.

Stieltjes [**368**] proved the following theorem:

**Theorem 15.1.3** (Stieltjes' Theorem). *Suppose that $\mu$ is a measure on $[0, \infty)$ with $\int |x|^n \, d\mu < \infty$ for all $n$ and for no other measure, $\nu$, on $[0, \infty)$*

*is* $\int x^n\,d\mu = \int x^n\,d\nu$ *for* $n = 0, 1, 2, \dots$. *For* $z \in \mathbb{C} \setminus (-\infty, 0]$, *let*

$$f(z) = \int \frac{d\mu(t)}{1+zt} \tag{15.1.19}$$

*Then*

(a) *In any sector* $\{z \mid |\arg(z)| < \pi - \varepsilon\}$, *as* $|z| \downarrow 0$,

$$f(z) = \sum_{n=0}^{\infty} a_n z^n, \qquad a_n = (-1)^n \int t^n\,d\mu(t) \tag{15.1.20}$$

(b) *For each* $j \in \mathbb{Z}$, $f^{[N,N+j]}(z) \to f(z)$ *uniformly on compact subsets of* $\mathbb{C} \setminus (-\infty, 0]$.

**Remarks.** 1. If $d\mu(t) = e^{-t}\,dt$, the uniqueness assumption holds (see Problem 2 of Section 5.6 of Part 1). So even though the Taylor series is everywhere divergent, the diagonal Padé approximants converge.

2. $f^{[N,N]}(z)$ is associated with continued fractions; see Section 7.5 of Part 2A.

To describe another summability technique, suppose that

$$|a_n| \le CR^{-n} n! \tag{15.1.21}$$

for some $C$ and $R > 0$. Then $B(z) = \sum_{n=0}^{\infty} \frac{a_n}{n!} z^n$ converges in the disc of radius $R$. Suppose it has an analytic continuation to a neighborhood of $[0, \infty)$ and that for $z \in (0, \infty)$,

$$f(z) = \int_0^{\infty} B(xz) e^{-x}\,dx \tag{15.1.22}$$

is an absolutely convergent integral. Since $\int_0^\infty (xz)^n e^{-x}\,dx = n! z^n$, formally $f(z) = \sum_{n=0}^{\infty} a_n z^n$, and this might make sense even if the series is divergent. $f$ is called the *Borel sum* of $\sum_{n=0}^{\infty} a_n z^n$ (after Borel [**62**]) and $B$ the *Borel transform* of $f$.

Watson [**403**] (see Nevanlinna [**289**] for a stronger result) has proven the following theorem:

**Theorem 15.1.4** (Watson's Theorem). *Let* $f(z)$ *be analytic in* $\Omega \equiv \{z \mid |\arg(z)| < \frac{\pi}{2} + \varepsilon,\ 0 < |z| < \rho\}$ *for some* $\varepsilon > 0$, *and for some series* $\{a_n\}_{n=0}^{\infty}$, *one has in* $\Omega$ *that*

$$\left| f(z) - \sum_{n=0}^{N} a_n z^n \right| \le CR^{-N-1}(N+1)!\,|z|^{N+1} \tag{15.1.23}$$

*for* $C, R$ *fixed and* $N = 0, 1, 2, \dots$. *Then the Borel transform has an analytic continuation to a neighborhood of* $[0, \infty)$, *and for* $z$ *with* $\operatorname{Re}(z^{-1}) > R^{-1}$, *one has*

$$f(z) = z^{-1} \int_0^\infty B(x) e^{-x/z}\, dx \tag{15.1.24}$$

*where the integral is absolutely convergent for such $z$.*

**Remarks.** 1. If $z$ is real, with $|z| < R$, (15.1.22) holds by scaling.

2. If $\mu$ is a measure on $[0, \infty)$ whose moments obey $\int x^n\, d\mu(x) \le CR^{-n} n!$, then (15.1.23) holds, so $f$ given by (15.1.19) can be recovered by Borel summation. This includes Example 15.1.1 where $B(z) = (1+z)^{-1}$.

These ideas have been applied to various problems of interest in quantum mechanics. If $E_1(\beta)$ is the smallest eigenvalue of $-\frac{d^2}{dx^2} + x^2 + \beta x^4$, then the formal perturbation series $E_1(\beta) = \sum_{n=0}^\infty a_n \beta^n$ is only asymptotic and is known to diverge. Using in part results of Simon [**351**], Loeffel et al. [**266**] proved it Padé summable, and Graffi et al. [**157**] Borel summable; see Simon [**352**] for a review. Borel summability is even known for some objects in certain quantum field theories; see, for example, the discussion in Glimm–Jaffe [**150**].

### Problems

1. (Incomplete gamma function) (a) For all $x \in \mathbb{C}$, $x > 0$, define

$$\Gamma(s, x) = \int_x^\infty t^{s-1} e^{-t}\, dt \tag{15.1.25}$$

Prove that

$$\Gamma(s, x) = e^{-x} x^{s-1} + (s-1)\Gamma(s-1, x) \tag{15.1.26}$$

(*Hint*: Integration by parts.)

(b) Prove that $\Gamma(s, x) \le x^{s-1} e^{-x}$ if $s < 1$.

(c) As $x \to \infty$, prove that for any $s \in \mathbb{C}$,

$$\Gamma(s, x) \sim e^{-x} \sum_{r=1}^\infty (s-1) \dots (s-r+1) x^{s-r} \tag{15.1.27}$$

where $(s-1)\dots(s-r+1) = 1$ if $r = 1$. (*Note*: This is asymptotic, not convergent.)

2. (Complementary error function) Let $\mathrm{Erfc}(T)$ be defined by

$$\mathrm{Erfc}(T) = \frac{2}{\sqrt{\pi}} \int_T^\infty e^{-u^2}\, du \tag{15.1.28}$$

(*Note*: Some authors drop the $2/\sqrt{\pi}$ factor in front.)

(a) In terms of $\Gamma$ given by (15.1.25), prove that

$$\mathrm{Erfc}(T) = \frac{1}{\sqrt{\pi}} \Gamma(\tfrac{1}{2} T^2) \tag{15.1.29}$$

(b) As $T \to \infty$, prove that

$$\operatorname{Erfc}(T) = \frac{1}{\pi} e^{-T^2} \sum_{r=1}^{\infty} \Gamma(r - \tfrac{1}{2}) \frac{(-1)^{r-1}}{T^{2r+1}} \tag{15.1.30}$$

(*Hint*: Use Euler reflection.)

(c) If $S_N(T)$ is obtained by using $\sum_{r=1}^{N}$ in (15.1.30), prove that for $n = 1, 2, \dots,$

$$S_{2n}(T) \le \operatorname{Erfc}(T) \le S_{2n+1}(T)$$

(d) Using Mathematica or Maple, find the optimal upper and lower bounds obtained from $S_n(T)$ for $2Te^{T^2} \operatorname{Erfc}(T)$ for $T = 3, 4, 5, 10$.

3. For $\operatorname{Re} s > 0$, define

$$F(x,s) = \lim_{R\to\infty} \int_x^R t^{-s} e^{it}\, dt \tag{15.1.31}$$

(a) Prove that

$$F(x,s) = \frac{ie^{ix}}{x^s} - isF(x, s+1) \tag{15.1.32}$$

(*Hint*: Integration by parts.)

(b) If $0 < \operatorname{Re} s \le 1$, conclude the limit in (15.1.31) exists.

(c) For $\operatorname{Re} s > 1$, $x > 1$, prove that

$$|F(x,s)| \le C_s x^{-\operatorname{Re} s} \tag{15.1.33}$$

(d) As $x \to \infty$, prove that for $s$ fixed,

$$F(x,s) \sim \frac{ie^{ix}}{x^s} \sum_{r=0}^{\infty} \frac{s(s+1)\dots(s+r-1)}{(ix)^s} \tag{15.1.34}$$

(*Note*: This series is not convergent.)

4. Recall the Fresnel integrals, $C(x)$ and $S(x)$, are given by (5.7.58) of Part 2A and that as $x \to \infty$, they have limits which are $\sqrt{\pi}/(2\sqrt{2})$.

(a) In terms of the function $F$ of Problem 3, prove that

$$C(x) = \frac{\sqrt{\pi}}{2\sqrt{2}} - \operatorname{Re} F(x^2, \tfrac{1}{2}), \quad S(x) = \frac{\sqrt{\pi}}{2\sqrt{2}} - \operatorname{Im} F(x^2, \tfrac{1}{2}) \tag{15.1.35}$$

(b) Find complete asymptotic series for $C(x)$ and $S(x)$ as $x \to \infty$.

## 15.2. Laplace's Method: Gaussian Approximation and Watson's Lemma

We turn to the analysis of (15.0.1)

$$Q(s) = \int f(x)e^{-sg(x)}\, d\mu(x) \tag{15.2.1}$$

as $s \to \infty$. In this section, we'll mainly consider $g$ real-valued and $f$ nonnegative. Here is the analysis of the leading term in great generality:

**Theorem 15.2.1.** *Let $\mu$ be a Baire measure on a compact Hausdorff space, $X$, with $\mu(A) > 0$ for any nonempty open set, $A$. Let $f, g$ be real-valued continuous functions on $X$ with $f \geq 0$ and so that $f(x_0) \neq 0$ for some $x_0$ with $g(x_0) = \inf_{y\in X} g(y)$. Then*

$$\lim_{s\to\infty} \frac{1}{s} \log(Q(s)) = -\inf_{y\in X} g(y) \tag{15.2.2}$$

**Remarks.** 1. If $\mu(A) > 0$ for all open $A$ is not assumed, one lets $X_0 = \{x \mid \mu(N) > 0$ for all neighborhoods, $N$, of $x\}$, which is a closed set (see Problem 1) and (15.2.2) holds (if, say, $f$ is everywhere nonnegative) with $\inf_{y\in X}$ replaced by $\inf_{y\in X_0}$.

2. There are extensions to locally compact spaces (see Problem 2).

**Proof.** If $g_- = \inf_{y\in X} g(x)$, then $g(x) \geq g_-$ implies

$$Q(s) \leq \|f\|_\infty e^{-sg_-}\mu(X) \tag{15.2.3}$$

from which

$$\limsup \frac{1}{s} \log(Q(s)) \leq -g_- \tag{15.2.4}$$

follows.

On the other hand, if $g(x_0) = g_-$, $f(x_0) \neq 0$ and $\varepsilon > 0$. Let $A_\alpha = \{x \mid g(x) \leq g_- + \varepsilon;\ f(x) \geq \frac{1}{2} f(x_0)\}$, which is a nonempty open set. Then

$$Q(s) \geq \tfrac{1}{2}\, f(x_0)\mu(A_\alpha)e^{-s(g_-+\varepsilon)} \tag{15.2.5}$$

so

$$\liminf \frac{1}{s} \log(Q(s)) \geq -(g_- + \varepsilon)$$

Since $\varepsilon$ is arbitrary, (15.2.2) follows. □

For going to higher order in the one-dimensional case, there are two approaches: the Gaussian approximation and Watson's lemma. We'll discuss both—not because either is deficient, but because both have advantages in extensions; in particular, the Gaussian approximation has generalizations to higher, indeed, to infinite dimensions (see the Notes), while Watson's lemma has effortless extension to complex $s$ and to situations of a half-line with maxima on the edges. Here is the full leading order in finite dimensions.

**Theorem 15.2.2.** *Let $f(x), g(x)$ be real-valued functions on $\mathbb{R}^\nu$ so that*

$$g_- = \inf_{x\in\mathbb{R}^\nu} g(x) > -\infty \tag{15.2.6}$$

*and $f(x) \geq 0$ for all $x$. Suppose there is a unique $x_0 \in \mathbb{R}^\nu$ so that $g(x_0) = g_-$, that for some $R > 0$,*

$$g_R \equiv \inf_{|x|\geq R} g(x) > g_- \tag{15.2.7}$$

*that for some $\alpha$,*

$$\int e^{-\alpha g(x)} f(x)\, dx < \infty \tag{15.2.8}$$

*that $f(x_0) \neq 0$, that $f$ is continuous near $x_0$, and that $g$ is $C^2$ near $x_0$ with*

$$A_{ij} = \frac{\partial^2 g(x_0)}{\partial x_i \partial x_j} \tag{15.2.9}$$

*strictly positive definition. Then with $d\mu = d^\nu x$,*

$$\lim_{s\to\infty} s^{\nu/2} e^{sg_-} Q(s) = (2\pi)^{\nu/2} f(x_0) \det(A)^{-1/2} \tag{15.2.10}$$

**Proof.** Without loss of generality, we can suppose $x_0 = 0$. The hypotheses then imply that (15.2.7) holds for any $R$, so picking $R$ small, we can suppose $f$ is continuous (and so bounded) in $\{x \mid |x| < R\}$.

By the hypotheses, for $C < \infty$,

$$\left| \int_{|x|>R} e^{-sg(x)} f(x) \right| \leq Ce^{-(s-\alpha)g_R} \tag{15.2.11}$$

so since $g_R > g_-$ and (15.2.2) holds, it suffices to prove (15.2.3) with $f$ replaced by its restriction to $\{x \mid |x| < R\}$.

By Taylor's theorem with remainder, we have that, by shrinking $R$ if necessary, we have in $\{x \mid |x| \leq R\}$ that

$$g(x) \geq g_- + c|x|^2 \tag{15.2.12}$$

for some $c$.

Changing variables from $x$ to $y = s^{1/2}x$, we see that

$$s^{\nu/2} e^{sg_-} Q(s) = \int f(s^{-1/2}y) \exp[-s(g(s^{-1/2}y) - g_-)]\, d^\nu y \tag{15.2.13}$$

By (15.2.12) and $f$ bounded, we see the integrand is bounded uniformly in $s > 1$ by $\|f\|_\infty e^{-c|x|^2}$. By Taylor's theorem with remainder, the integrand converges to $f(x_0)\exp(-\frac{1}{2}\langle y, Ay\rangle)$. By the dominated convergence theorem, the integral converges to the right side of (15.2.10). □

The same proof works for complex-valued $g$ and $f$ so long as the $\operatorname{Re} g$ dominates $\operatorname{Im} g$ near the minimum. To compare with the real-valued case, we'll write $g + ih$ with $g, h$ real-valued.

**Theorem 15.2.3.** *Let $g, f$ be as in Theorem 15.2.2 with the change that $f$ can be complex-valued and (15.2.8) holds with $|f(x)|$ in place of $f$. Let $h$ be real-valued and continuous near $x_0$ and let*

$$\widetilde{Q}(s) = \int f(x) e^{-s(g(x)+ih(x))}\, d^\nu x \tag{15.2.14}$$

*Suppose $A$, given by (15.2.9), is still strictly positive definite, and as $x \to x_0$,*

$$h(x) - h(x_0) = o(|x - x_0|^2) \tag{15.2.15}$$

*Then*

$$\lim_{s\to\infty} s^{\nu/2} e^{sg_-} e^{ish(x_0)} \widetilde{Q}(s) = \text{RHS of (15.2.10)} \tag{15.2.16}$$

**Proof.** (15.2.12) holds with $sg(x)$ replaced by $s(g(x)+ih(x))$. In (15.2.13), we have an $is(h(s^{-1/2}y) - h(0))$, but by (15.2.15), this goes to 0. $\square$

**Remark.** We'll call this theorem *Laplace's method with complex argument.*

To go beyond leading order, we want to reduce to the Gaussian case. To have a result we can use also in the next section, we'll only require that $A$ be nondegenerate:

**Theorem 15.2.4** (Morse Lemma). *Let $g$ be a function $C^\infty$ in a neighborhood of 0 in $\mathbb{R}^\nu$ so $g(0) = 0$, $\vec{\nabla} g(0) = 0$, and so that $A_{ij}$, given by (15.2.9) (with $x_0 = 0$), is invertible. Then there exists a $C^\infty$ change of variables $x \mapsto y(x)$ near $x = 0$ so that*

$$y(x) = x + O(x^2) \tag{15.2.17}$$

*and if $y \mapsto x(y)$ is the inverse map, then*

$$g(x(y)) = \tfrac{1}{2} \sum_{i,j=1}^{\nu} A_{ij} y_i y_j \tag{15.2.18}$$

**Remark.** By basic linear algebra (diagonalization plus scaling; see Theorem 1.7.6 of Part 2A), for any invertible $A$, there is an invertible $T$ so $T^t A T$ is diagonal with elements $\pm 1$. By letting $y = Tz$, we get

$$g(x(Tz)) = z_1^2 + \cdots + z_k^2 - z_{k+1}^2 - \cdots - z_\nu^2 \tag{15.2.19}$$

which is the way it is usually expressed. We write it as (15.2.18) since it will have (15.2.17) which makes the proof of the next theorem somewhat simpler. But we will use (15.2.19) in the next section.

**Proof.** We can make an orthogonal change of coordinates, which we'll still call $\{x_j\}$, so that $A_{ij} = \alpha_j \delta_{ij}$. $\frac{\partial g}{\partial x_\nu}(\vec{0}, 0) = 0$ and $\frac{\partial^2 g}{\partial^2 x_\nu} = \alpha_\nu \neq 0$. So by the implicit function theorem (Theorem 1.4.2 of Part 2A), there is a smooth function $w$ taking a neighborhood of zero in $\mathbb{R}^{\nu-1}$ to $\mathbb{R}$ with $w(0) = 0$ so that

$$\frac{\partial g}{\partial x_\nu}(x_1, \dots, x_{\nu-1}, w(x_1, \dots, x_\nu)) = 0 \tag{15.2.20}$$

and because $\frac{\partial g}{\partial x_\nu \partial x_j}(0) = 0$ for $j \neq \nu$, but $\frac{\partial^2 g}{\partial x_\nu^2}(0) \neq 0$, we have $\nabla w(0) = 0$, so

$$w(x_1, \dots, x_{\nu-1}) = O(|x_1|^2 + \dots + |x_{\nu-1}|^2) \tag{15.2.21}$$

It follows, by Taylor's theorem with remainder, that near $x = 0$,

$$g(x_1, \dots, x_\nu) = g(x_1, \dots, x_{\nu-1}, w(x_1, \dots, x_{\nu-1})) + \tfrac{1}{2}\, h(x)(x_\nu - w)^2 \tag{15.2.22}$$

where $h$ is smooth near $x = 0$ and

$$h(0) = \alpha_\nu \tag{15.2.23}$$

So we define

$$\tilde{y}_\nu(x_1, \dots, x_\nu) = \sqrt{\frac{h(x)}{\alpha_\nu}}\,(x_\nu - w) \tag{15.2.24}$$

$$\tilde{y}_j(x_1, \dots, x_\nu) = x_j, \qquad j = 1, \dots, \nu - 1 \tag{15.2.25}$$

$$\tilde{g}(x_1, \dots, x_{\nu-1}) = g(x_1, \dots, x_{\nu-1}, w(x_1, \dots, x_{\nu-1})) \tag{15.2.26}$$

By (15.2.21), (15.2.22), and (15.2.23),

$$\tilde{y} = x + O(x^2), \quad g(x_1, \dots, x_\nu) = \tilde{g}(x_1, \dots, x_{\nu-1}) + \tfrac{1}{2}\, \alpha_\nu y_\nu^2 \tag{15.2.27}$$

and

$$\frac{\partial \tilde{g}}{\partial x_i \partial x_j} = \alpha_{ij}, \qquad i, j = 1, \dots, \nu - 1 \tag{15.2.28}$$

If $\nu = 1$, this proves the result. If we have the result for $\mathbb{R}^{\nu-1}$, we can apply it to $\tilde{g}$, and so get the result for $g$. Thus, by induction, the result holds. □

To avoid some technical issues, we'll do asymptotics to all orders only in the one-dimensional case.

**Theorem 15.2.5.** *Let $f(x), g(x)$ be real-valued functions on $\mathbb{R}^\nu$ so that* (15.2.6) *holds, $f(x) \geq 0$, and there is a unique $x_0 \in \mathbb{R}^\nu$ so that $g(x_0) = g_-$, that* (15.2.7) *and* (15.2.8) *hold, and that $A_{ij}$ given by* (15.2.9) *is strictly positive, $f(x_0) \neq 0$ and $f$ and $g$ are $C^\infty$ near $x_0$. Then with $d\mu = d^\nu x$,*

$$s^{\nu/2} e^{sg_-} Q(s) \sim (2\pi)^{\nu/2} f(x_0) \det(A)^{-1/2} \left(1 + \sum_{j=1}^\infty a_j s^{-j}\right) \tag{15.2.29}$$

*as an asymptotic series as $s \to \infty$.*

**Remark.** There are explicit formulae for $a_j$ in terms of derivatives of $f$ and $g$ at $x_0$, but they are quite complicated. For example (Problem 3), if $\nu = 1$, $f \equiv 1$,

$$a_1 = -\frac{1}{8}\frac{d_4}{(d_2)^2} + \frac{5}{24}\frac{(d_3)^2}{(d_2)^3}; \qquad d_k = g^{(k)}(x_0) \tag{15.2.30}$$

**Proof.** Without loss, suppose $x_0 = 0$. By the Morse lemma applied to $g - g_-$ and dropping the contribution to 0 away from a neighborhood of 0 which is negligible compared to $e^{sg_-}$,

$$\text{LHS of (15.2.29)} = s^{\nu/2} \int_N \exp\biggl(-\frac{s}{2}\sum A_{ij}y_iy_j\biggr) h(y)\, d^\nu y \tag{15.2.31}$$

where

$$h(y) = f(x(y))\biggl(\frac{\partial x}{\partial y}\biggr)^{-1}(x(y)) \tag{15.2.32}$$

and $h$ is smooth since $f$ and $x$ are. Since $g(x) = x + O(x^2)$, $h(0) = f(0)$.

By scaling,

$$\begin{aligned}\text{LHS of (15.2.29)} &= \int_{s^{1/2}N} \exp\biggl(-\tfrac{1}{2}\sum A_{ij}w_iw_j\biggr) h(s^{-1/2}w)\, d^\nu w \\ &\sim (2\pi)^{\nu/2}\deg(A)^{-1/2}h(0) + \sum_{k=1}^\infty b_k s^{-k/2}\end{aligned}$$

where $b_k$ has a sum of derivatives of $h$ at 0 multiplied by integrals of $\int w_{j_1}\dots w_{j_k}\exp(-\frac{1}{2}\sum A_{ij}w_iw_j)\, d^\nu w$. If $k$ is odd, these integrals are zero, so $b_k = 0$ for $k$ odd. □

**Example 15.2.6** (Stirling's Formula). As noted in the introduction to this chapter (see (15.0.3)),

$$n! = n^{n+1}\int_0^\infty \exp(-ng(x))\, dx, \qquad g(x) = x - \log x \tag{15.2.33}$$

Thus, $x_0 = 1$, where $g'(x_0) = 0$ and

$$g(x_0) = 1, \quad g^{(2)}(x_0) = 1, \quad g^{(3)}(x_0) = -2, \quad g^{(4)}(x_0) = 6 \tag{15.2.34}$$

By Theorem 15.2.2 with $\nu = 1$, $A = \mathbb{1}$,

$$e^n n^{1/2}\int_0^\infty e^{-ng(x)}\, dx \to \sqrt{2\pi} \tag{15.2.35}$$

giving the leading Stirling's formula

$$\frac{n!}{n^{n+\frac{1}{2}}e^{-n}\sqrt{2\pi}} \to 1 \tag{15.2.36}$$

By Theorem 15.2.5, we have a complete asymptotic series in $1/n$. By (15.2.30), we have

$$a_1 = -\frac{1}{8}6 + \frac{5}{24}(-2)^2 = \frac{1}{12} \tag{15.2.37}$$

yielding the explicit leading correction.

In Problem 19 of Section 9.7 of Part 2A, using only the leading asymptotics, (15.2.36), you deduced an integral formula for the log of the ratio. In Problem 4, you'll use that to get complete asymptotics. □

That completes the discussion of the Gaussian approximation. We turn to the other approach using Watson's lemma. We start with the case of $s$ real:

**Theorem 15.2.7** (Watson's Lemma—Real Form). *Let $f$ be a Borel function on $(0,\infty)$ and define*

$$Q(s) = \int_0^\infty e^{-x^\beta s} f(x)\, dx \tag{15.2.38}$$

*for $\beta$ fixed and nonnegative, and suppose for some $C > 0$, $s_0$, and $\varepsilon > 0$, we have for all $x > 0$,*

$$|f(x)| \le C(x^{-(1-\varepsilon)} + e^{x^\beta s_0}) \tag{15.2.39}$$

*Suppose that $f$ has an asymptotic expansion for $x \in (0,\infty)$ near $x = 0$,*

$$f(x) \sim \sum_{n=0}^\infty a_n x^{(n+\lambda-\mu)/\mu} \tag{15.2.40}$$

*for $\lambda, \mu > 0$. Then on $(s_0,\infty)$ near $s = \infty$, one has the asymptotic expansion*

$$Q(s) \sim \sum_{n=0}^\infty a_n \beta^{-1} \Gamma\left(\frac{n+\lambda}{\beta\mu}\right) s^{-(n+\lambda)/\beta\mu} \tag{15.2.41}$$

**Remarks.** 1. The most common case is $\mu = \lambda = 1$.

2. By (15.2.39), $Q$ is given by a convergent integral if $\delta > s_0$. In many cases, (15.2.39) holds for each $s_0 > 0$ for an $s_0$-dependent constant, $C_{s_0}$ (e.g., $|f(x)| \le C(|x|^{-(1-\varepsilon)} + |x|^N)$) and $Q$ is defined on $(0,\infty)$.

3. The standard form of Watson's lemma has $\beta = 1$; but in applications, one often changes variables to effectively turn the general $\beta$ case to $\beta = 1$.

**Proof.** By a direct change of variables from $x$ to $t = sx^\beta$, we have that

$$Q(s) = \beta^{-1} s^{-1/\beta} \int_0^\infty e^{-t} f\left(\left(\frac{t}{s}\right)^{1/\beta}\right) t^{-1+1/\beta}\, dt \tag{15.2.42}$$

For each $N = 1, 2, \ldots$, define

$$R_N(x) = f(x) - \sum_{n=0}^{N-1} a_n x^{(n+\lambda-\mu)/\mu} \tag{15.2.43}$$

and $Q_N$ the integral (15.2.42) with $f$ replaced by $R_N$. Then for any $R < \infty$, we have

$$|R_N(x)| \le C_{N,R} x^{(N+\lambda-\mu)/\mu}, \qquad x \in [0, R) \tag{15.2.44}$$

and, by (15.2.39), for any $R$ and $N$,

$$|R_N(x)| \le \widetilde{C}_{N,R} e^{x^\beta s_0}, \qquad x \in [R, \infty) \tag{15.2.45}$$

Since $e^{x^\beta s_0} \ge 1$ on $[0,1]$ and $x^{(N+\lambda-\mu)/\mu} \ge 1$ on $[1, \infty)$ if $N \ge \lambda - \mu$, we see

$$|R_N(x)| \le \max(C_{N,1}, \widetilde{C}_{N,1}) x^{(N+\lambda-\mu)/\mu} e^{x^\beta s_0}, \quad x \in (0, \infty),\ N \ge \lambda - \mu \tag{15.2.46}$$

Plugging this into the formula, (15.2.42), for $Q_N(s)$, we see for $N \ge (\lambda - \mu)$,

$$|Q_N(s)| \le C s^{-(N+\lambda)/\mu\beta} \int_0^\infty e^{-t} e^{ts_0/s} t^{-1+(N+\lambda)/\beta\mu}\, dt \tag{15.2.47}$$

The integral is uniformly bounded for $s \in [2s_0, \infty)$. This plus the gamma integrals we get from putting the first $N$ terms of the asymptotic series into (15.2.42) gives the required estimate, proving the series for $Q(s)$ is asymptotic at $s = \infty$. □

Actually, the argument applies to complex $s$:

**Theorem 15.2.8** (Watson's Lemma—Complex Form). *Under the hypotheses of Theorem 15.2.7, $Q(s)$ is analytic in $\{s \mid \operatorname{Re} s > s_0\}$ and (15.2.41) is valid in each sector $|\arg(s)| \le \frac{\pi}{2} - \delta$.*

**Proof.** We let $s = |s|e^{-\theta(s)}$ and scale by $|s|$ to get

$$Q(s) = \beta^{-1}|s|^{-1/\beta} \int_0^\infty e^{-t\exp(i\theta(s))} f\left(\left(\frac{t}{|s|}\right)^{1/\beta}\right) t^{-1+\beta}\, dt \tag{15.2.48}$$

Plugging in (15.2.46) leads to an error bound

$$|Q_N(s)| \le |s|^{-(N+\lambda)/\mu\beta} \int_0^\infty e^{-t\cos(\theta(s))} e^{ts_0/|s|} t^{-1+(N+\lambda)/\beta}\, dt \tag{15.2.49}$$

If $\cos(\theta(s)) - s_0/|s| > 0$, that is, $\operatorname{Re} s > s_0$, the integral converges and $Q(s)$ is defined as analytic, and if $\operatorname{Re} s \in [2s_0, \infty)$, we get the required estimate. In the terms from the finite asymptotic series, we can use analyticity to move the contour to effectively do a scaling by $s$, not $|s|$, and find the required terms for the asymptotic series for $Q(s)$ when $\cos(\theta(s))$ is bounded strictly away from 0. □

If $f$ is analytic, we can say more:

**Theorem 15.2.9** (Watson's Lemma—Analytic Form). *Let the hypotheses of Theorem 15.2.7 hold and suppose for some $\gamma_\pm > 0$, $f(x)$ has an analytic continuation to the sector $\{x \mid -\gamma_- < \arg(x) < \gamma_+\}$ and that for each $\delta \in (0, \min(\gamma_-, \gamma_+))$, (15.2.40) holds uniformly in the sector*

$$S_\delta = \{x \mid -\gamma_- + \delta \le \arg(x) \le \gamma_+ - \delta\} \tag{15.2.50}$$

*and that for an $s_0$ (depending on $\delta$), (15.2.39) holds in $S_\delta$. Then $Q(s)$ has an analytic continuation to*

$$\bigcup_{-\gamma_- < \gamma < \gamma_+} \left\{ s \,\middle|\, -\frac{\pi}{2} < \arg(s) - \beta\gamma < \frac{\pi}{2};\ |s| \ge A_\delta \right\} \tag{15.2.51}$$

*where $A_\delta$ depends on the $s_0$-dependence of $\delta$ but $A_\delta$ is finite on $(-\gamma_-, \gamma_+)$ and may be taken continuous. Moreover, uniformly on each sector with $\delta$ fixed, (15.2.41) is valid as $s \to \infty$.*

**Remark.** If, in fact, there is a constant for any $s_0$, and all $\delta$, then $Q(s)$ is analytic in $\{s \mid -\frac{\pi}{2} - \frac{\gamma_-}{\beta} < \arg(s) < \frac{\pi}{2} + \frac{\gamma_+}{\beta}\}$ and the asymptotic series is valid in each sector with $\arg(s) \in [-\frac{\pi}{2} - \frac{\gamma_-}{\beta} + \delta, \frac{\pi}{2} - \frac{\gamma_+}{\beta} - \delta]$ for each $\delta > 0$ and small.

**Proof.** By the assumed bounds and analyticity, one can rotate the contour for $s$ real and large and then modify $\arg(s)$ (perhaps in several small steps) to see that for each $\gamma \in (-\gamma_-, \gamma_+)$, we have

$$Q(se^{-i\beta\gamma}) = \int_0^\infty e^{x^\beta s} f(xe^{i\gamma})\, dx \tag{15.2.52}$$

Then we apply the complex Watson lemma to this function. □

We now turn to applying Watson's lemma to a variety of special functions. In dealing with these examples, there is something called *Stokes' phenomenon* that is illustrated in a case that is trivial because it is so explicit.

**Example 15.2.10** (Stokes' Phenomenon). Let $f(z) = \cosh(z)$. Then the leading exponential order on $\mathbb{H}_+$ is

$$f(z) \sim \tfrac{1}{2} e^z (1 + O(|z|^{-n})), \qquad \text{all } n \tag{15.2.53}$$

uniformly in each $\{z \mid |\arg(z)| \le \frac{\pi}{2} - \delta\}$ for any $\delta > 0$, and on $-\mathbb{H}_+$,

$$f(z) \sim \tfrac{1}{2} e^{-z} (1 + O(|z|^{-n})), \qquad \text{all } n \tag{15.2.54}$$

uniformly in each $\{z \mid \frac{\pi}{2} + \delta < \arg(z) < \frac{3\pi}{2} - \delta\}$. The asymptotic form is different, in fact, discontinuous, as one crosses $\arg(z) = \frac{\pi}{2}, \frac{3\pi}{2}$ even though $f$ is continuous. These lines where the asymptotics shift are called *Stokes lines* and the notion of variable asymptotics in different regions is called *Stokes' phenomenon.* Of course, it is rather trivial in this case but is still illustrated. It is also obvious that using $f(z) \sim \frac{1}{2}e^z + \frac{1}{2}e^{-z}$ yields an asymptotic formula in $\mathbb{H}_+$ which has exponentially negligible terms that can become dominant as one crosses a Stokes line. □

**Example 15.2.11** (Bessel Functions of Imaginary Argument). We start with $K_\alpha$ using (14.7.20). Letting $x = \cosh t - 1$ so $dx = (\sinh t)\,dt = \sqrt{x(x+2)}\,dt$, we see

$$K_\alpha(z) = e^{-z} \int_0^\infty e^{-xz} \frac{\cosh(\alpha + t(x))}{\sqrt{x(x+2)}}\,dx \tag{15.2.55}$$

$$= e^{-z} \int_0^\infty \frac{e^{-xz}}{\sqrt{x}} \left( \frac{1}{\sqrt{2}} + O(x) \right) \tag{15.2.56}$$

$$= \frac{\Gamma(\frac{1}{2})}{\sqrt{2}} \frac{e^{-z}}{\sqrt{z}} \left( 1 + O\left(\frac{1}{z}\right) \right)$$

by Watson's lemma. $\Gamma(\frac{1}{2})$ is, of course, $\sqrt{\pi}$, so

$$K_\alpha(z) \sim \sqrt{\frac{\pi}{2z}}\, e^{-z} \left( 1 + \sum_{n=1}^\infty \frac{a_n(\alpha)}{z^n} \right) \tag{15.2.57}$$

A detailed analysis (see the Notes) shows that

$$a_n(\alpha) = \frac{2^{-n}}{n!} \frac{\Gamma(\frac{1}{2}) + \alpha + n)}{\Gamma(\frac{1}{2} + \alpha - n)} \tag{15.2.58}$$

$$\equiv 2^{-n} h_n(\alpha)$$

and analysis of the analytic Watson lemma shows this works in $|\arg(z)| < \frac{3\pi}{2} - \delta$. The numbers $h_n(\alpha)$ are called *Hankel's symbol*; they arise in many asymptotic formulae for Bessel functions.

By taking $\frac{d}{dz}$ in (15.2.55) and then using Watson's lemma, one also gets asymptotic formulae for $K'_\alpha(z)$ that are equivalent to term-by-term derivation of (15.2.57) (by Leibniz's rule, several terms combine to get the coefficient of $z^{-n-1/2}e^{-z}$).

Notice the region of validity $|\arg(z)| < \frac{3\pi}{2}$ (see the Notes) includes $\arg(z) = \pm i\frac{\pi}{2}$, so the real axis for $H_\nu^{(1)}(z)$. We thus see, by (14.5.37) and (15.2.57) that (14.5.26) holds, thus so do (14.5.23) and (14.5.24). We can also thereby (by going to $\arg(z) = \pm\pi$ in (15.2.57)) obtain asymptotics of $I_\alpha$, but we show how to get them for $\alpha = n$ from an alternate representation.

As for $I_n$, we use (14.7.21) and change variables to $x = 1 - \cos\theta$, so

$$I_n(z) = \frac{e^z}{\pi} \int_0^2 e^{-zx} T_n(1-x) \frac{dx}{\sqrt{x(2-x)}} \tag{15.2.59}$$

where $T_n$ is a Chebyshev polynomial. As in the analysis of $K_\alpha$, this leads to

$$I_n(x) \sim \sqrt{\frac{1}{2\pi x}}\, e^x \left(1 + O\left(\frac{1}{x}\right)\right) \tag{15.2.60}$$

valid for $|\arg(x)| < \frac{\pi}{2} - \delta$. □

**Example 15.2.12** (Airy Functions). (14.5.46) is in a form that one can directly apply Watson's lemma to get asymptotics of $Ai(z)$ and $Ai'(z)$ in $|\arg(z)| < \frac{2}{3}\pi - \delta$. Since the exponent has $z^{1/2}$, the gamma integral gives a $z^{-1/4}$ rather than $z^{-1/2}$ as in the Bessel case. (14.5.50) and (14.5.51) result. □

**Notes and Historical Remarks.** Laplace's method was developed by Pierre-Simon Laplace (1749–1827) in his work on probability theory. A special case appeared in 1774 [**250**] and a fuller development eight years later [**251**] and in his monumental book on probability theory [**253**].

Laplace was raised in a prosperous family in Normandy. In 1768, he was taken as a student by d'Alembert when he came to Paris (d'Alembert later had problems coping with Laplace's fame). After several failed attempts at election to the French Academy, he succeeded in 1773. His later fame was such that Napoleon appointed him Minister of the Interior, but shortly after dismissed him, complaining that he was an incompetent administrator. After the Bourbon restoration, Laplace was made a marquis.

Laplace's fame rested most of all on his work in celestial mechanics where he not only gave a mathematically polished presentation but contributed important physical ideas like the nebular hypothesis (origin of the solar system), gravitational collapse, and black holes. His mathematical work includes potential theory (Laplace's equation, spherical harmonics) and probability theory (generating functions, least squares, asymptotics via Laplace's method, central limit theorem; see the discussion on Section 7.3 of Part 1).

Important early followup on Laplace's work was by Cauchy [**77**], who provided detailed proofs, and a posthumous paper by Abel [**2**], who applied the method to several dimensions. The Morse lemma is from Morse [**281**].

Laplace's method in infinite dimensions is the basis of subtle estimates in stochastic processes, going under the name the method of *large deviations.* An early example was Cramèr [**93**], with the fundamentals by Varadhan [**391**] and in a series by Donsker–Varadhan [**107**]. The key is a tradeoff between volumes in function space ("entropy") and the exponent ("energy")—so an analogy to ideas developed by Lanford, Robinson, and Ruelle in statistical mechanics [**246, 331**]. For books and surveys on large deviations, see [**58, 99, 114, 134, 138, 375, 392, 393**].

Numerous nineteenth-century calculations of asymptotics of integrals had reductions to gamma integrals, but a general result with error estimates only became possible after Poincaré's formal definition of asymptotic series. The idea behind Watson's lemma was not new, but the precise general formulation in 1918 by Watson [**404**] gives the result its name. Actually, Barnes [**27**] essentially had the result a dozen years earlier. The complex versions go back at least to Doetsch [**106**] in 1950.

Stokes' phenomenon is named after George Gabriel Stokes (1819–1903), known for his work in optics and in fluid dynamics. In 1857, he noted [**371**] the phenomenon in the case of asymptotics of Airy functions. Because the transitions occur at edges of sectors, one refers to Stokes lines. There is some confusion in the physics literature because there is also an unrelated spectral phenomenon involving line spectrum associated with Stokes, also called Stokes lines. As we'll see (see the Notes to Section 15.5), within WKB theory, there are natural dividing curves that are asymptotic to the Stokes lines, called *Stokes curves.*

It is interesting that the validity of (15.2.57) into $|\arg(z)| < \frac{3\pi}{2} - \delta$ includes a second sheet ($K_\alpha$ is multisheeted if $\alpha \notin \mathbb{Z}$, infinite-sheeted if $\alpha$ is irrational because of the Euler indices of Bessel's equation at the regular point, $z = 0$). At $\arg(z) = \pm\frac{3\pi}{2}$, we are essentially on the real axis for an analytic continuation of $H_\alpha^{(1)}$, and since $H_\alpha^{(2)}$ mixes in, the asymptotics changes.

For the calculations of the asymptotics of $K_\alpha(z)$ to all orders, see, for example, Temme [**381**, Sect. 9.7].

**Problems**

1. Let $\mu$ be a Baire measure on a compact Hausdorff space, $X$, and let

$$X_0 = \{x \mid \mu(N) > 0 \text{ all Baire neighborhoods, } N \text{ of } x\}$$

Let $x_\alpha \to x$ be a net with all $x_\alpha \in X_0$. If $N$ is a Baire neighborhood of $x$, prove it is a Baire neighborhood of some $x_\alpha$ and conclude that $x \in X_0$. Conclude that $X_0$ is closed.

2. Extend Theorem 15.2.1 to the case where $X$ is locally and $\sigma$-compact, $d\mu$ is a Baire measure so that for some $s_0$, $fe^{-s_0 g} \in L^1(d\mu)$ and where $g$ is bounded from below.

3. Verify (15.2.30).

4. (a) For $|t| < 2\pi$, prove that

$$(e^t - 1)^{-1} - t^{-1} + \frac{1}{2} = \sum_{m=1}^{\infty} \frac{B_{2m}}{(2m)!}\, t^{2m-1}$$

where the $B_m$ are the Bernoulli numbers given by (3.1.48) of Part 2A (recall $B_0 = 1$, $B_1 = -\frac{1}{2}$, $B_{2m+1} = 0$ for $m \geq 1$).

(b) Using Binet's formula (9.7.40) of Part 2A (as proven in Problem 19 of Section 9.7 of Part 2A), prove the full Stirling approximation for $\log \Gamma$. By applying the analytic Watson lemma, show this holds in any sector $\{z \mid |\arg(z)| < \pi - \varepsilon\}$.

**Remark.** Whittaker–Wason [**411**] used both of Binet's formulae, (9.7.92) and (9.7.40) of Part 2A, to get the full Stirling formula. They also have an interesting direct proof of (9.7.40) of Part 2A.

5. Verify the details of Example 15.2.12.

6. The *logarithmic integral*, $\mathrm{li}(x)$, is defined for $x \in (0,1) \cup (1,\infty)$ by

$$\mathrm{li}(x) = \mathrm{pv} \int_0^x \frac{dt}{\log t} \tag{15.2.61}$$

where the integral is a principal value integral if $x > 1$. The purpose of this problem is for the reader to prove that as $x \to \infty$,

$$\mathrm{li}(x) \sim \frac{x}{\log x} \sum_{k=0}^{\infty} \frac{k!}{(\log x)^k} \tag{15.2.62}$$

(a) Letting $x = e^y$, $t = e^{y-u}$, prove that

$$\mathrm{li}(e^y) = e^y \, \mathrm{pv} \int_0^\infty \frac{e^{-u}}{y-u} \, du \tag{15.2.63}$$

(b) Prove that

$$e^{-y} \, \mathrm{li}(e^y) = \mathrm{pv} \int_0^\infty \frac{e^{-ys}}{1-s} \, ds \tag{15.2.64}$$

(c) Prove (15.2.62) by extending Watson's lemma to allow the pv.

**Remark.** li enters in some treatments of the prime number theorem.

7. Find the large $z$ behavior of

$$G(z) = \int_0^\infty \exp(zt - t^\alpha) \, dt$$

for $\alpha$ fixed and real with $\alpha > 1$.

8. Following Riemann [**328**], you will analyze the hypergeometric integral

$$I_n(a,b,c) = \int_0^1 t^{a+n}(1-t)^{b+n}(1-xt)^{c-n} \, dt \tag{15.2.65}$$

as $n \to \infty$, where $0 < x < 1$, $a > -1$, $b > -1$, $c$ real.

(a) Let $f(t) = \log[\frac{x(1-t)}{1-xt}]$. Prove that $f''(t) < 0$ on $[0,1]$ and that $f'(t)$ has exactly one zero, $t_0$, in $[0,1]$.

(b) In terms of $t_0$ and $f''(t_0)$, find the leading asymptotics of $I_n$ as $n \to \infty$.

## 15.3. The Method of Stationary Phase

In this section, we'll consider integrals of the form

$$S(s) = \int e^{isg(x)} f(x)\, d^\nu x \tag{15.3.1}$$

for $f, g$ real-valued, as $s \to \infty$. Here the issues are more subtle than in the real case of Section 15.2 because the issue is cancellations rather than brute force comparison of pieces. The model, of course, is the Fourier transform, $g(x) = -k_0 \cdot x$ for a unit vector $k_0$. If $f \in \mathcal{S}(\mathbb{R}^\nu)$, we already know $S(s) = o(s^{-n})$ for all $n$. This, of course, comes from

$$(is)^n S(s) = (2\pi)^{\nu/2} k_0 \cdot \widehat{\nabla_x f}\ (sk_0)$$

suggesting one key to analyzing (15.3.1) will be integration by parts. That this is so is seen already in our choice to use $\mathbb{R}^\nu$ and $d^\nu x$ rather than a general locally compact space.

Our first theme will be that the only points, $x$, that matter to order $s^{-N}$ are those with $\vec{\nabla} g = 0$ (hence, "stationary phase") in that those points where $\vec{\nabla} g \neq 0$ contribute $o(s^{-N})$ for all $N$. We'll illustrate this in the case

$$\varphi(\vec{x}, s) = (2\pi)^{-\nu/2} \int \exp(i\vec{k} \cdot \vec{x} - isE(\vec{k}))\widehat{\varphi}_0(k)\, d^\nu k, \tag{15.3.2}$$

which, as discussed in the Notes, describes the quantum mechanics of a particle with energy $E(\vec{k})$ ($k$ is momentum, so the usual case is $E(k) = (\vec{k})^2/2m$) at time $s$. We'll prove that $|\varphi(\vec{x}, s)|$ is $O((|x| + s + 1)^{-N})$ as $s \to \infty$ in $\{x \mid (x,s) \notin C\}$, where $C$ is the classically allowed region $\{(x,s) \mid \frac{\vec{x}}{s} \in V \equiv \{\vec{v} \mid \vec{v} \in \nabla E(\vec{k}),\, k \in \operatorname{supp}(\widehat{\varphi}_0)\}\}$. As discussed in the Notes and Problems, such estimates are important in quantum scattering theory.

The second theme will be a one-dimensional analog of Watson's lemma, but with global hypotheses needed, which will allow calculation of the contributions of stationary phase points. Our final theme will involve decay of integrals of the form (15.3.1), where each point with $\nabla g = 0$ has nongenerative second derivatives. We'll apply this to prove decay of Fourier transforms of measures on curved hypersurfaces.

Ironically, in the shift from real to pure imaginary exponent, we'll have ideas that are almost entirely real-analysis based, not depending on analyticity. Still it fits into the subject of this chapter and, as seen, for example, in Section 6.9 of Part 1, describes ideas central to large parts of analysis, especially the theory of PDEs.

The following illustrates the integration-by-parts machine:

**Theorem 15.3.1.** *Let $f$ be a $C^\infty$ function of compact support, $K$. Suppose that $g$ is $C^\infty$ with $\vec{\nabla} g \neq 0$ on $K$. Then $S(s)$, given by (15.3.1), is $O(s^{-n})$ for all $n$.*

**Proof.** For $k = 1, \dots, \nu$, let $U_k = \{x \mid \frac{\partial g}{\partial x_k} \neq 0\}$. By $\vec{\nabla} g \neq 0$ on $K$, $\{U_k\}_{k=1}^{\nu}$ cover $K$, so let $j_1, \dots, j_\nu$ be a partition of unity (see Theorem 1.4.6 of Part 2A) with $\operatorname{supp}(j_k) \subset U_k$. Let

$$S_k = \int f(x) j_k(x) e^{isg(x)}\, d^\nu x \tag{15.3.3}$$

On $\operatorname{supp}(f j_k)$, $\frac{\partial g}{\partial x_j} \neq 0$, so if we control the special case where $\frac{\partial g}{\partial x_1} \neq 0$ on $K$, we obtain the general case.

In that case,

$$\left( \frac{1}{is \frac{\partial g}{\partial x_1}} \frac{\partial}{\partial x_1} \right)^n e^{isg} = e^{isg} \tag{15.3.4}$$

so integrating by parts $n$ times,

$$\begin{aligned} |s^n S(s)| &= \left| \int e^{isg(x)} \left( \frac{\partial}{\partial x_1} \frac{1}{\frac{\partial g}{\partial x_1}} \right)^n f\, d^\nu x \right| \\ &\leq \int \left| \left( \frac{\partial}{\partial x_1} \frac{1}{\frac{\partial g}{\partial x_1}} \right)^n f \right| d^\nu x \end{aligned} \tag{15.3.5}$$

is finite since $f$ and $g$ are $C^\infty$ and $\frac{\partial g}{\partial x_1}$ is bounded away from zero on $K$. □

**Theorem 15.3.2.** *Let $\varphi$ be given by (15.3.2) where $\widehat{\varphi}_0$ has support as a compact set $K$ in $\mathbb{R}^\nu$ and where $\varphi_1$ and $E$ are $C^\infty$ functions. Let $U$ be an open set in $\mathbb{R}^\nu$ with*

$$U \supset \{\vec{v} = \vec{\nabla} E(\vec{k}) \mid \vec{k} \in K\} \tag{15.3.6}$$

*Then for any $n = 0, 1, 2, \dots$, there exists a $C_{n,U}$ so that*

$$|\varphi(\vec{x}, t)| \leq C_{n,U} (1 + |\vec{x}| + |t|)^{-n} \quad \textit{if } \frac{\vec{x}}{t} \notin U \tag{15.3.7}$$

**Proof.** Let $\vec{x} = s\vec{v}$ and write (15.3.2) as

$$\varphi(s\vec{v}, s) = (2\pi)^{-\nu/2} \int \exp(is[\vec{k} \cdot \vec{v} - E(\vec{k})]) \widehat{\varphi}_0(k)\, d^\nu k \tag{15.3.8}$$

(15.3.6) implies $\vec{\nabla}_k [\vec{k}\vec{v} - E(\vec{k})] \neq 0$ for all $\vec{k} \in K$, $\vec{v} \notin U$. Theorem 15.3.1 implies the result for fixed $\vec{v}$. We leave it to the reader (Problem 1) to check that the constant can be chosen $\vec{v}$-independent over $\vec{v} \in \mathbb{R}^\nu \setminus U$. □

That concludes what we want to say about points of nonstationary phase and we turn in one dimension to asymptotics when there are points of stationary phase. We'll later have results on leading decay in higher dimension.

There is an analog of Watson's lemma, but unlike the basic version of that lemma, it requires global smoothness hypotheses on $f$:

**Theorem 15.3.3.** *Let $\alpha > -1$, $f \in \mathcal{S}(\mathbb{R})$, and define*

$$R_\alpha(f)(s) \equiv \int_0^\infty x^\alpha e^{ixs} f(x)\,dx \tag{15.3.9}$$

*Then on $\mathbb{R}$, at $s = \pm\infty$, we have*

$$R_\alpha(f)(s) \sim \sum_{n=0}^\infty \frac{f^{(n)}(0)}{n!}\,\Gamma(n+1+\alpha)(-is)^{-(1+\alpha+n)} \tag{15.3.10}$$

**Remark.** The proof we give also yields a proof of Watson's lemma, but only if $f \in \mathcal{S}$.

We first need

**Lemma 15.3.4.** *If $m \in \mathbb{N}$ and $m < \alpha$, then*

$$\lim_{s\to\pm\infty} s^m R_\alpha(f)(s) = 0 \tag{15.3.11}$$

**Proof.** We use a simple integration by parts. Write

$$e^{ixs} = \left(\frac{1}{s}\,\frac{d}{dx}\right)^m e^{ixs} \tag{15.3.12}$$

When we integrate by parts, there is no boundary term at $\infty$ since $f \in \mathcal{S}$ and none at zero since

$$\left(\frac{d}{dx}\right)^\ell (x^\alpha f)\bigg|_{x=0} = 0, \qquad \ell = 0, \dots, m$$

since $x^{\alpha-j} f^{(k)}\big|_{x=0} = 0$ if $j < \alpha$. Thus,

$$(-is)^m R_\alpha(f)(s) = \int_0^\infty e^{ixs}\left(\frac{d}{dx}\right)^m (x^\alpha f(x))\,dx$$

goes to zero by the Riemann–Lebesgue lemma (see Theorem 6.5.3 of Part 1). □

**Proof of Theorem 15.3.3.** We begin with the special case

$$f_0(x) = e^{-x} \tag{15.3.13}$$

Then by rotating the contour so $(1-is)z$ is real and then scaling by $|1-is|^{-1}$, we see

$$R_\alpha(f_0)(s) = \Gamma(1+\alpha)(1-is)^{-1-\alpha} \tag{15.3.14}$$

$$= (-is)^{-1-\alpha}\Gamma(1+\alpha)\left(1-\frac{1}{is}\right)^{-1-\alpha}$$

$$= \Gamma(1+\alpha)\sum_{j=0}^{\infty}\frac{(-1-\alpha)\dots(-j-\alpha)}{j!}(-is)^{-1-\alpha-j} \tag{15.3.15}$$

$$= \sum_{j=0}^{\infty}\frac{\Gamma(j+\alpha+1)}{j!}(-1)^j(-is)^{-1-\alpha-j} \tag{15.3.16}$$

In (15.3.15), the coefficient is interpreted as 1 if $j = 0$, and we use the binomial theorem. To get (15.3.16), we use $(-1-\alpha)\dots(-j-\alpha)\Gamma(1+\alpha) = (-1)^j\Gamma(j+1+\alpha)$ (since $\Gamma(z+1) = z\Gamma(z)$).

Since $f_0^{(j)}(0) = (-1)^j$, (15.3.16) is just (15.3.10) for $f_0$. Now let

$$f_m(x) = x^m e^{-x} \tag{15.3.17}$$

Since $R_\alpha(f_m) = R_{\alpha+m}(f_0)$, we get (15.3.10) for $f_m$.

Given any $h \in \mathcal{S}$ and $\ell = 1, 2, \dots$, we can find $\{c_j\}_{j=0}^{\infty}$ and $h_\ell \in \mathcal{S}$ so that

$$h(x) = \sum_{j=0}^{\ell-1} c_j f_j + x^\ell h_\ell(x) \tag{15.3.18}$$

Since $S_\alpha(x^\ell h_\ell) = S_{\alpha+\ell}(h)$, we have, by the lemma, that for any $\alpha$ and $m$, there is an $\ell$ with $s^m S_\alpha(x^\ell h_\ell) \to 0$. Thus, (15.3.10) for all $f_m$ implies it for all $h \in \mathcal{S}$. $\square$

The same method shows the following (see Problem 4):

**Theorem 15.3.5.** *Let $\alpha > -1$, $\beta, \mu \in (0, \infty)$, and $f \in \mathcal{S}(\mathbb{R})$. Define*

$$R_{\alpha,\beta,\mu}(f)(s) = \int_0^\infty x^\alpha e^{ix^\beta s} f(x^\mu)\, dx \tag{15.3.19}$$

*Then on $\mathbb{R}$ at $s = \pm\infty$, we have*

$$R_{\alpha,\beta,\mu}(f)(s) \sim \beta^{-1}\sum_{n=0}^{\infty}\frac{f^{(n)}(0)}{n!}\Gamma\left(\frac{\alpha+n\mu+1}{\beta}\right)(-is)^{-(\alpha+n\mu+1)/\beta} \tag{15.3.20}$$

**Example 15.3.6.** Let $f \in \mathcal{S}$ and consider

$$\widetilde{R}(f)(s) = \int_{-\infty}^{\infty} e^{ix^2 s} f(x)\, dx \tag{15.3.21}$$

Define $h \in \mathcal{S}$ by

$$f(x) + f(-x) = 2h(x^2) \tag{15.3.22}$$

(Problem 2 asks you to prove that $h \in \mathcal{S}$.) Then

$$\widetilde{R}(f) = \int_0^\infty e^{ix^2 s}(f(x) + f(-x))\,dx$$

$$= R_{-1/2}(h)(s) \tag{15.3.23}$$

$$\sim \frac{f(0)\sqrt{\pi}}{\sqrt{-is}} + \sum_{k=1}^\infty a_k s^{k-1/2} \tag{15.3.24}$$

□

As in Section 15.2, this allows, by a local change of variables, to obtain an asymptotic series for $S(s)$, given by (15.3.1) in the case $\nu = 1$, if $g$ has finitely many points where $g'(x_j) = 0$ so long as $g''(x_j) \neq 0$ for all $j$. By using other values of $\alpha$, one can even consider the case where some $g''(x_j) = 0$ so long as some derivative is nonzero (see below).

We note also that there is no problem dealing with finite intervals, that is,

$$R^\sharp(f,g)(s) = \int_a^b e^{ig(x)s} f(x)\,dx \tag{15.3.25}$$

If $g$ has a single point, $x_0$, with $g'(x) = 0$ in $(a,b)$ with $g''(x_0) \neq 0$, and $g'(a) \neq 0 \neq g'(b)$, we can use a local coordinate change near $x_0$ and $f(x) = f(x)h(x) + f(x)(1-h(x))$, with $h$ supported in a small neighborhood of $x_0$, and $h \equiv 1$ near $x_0$. The leading term comes from an integral, now of the form (15.3.21); the others from integrals like $\int_a^c e^{ig(n)s} f(n)h(n)$, where $g'$ is bounded strictly away from zero on $[a,c]$. An integration by parts with boundary term shows this is $O(s^{-1})$. Thus,

$$R^\sharp(f)(s) = \frac{f(x_0)e^{ig(x_0)s}\sqrt{\pi}}{\sqrt{-ig''(x_0)s}} + O(s^{-1}) \tag{15.3.26}$$

**Example 15.3.7.** ($J_\alpha(x)$ as $\alpha, x \to \infty$ with $\alpha/x$ fixed in $(1,\infty)$). If $x$ is fixed and $\alpha \to \infty$ along $(0,\infty)$, then by the power series (14.5.6) becomes

$$\Gamma(\alpha+1)J_\alpha(x)\left(\frac{2}{x}\right)^{-\alpha} = 1 - \sum_{n=1}^\infty \frac{(-1)^n}{n!}\,\frac{1}{(\alpha+1)\dots(\alpha+n-1)}\left(\frac{x}{2}\right)^{2n} \tag{15.3.27}$$

so $J_\alpha(x) \sim \exp(-\alpha\log\alpha)$ by Stirling's formula, and using the full Stirling formula, we get a complete asymptotic series for $\log J_\alpha(x)$.

What is more subtle is $J_\alpha(\alpha y)$, where $y$ is fixed in $(0,\infty)$ and $\alpha \to \infty$ in $(0,\infty)$. It is also relevant to some problems in optics. Here we'll treat the case $y \in (1,\infty)$; in the next example, $y = 1$; and, using different methods,

$y \in (0,1)$ in Example 15.4.5. We'll obtain the formula of Nicholson and Rayleigh that for $\beta \in (0, \frac{\pi}{2})$,

$$J_\alpha(\alpha \sec\beta) = \left(\frac{2}{\pi\alpha\tan\beta}\right)^{1/2} \cos(\nu(\beta - \tan\beta) + \tfrac{1}{4}\pi) + O\left(\frac{1}{\alpha}\right) \tag{15.3.28}$$

We'll use (14.7.12). The second term is bounded by (since $(\sec\beta)(\sinh x) > 0$)

$$\frac{1}{\pi}\int_0^\infty e^{-\alpha x}\,dx = \frac{1}{\pi\alpha} \tag{15.3.29}$$

and so is $O(\alpha^{-1})$. The first term is the real part of $R^\sharp(f,g)(\alpha)$, where

$$a = 0, \quad b = \pi, \quad f(\theta) = \frac{1}{\pi}, \quad g(\theta) = \theta - \sec\beta\sin\theta \tag{15.3.30}$$

We have $g'(\theta) = 1 - \cos\theta/\cos\beta$, so the point of stationary phase is $\theta_0 = \beta$, $g''(\theta_0) = \tan\beta$ and so, by (15.3.26), $R^\sharp = (2/\pi\alpha\tan\beta)^{1/2}\exp(i\nu(\beta - \tan\beta) + \frac{1}{4}\pi i)$ (the $\frac{1}{4}\pi i$ comes from $1/\sqrt{-i}$). Taking real parts gives (15.3.28). $\square$

**Example 15.3.8.** ($J_\alpha(\alpha)$ as $\alpha \to \infty$ along $(0,\infty)$). We follow the analysis of the last example, but now $g(\theta) = \theta - \sin\theta$ has $g'(0) = 0$, $g''(0) = 0$, $g'''(0) = 1$. The point of stationary phase is at an endpoint (which has little impact on our analysis) and is degenerate, which does have impact. After a local change of variables, we get

$$J_\alpha(\alpha) = O(\alpha^{-1}) + \frac{1}{\pi}\operatorname{Re}\left(R_{0,\frac{1}{3},1}(f)\left(\frac{\alpha}{6}\right)\right) \tag{15.3.31}$$

with $R$ given by (15.3.19) and $f$ a Jacobian of smooth change of variables so $g(\theta) = x(\theta)^{1/3}$. In particular, $f(0) = 1$. By (15.3.20),

$$R_{0,\frac{1}{3},1}(f)\left(\frac{\alpha}{6}\right) = \frac{1}{3}\Gamma\left(\frac{1}{3}\right)\left(\frac{\alpha}{6}\right)^{-1/3} e^{\frac{1}{6}\pi i}(1 + O(\alpha^{-1/3})) \tag{15.3.32}$$

Since $\cos(\pi/6) = \sqrt{3}/2$, we get

$$J_\alpha(\alpha) = \frac{\Gamma(\frac{1}{3})}{2^{2/3}3^{1/6}\pi}\alpha^{-1/3} + O(\alpha^{-2/3}) \tag{15.3.33}$$

Watson [**405**] has shown that the error is of order $\alpha^{-5/3}$. $\square$

In higher dimensions, we'll settle for any upper bound at nondegenerate stationary phase points. We're heading towards a proof of

**Theorem 15.3.9.** *Let $S(s)$ have the form* (15.3.1) *where $f \in C_0^\infty(\mathbb{R}^\nu)$ and $g$ is $C^\infty$. Suppose that $(\nabla g)(y) = 0$ at only finitely many points $y_1, \dots, y_k$,* *and at each* $y_j$, $\frac{\partial^2 g}{\partial x_i \partial x_j}$ *is an invertible matrix. Then*

$$S(s) = \sum_{j=1}^{k} e^{isg(y_j)} T_j(s) \tag{15.3.34}$$

*where for all* $\ell = 0, 1, 2, \dots; j = 1, \dots, k$,

$$\left| \frac{d^\ell}{ds^\ell} T_j(s) \right| \le C_{j\ell}(1 + |s|)^{-\frac{1}{2}\nu - \ell} \tag{15.3.35}$$

By a localization, we can suppose $k = 1$ and then that $y_1 = 0$ and $g(y_1) = 0$. Also by changing variables using the Morse lemma in the form (15.2.19), we can suppose

$$g(x) = x_1^2 + \cdots + x_k^2 - x_{k+1}^2 - \cdots - x_\nu^2 \tag{15.3.36}$$

We'll use induction in $\nu$ but this will require us to allow $s$-dependence of $f$.

**Definition.** Let $f(x, s)$ be $C^\infty$ defined on $\mathbb{R}^{\nu+1}$ with $f(x, s) = 0$ for $x \notin K$ and $K$ some $s$-independent compact set. We say $f$ is a *symbol* (of order 0) if and only if

$$\left| \frac{\partial^{|\alpha|}}{\partial x_1^{\alpha_1} \dots \partial x_\nu^{\alpha_\nu}} \frac{\partial^\ell}{\partial s^\ell} f(x, s) \right| \le C_{\alpha,\ell}(1 + |s|)^{-\ell} \tag{15.3.37}$$

for $\ell = 0, 1, 2, \dots$ and every multi-index $\alpha$ (see Section 4.3 of Part 3).

**Lemma 15.3.10** (van der Corput's Lemma)**.** *Let* $f \in C_0^\infty(\mathbb{R})$. *Then for* $\alpha \ge 0$,

$$g(s) = \int y^\alpha e^{isy^2} f(y)\, dy \tag{15.3.38}$$

*obeys*

$$|g(s)| \le C_\alpha s^{-(1+\alpha)/2} \tag{15.3.39}$$

*where* $C_\alpha$ *depends only on the* $\|\cdot\|_\infty$ *of the* $f^{(k)}$ *and* $R$, *where* $\operatorname{supp}(f) \subset [0, R]$.

**Proof.** Write $g(s) = g_1(s) + g_2(s)$, where

$$g_1(s) = \int y^\alpha \varphi(ys^{1/2}) e^{isy^2} f(y)\, dy, \qquad g_2 = g - g_1 \tag{15.3.40}$$

and $\varphi(y)$ is $C^\infty$ with $0 \le \varphi \le 1$, $\varphi = 1$ on $[0, 1]$, $\varphi = 0$ on $[2, \infty)$. Clearly,

$$|g_1(s)| \le \|f\|_\infty \int_0^{2s^{-1/2}} y^\alpha\, dy = (\alpha + 1)^{-1}(2s^{-1/2})^{(\alpha+1)} \|f\|_\infty \tag{15.3.41}$$

In $g_2$, change variables from $y$ to $x = y^2$ and get

$$g_2(s) = \int e^{isx} h(x) \big(1 - \rho(\sqrt{xs})\big)\, dx \tag{15.3.42}$$

where

$$h(x) = \tfrac{1}{2}\, x^{\frac{\alpha}{2}-\frac{1}{2}} f\big(\sqrt{x}\big) \tag{15.3.43}$$

Write $e^{isx} = ((is)\frac{d}{dx})^N(e^{isx})$ and integrate by parts $N$ times. In the integral, $x$ lies between $s^{-1}$ and $R^2$. $\ell$-th derivatives of $h$ are bounded by $C(1 + x^{-(\ell-\frac{\alpha}{2}+\frac{1}{2})})$ and $(N-\ell)$-th derivatives of $(1-\rho(\sqrt{xs}))$ by $(1+s^{N-\ell})$. Thus, using the fact that $\int_{s^{-1}}^{R^2} x^\beta\, dx \le C(1+s^{-(\beta+1)})$, we have for $s \ge 1$ that

$$|g_2(s)| \le Cs^{-N}(1+s^{N-\frac{\alpha}{2}-\frac{1}{2}}) = 2Cs^{-(1+\alpha)/2} \tag{15.3.44}$$

if $N > (1+\alpha)/2$. □

**Lemma 15.3.11.** *Let $f$ be a symbol on $\mathbb{R}^{\nu+1}$. Define $g(x_1, \dots, x_{\nu-1}, s) = (1+|s|^2)^{1/4} \int e^{isx_\nu^2} f(x_1, \dots, x_\mu, s)\, dx_\nu$. Then $g$ is a symbol (of order 0) on $\mathbb{R}^\nu$.*

**Proof.** This is immediate from van de Corput's lemma: $s$-derivatives of order $\ell$ have a sum of terms with $x_\nu^{2k}$ and $\frac{\partial^{\ell-k}}{\partial s^{\ell-k}}$. □

**Proof of Theorem 15.3.9.** As noted, we can suppose $k = 1$, $y_1 = 0$, and $g(0) = 0$ and the $g$ has the form (15.3.36). In that case, we use Lemma 15.3.11 for successive integrations and see inductively that

$$(1+|s|^2)^{m/4} \int (e^{\sum_{j=\nu+1-m}^{\nu} \pm isx_j^2}) f(x_1, \dots, x_\nu)\, dx_{\nu+1-m} \dots dx_\nu \tag{15.3.45}$$

is a symbol. So taking $m = \nu$, $(1+|s|^2)^{\nu/4} S(s)$ is a symbol, which implies (15.3.35). □

**Example 15.3.12** (Fourier Transform of Spherical Measures). In $\mathbb{R}^\nu$, let $S^{\nu-1}$ be the unit sphere, $\{\vec{x} \mid |x| = 1\}$ and let $\tau_\nu$ be the unique normalized measure with support $S^{\nu-1}$ which is rotationally invariant (up to normalization, the measure generated by the Riemann metric on $S^{\nu-1}$ induced by the Euclidean metric on $\mathbb{R}^\nu$). Then its Fourier transform obeys

$$\widehat{\tau}_\nu(\vec{k}) = \frac{\Gamma(\frac{\nu}{2})}{2(\pi)^{\nu/2}}\, |k|^{-(\nu-2)/2} J_{(\nu-2)/2}(|k|) \tag{15.3.46}$$

For $\nu = 2$, this follows from (3.7.22) of Part 2A and more generally from the analysis of $\nu$-dimensional rotationally invariant Fourier transforms (Theorem 3.5.13 of Part 3). Notice that as $|k| \to \infty$, by (14.5.23), $|\tau_\nu(\vec{k})| \le C|k|^{-(\nu-1)/2}$ (and no power is better). This might be surprising at first. If $\sigma$ is a measure concentrated on $\{\vec{x} \mid x_\nu = 0\}$ of the form $f(x_1, \dots, x_{\nu-1})\, d^{\nu-1}x$ with $f \in C_0^\infty(\mathbb{R}^{\nu-1})$, then $\widehat{\sigma}$ has no decay at all at $k_\nu \to \infty$. This example will play a starring role in Section 6.8 of Part 3 where we discuss properties of the Fourier transform restricted to curved surfaces. □

The potential surprise is that curvature of $\mathrm{supp}(\tau)$ implies decay. We'll explore that in general. A *hypersurface* in $\mathbb{R}^\nu$ is a connected subset, $S$, so that for every $x_0 \in S$, there is a function $f_{x_0}$ defined near $x_0$ with $f(x_0) = 0$, $\nabla f_{x_0}(x_0) \neq 0$ so that for some $\varepsilon$, $S \cap \{x \mid |x - x_0| < \varepsilon\} = \{x \mid f(x) = 0, |x - x_0| < \varepsilon\}$. Since the gradient is nonzero, locally we can write

$$S = \{x \mid x = (x_1, \dots, x_{j-1}, h(x_1, \dots, x_{j-1}, x_{j+1}, \dots, x_\nu), x_{j+1}, \dots, x_\nu) \tag{15.3.47}$$

We say $S$ is *curved* at $x_0$ if $\frac{\partial^2 h}{\partial x_i \partial x_\ell}\Big|_{i,\ell \in \{1,\dots,j-1,j+1,\dots,\nu\}}$ is invertible. In fact, the total curvature at $x_0$ in the metric induced by the Euclidean metric on $\mathbb{R}^2$ is $(1 + |\nabla h|^2)^{-(\nu+1)/2} \det(\frac{\partial^2 h}{\partial x_i \partial x_j})$, so this condition is saying the total curvature is nonzero and is independent of the coordinate picked (so long as $(\nabla f)_j(x_0) \neq 0$).

Locally, the surface measure, $d\sigma$, induced by the metric is given by

$$d\sigma(x) = \sqrt{1 + |\nabla h(x)|^2}\, dx_1 dx_2 \dots dx_{j-1} dx_{j+1} \dots dx_\nu$$

The normal to $y_0 \in S$ is $(\vec{\nabla} f)(y_0)/|\vec{\nabla} f|$, so since

$$f = x_j - h(x_1, \dots, x_{j-1}, x_{j+1}, \dots, x_\nu)$$

we see (if $j = \nu$ for simplicity) that

$$\vec{n} = \frac{(-\vec{\nabla}_{\nu-1} h, 1)}{(1 + |\vec{\nabla} h|^2)^{1/2}} \tag{15.3.48}$$

**Theorem 15.3.13.** *Let $d\tau = g\, d\sigma$, where $g \in C_0^\infty(\mathbb{R}^\nu)$, and $d\sigma$ is the surface measure on a hypersurface which is everywhere curved. Then the Fourier transform obeys*

$$|\widehat{\tau}(\vec{k})| \le C|k|^{-(\nu-1)/2} \tag{15.3.49}$$

**Remarks.** 1. The proof and the full form of Theorem 15.3.9 actually show (see Problem 5) that for any unit vector, $\vec{n} \in S^{\nu-1}$, if $y_1, \dots, y_\ell \in S \cap \mathrm{supp}(g)$ are precisely those points where $\pm\vec{n}$ is the normal to $S$ at $y_k$, then

$$\widehat{\tau}(\kappa\vec{n}) = \sum_{k=1}^{\ell} T_k(\kappa) e^{-i\kappa\vec{n}\cdot y_k} \tag{15.3.50}$$

where, in $\kappa > 1$,

$$\frac{d^\ell T_k(\kappa)}{d\kappa} \le C_{\ell,\vec{n}} (1 + \kappa)^{-\frac{1}{2}(\nu-1)-\ell} \tag{15.3.51}$$

2. The method of stationary phase when there are no points of stationary phase also shows that if $U$ is a neighborhood of $\{\vec{n} \in S^{\nu-1} \mid \vec{n}$ is normal to some point $S \cap \mathrm{supp}(g)\}$, then for all $\vec{k}$ with $\vec{k}/|\vec{k}| \notin U$, $|\vec{k}| > 1$, one has $|\widehat{\tau}(\vec{k})| \le C_N(|k|^{-N}$ for all $N$.

**Proof.** We prove (15.3.49) for $\vec{k} = \kappa\vec{n}$ for fixed $\vec{n}$ in $S^{\nu-1}$ and leave the uniformity in $\vec{n}$ to the reader (Problem 6). By using a partition of unity, we can suppose a single $j$ and $h$ work on all of $S$. We'll suppose $j = \nu$. then with $\tilde{g}(x_1, \dots, x_{\nu-1}) = g(x_1, \dots, x_{\nu-1}, h(x_1, \dots, x_\nu))(1 + |\nabla h|^2)^{1/2}$, we have

$$\widehat{\tau}(\kappa\vec{n}) = (2\pi)^{-\nu/2} \int \exp(-i\kappa(n_1 x_1 + \cdots + n_{\nu-1}x_{\nu-1} + n_\nu h(x_1, \dots, x_{\nu-1})) \tilde{g}(x_1, \dots, x_{\nu-1}) d^{\nu-1}x \tag{15.3.52}$$

The points of stationary phase are precisely the points where $\vec{n}$ has the form (15.3.48). The second derivatives of the exponential are those of $h$, so the nonzero curvature condition is precisely what is needed for the points of stationary phase to be nondegenerate, and (15.3.35) for $\ell = 0$ implies (15.3.49). $\square$

**Notes and Historical Remarks.** The use of stationary phase ideas in a special context (integral representations of Airy functions) goes back to Stokes in 1850 [**370**]. General theories were first formulated by Kelvin [**226**] and Poincaré [**318**].

The method is central to many aspects of modern analysis, especially PDEs and to geometric analysis. Two classics of the modern approach are Hörmander [**198, 199**]. The first studies general integral operators $h \to S(\,\cdot\,, fh)$ for smooth $g$ and $f$ and the latter to quantum scattering theory (somewhat earlier, other authors used stationary phase in scattering theory).

For books that discuss the method and its myriad applications, see [**110, 165, 170, 200, 268, 275, 284, 362**].

(15.3.2) describes "free" quantum dynamics in units with $\hbar = 1$. If $\hbar$ is restored, $\vec{k}\cdot\vec{x}$ becomes $\vec{k}\cdot\vec{x}/\hbar$ and below $\frac{1}{i}\nabla$ becomes $\frac{\hbar}{i}\nabla$, so $-\frac{\Delta}{2m}$ becomes $-\frac{\hbar^2}{2m}\Delta$ and $[p,x] = -i$ becomes $[p,x] = -i\hbar$.

If $E$ is polynomial in $\vec{k}$, then $\varphi(\vec{x}, s)$ solves

$$i\,\frac{\partial}{\partial s}\,\varphi = E\biggl(\frac{1}{i}\,\vec{\nabla}\biggr)\varphi \tag{15.3.53}$$

as we saw in our discussion of parabolic equations in Section 6.9 of Part 1. In particular, if $E(\vec{k}) = \frac{|\vec{k}|^2}{2m}$, then (15.3.53) becomes

$$i\,\frac{\partial}{\partial s}\,\varphi = -\frac{1}{2m}\,\Delta\varphi \equiv H_0\varphi \tag{15.3.54}$$

$H$ is then given by

$$H = H_0 + V \tag{15.3.55}$$

The meaning of such an unbounded operator will be discussed in Chapter 7 of Part 4.

Scattering theory begins by finding functions $\varphi_\pm \in L^2$ for any given $\varphi_0 \in L^2$, so $e^{-itH}\varphi_\pm - e^{-itH}\varphi_0 \to 0$ as $t \to \mp\infty$. This is equivalent to the existence of the limit

$$\varphi_\pm = \lim_{t\to\mp\infty} e^{itH} e^{-itH_0}\varphi_0 \tag{15.3.56}$$

By a simple argument, it suffices to prove this for a dense set of $\varphi_0$. In Problem 3, the reader will prove, using (15.3.7), that this limit exists if $\widehat{\varphi}_0 \in C_0^\infty(\mathbb{R}^\nu \setminus \{0\})$ and $V$ obeys

$$|V(x)| \le C(1+|x|)^{-1-\varepsilon} \tag{15.3.57}$$

for some $C < \infty$ and $\varepsilon > 0$.

Remarkably, Enss [**116**] discovered that (15.3.7) can be used to prove much more, namely, what is known as asymptotic completeness and absence of singular spectrum. For discussions of scattering theory, see Reed–Simon [**325**], which includes a discussion of Enss theory. For more on that subject, see Perry [**313**].

Examples 15.3.7 and 15.3.8 were first studied by Nicholson [**300**] and Rayleigh [**324**], motivated by problems in physics. Langer [**249**] developed these ideas further to get expansions uniform in $x/\alpha$ (see also [**306, 87**]); they are discussed in Copson [**91**, Ch. 10].

Van der Corput's lemma was found by him in 1921 [**389**] for applications in number theory. You'll note that Theorem 15.3.3 is closely related to van der Corput's lemma.

The idea that the Fourier transform of a measure on a general curved surface decays goes back at least to Hlawka [**195**]. That the decay of the Fourier transform of the surface measure of a sphere has cosine-type oscillations is connected with the fact that there are two points of stationary phase and their oscillations sometimes cancel.

Hlawka considered the decay of the boundary measure in discussing lattice points in large balls and large multiples of convex bodies. For balls, he and, earlier, Landau [**243**] proved that if $N_\nu(\lambda) = \{j \in \mathbb{Z}^\nu \mid |j| \le \lambda\}$ and $\eta_\nu = |\{x \mid |x| \le 1\}|$, then

$$N(\lambda) = \eta_\nu\lambda^\nu + O(\lambda^{\nu-1-(\nu+1)^{-1}(\nu-1)}) \tag{15.3.58}$$

One might have expected an error of $\lambda^{\nu-1}$, but it is smaller. For a textbook presentation, see Sogge [**362**, §1.2]. We'll say more about this question when $\nu = 2$ in the Notes to Section 7.5 of Part 4.

### Problems

1. Carry through the estimates of $\varphi(s\vec{v}, s)$ in (15.3.8) to prove the bounds are uniform in $\vec{v} \in \mathbb{R}^\nu \setminus U$.

2. If $g$ is an even function in $\mathcal{S}(\mathbb{R})$, prove that $g(x) = h(x^2)$ for some $h \in \mathcal{S}(\mathbb{R})$.

3. (a) Let $H = H_0 + V$, with $V$ bounded, $H_0 = -\Delta$, and $\eta_t = e^{itH}e^{-itH_0}\varphi_0$. Suppose you know the formally correct $\frac{d\eta}{dt} = e^{itH}Ve^{-itH_0}\varphi_0$ is actually correct. Show that if

$$\int_{-\infty}^{\infty} \|Ve^{-itH_0}\varphi\|_2 \, dt < \infty \tag{15.3.59}$$

then $\lim_{t\to\mp\infty} e^{itH}e^{-itH_0}\varphi_0$ exists.

(b) If (15.3.57) holds, show that for any $c > 0$,

$$\int_{-\infty}^{\infty} \|\chi_{\{x||x|>c|t|\}}Ve^{-itH_0}\varphi_0\|_2 \, dt < \infty \tag{15.3.60}$$

(c) Let $\widehat{\varphi}_0 \in C_0^\infty(\mathbb{R}^\nu \setminus \{0\})$. Using (15.3.7), prove that for some $c > 0$,

$$\int_{-\infty}^{\infty} \|\chi_{\{x||x|<c|t|\}}Ve^{-itH_0}\varphi_0\|_2 \, dt < \infty$$

and conclude that, for such $\varphi_0$, $\lim_{t\to\mp\infty} \eta_t$ exists.

**Remark.** (a) is known as Cook's method after Cook [**88**].

4. This problem will prove Theorem 15.3.5.

(a) By changing variables from $x$ to $y = x^{1/\beta}$, prove it suffices to consider the case $\beta = 1$ (and henceforth do that).

(b) Prove that for any $\mu > 0$, $\alpha > m$, and $f \in \mathcal{S}(\mathbb{R})$, $x^\alpha f(x^\mu)$ is $C^\infty$ on $(0,\infty)$, and that $(\frac{d}{dx})^m(x^\alpha f(x^\mu))$ is in $L^1$.

(c) Prove an analog of Lemma 15.3.4 for $R_{\alpha,\beta=1,\mu}(f)$.

(d) Let $g \in \mathcal{S}(\mathbb{R})$ be identically 1 near $x = 0$. Let $f \in \mathcal{S}(\mathbb{R})$. Prove that for any $N = 0, 1, \dots$ and some $h_N \in \mathcal{S}(\mathbb{R})$, one has

$$f(x^\mu) = \sum_{n=0}^{N} \frac{f^{(n)}(0)}{n!} x^{\mu n} g(x) + x^{\mu(N+1)} h_N(x^\mu)$$

(e) Complete the proof of Theorem 15.3.5.

5. Verify (15.3.50) and (15.3.51).

6. Prove the uniformity-in-direction part of Theorem 15.3.13.

## 15.4. The Method of Steepest Descent

Thus far, with the exception of Theorem 15.2.9, the techniques of this chapter haven't been complex analytic. In this section where contour shifting will be critical, that will change. The technique has two standard names: "the method of steepest descent" (the most common) and also "the saddle-point method." While each name captures one aspect of the method, I do not think they capture the essence and would prefer "the method of constant phase." Alas, the usual names are firmly established, but the reader should think of the alternate name in understanding what is going on.

We'll consider integrals of the form (15.0.1) but now $f$ and $g$ are both complex-valued and analytic, and the integral is, in general, a contour integral. Included in applications are subtle points of the asymptotics of Bessel functions and also the Hardy–Ramanujan–Uspensky formula for the leading asymptotics of the partition function. The issues are illustrated in:

**Example 15.4.1.** Let

$$Q(s) = \int_{-\infty}^{\infty} \exp[-s(x^2 - 2ix + i - 1)]\, dx \tag{15.4.1}$$

If we just look at the absolute value of the integrand at the minimum of $x^2 - 1$ at $x = 0$, one might conjecture that $Q(s) \sim e^s$ and that is an upper bound for the leading behavior. It is true there is an oscillatory term, but so far, these have only been proven to be $O(s^{-N})$, so maybe they can't cancel the $e^s$. In fact, as the value at $k = 2s$ of the Fourier transform of an $s$-dependent Gaussian, one can compute $Q$ explicitly and see that

$$Q(s) = e^{-is}\left(\frac{\pi}{s}\right)^{1/2} \tag{15.4.2}$$

Thus, we see oscillations can cancel exponential growth! The easiest way to analyze this $Q$ (without doing an exact integral) is to recognize that

$$(x^2 - 2ix - 1) + i = (z - i)^2 + i\big|_{z=x} \tag{15.4.3}$$

and move the contour from $z = x$ $(-\infty < x < \infty)$ to $z = x + i$ $(-\infty < x < \infty)$, so

$$Q(s) = \int_{-\infty}^{\infty} \exp(-sx^2)\exp(-is)\, dx \tag{15.4.4}$$

where we can use Laplace's methods because up to a constant phase, the integrand is real with no oscillations! □

The point, of course, is that if we can deform our contour so that the phase of $e^{sg}$ is constant, that is, $\operatorname{Im} g$ is a constant, $\alpha$, then $e^{i\alpha s}Q(s)$ is an integral of the form (15.0.1) with $g$ replaced by $\operatorname{Re} g$, and we can use Laplace's method getting leading order behavior $e^{-s(\operatorname{Re} g(z_0)+i\alpha+i\beta)}$, where $z_0$ is the point where $\operatorname{Re} g(z_0)$ is minimal and $\beta$ comes from a potential change of variable from $dz$ to something real. If $\Gamma(s)$ is the deformed contour with $\Gamma(s_0) = z_0$ and $\Gamma$ is smooth near $s_0$, then $\operatorname{Im} g$ constant implies $\frac{d}{ds}\operatorname{Im} g(\Gamma(s)) = 0$ for $s$ near $s_0$. If $\operatorname{Re} g$ is minimized on $\Gamma$ at $s_0$, then

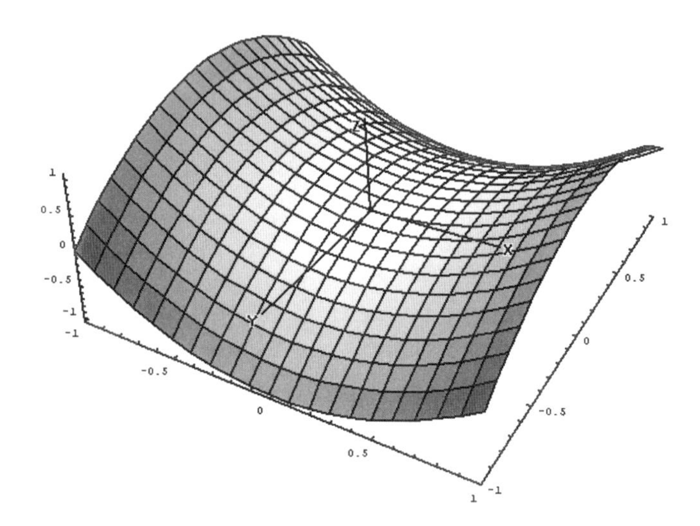


**Figure 15.4.1.** A saddle point (reproduced under GNU Free Documentation License from `http://commons.wikimedia.org/wiki/File:Saddle_point.png`).

$\frac{d}{ds} \operatorname{Re} g(\Gamma(s))\big|_{s=s_0} = 0$. Therefore (if $\Gamma'(s_0) \neq 0$), we have $g'(z_0) = 0$. Since $\operatorname{Re} g$ is harmonic, $z_0$ cannot be a maximum or minimum in two dimensions, and thus, if $g''(z_0) \neq 0$, $z_0$ is a saddle point of $\operatorname{Re} g(z)$ (see Figure 15.4.1).

Thus, one wants to deform the contour to a critical point for $\operatorname{Re} g$ (so, a saddle point). If $g'(z_0) = 0$ and $g''(z_0) \neq 0$, by changing variables to $w = e^{i\varphi}(z - z_0)$ for suitable $w$, we can arrange that $g(z(w)) = g(z_0) + cw^2 + O(w^3)$ for $c > 0$. If we ignore the $O(w^3)$ term and $w = x + iy$, then $\operatorname{Im} g(z(w)) = \operatorname{Im} g(z_0)$ if $x = 0$ or $y = 0$ and $\operatorname{Re} g(z(w)) = \operatorname{Re} g(z_0) + c(x^2 - y^2)$. So if we want a minimum, we take a curve with $y = 0$. It is easy to see then that the $x$-axis direction has $g$ increasing at the fastest rate, that is, if we look at $-g$, it is the direction of steepest descent! Thus, the tangent direction in general (i.e., restoring the $O(w^3)$ term) will be in that direction. If $e^{iw}$ is the direction of the contour, $\Gamma(s)$, of steepest descent, then $dz = e^{iw}\, ds$ near $z_0$, so we get a phase factor of $e^{iw}$.

Of course, we may not be able to arrange that $\operatorname{Im} g$ is constant on the contour, for example, if the original contour runs between finite points $z_1$ and $z_2$ and $\operatorname{Im} g(z_1) \neq \operatorname{Im} g(z_2)$! But so long as $\operatorname{Re} g(z)$ is smaller than $\operatorname{Re} g(z_0)$ on the part of the contour where $\operatorname{Im} g$ is not constant, the part of constant phase will dominate the integrals. We have thus proven the following:

**Theorem 15.4.2.** *Let $f, g$ be analytic in a region, $\Omega$, and let*

$$Q(s) = \int_{\Gamma_1} e^{-sg(z)} f(z)\, dz \tag{15.4.5}$$

*where either* $\Gamma_1$ *is finite or the integral converges absolutely if* $s > s_0$ *and* $\Gamma_1$ *is infinite. Let* $z_0 \in \Omega$ *be such that* $g(z_0) = 0$, $g''(z_0) \neq 0$. *Suppose the contour can be deformed to a new one* $\Gamma_2$ *through* $z_0$ *so that*

(i) *there is a small disk,* $D$, *centered at* $z_0$ *and contained in* $\Omega$ *so that* $\operatorname{Im} g$ *is constant on* $\Gamma_2 \cap D$.

(ii) *if* $\Gamma_2(s_0) = z_0$, *then*

$$\left.\frac{\partial^2 g(\Gamma_2(s))}{\partial s^2}\right|_{s=s_0} > 0 \tag{15.4.6}$$

(iii) $g(z_0) > g(z)$ *for all* $z \in \Gamma_2 \setminus D$.

(iv) $\Gamma_2$ *has a tangent direction* $e^{iw}$ *at* $z_0$.

*Then as* $s \to \infty$,

$$Q(s) \sim f(z_0) e^{-sg(z_0)} e^{iw} \sqrt{\frac{2\pi}{|g''(z_0)|s}} \left(1 + \sum_{j=1}^{\infty} a_j s^{-j}\right) \tag{15.4.7}$$

Before turning to examples, here are some remarks:

(1) It is often complicated to find the curve of constant phase. Typically, all that matters is that the curve goes through the saddle point, $z_0$, and be tangent at $z_0$ to the curve of constant phase (see Example 15.4.3). $g$ now has an imaginary part along $\Gamma$ but $\operatorname{Im} g = O((z-z_0)^3)$ because of the tangency condition and the Laplace method with complex argument (Theorem 15.2.3) is applicable. Some authors distinguish between "the method of steepest descent" (for when the curve is the exact constant phase one) and "the saddle-point method" (where it goes through the saddle with the right tangent direction). Since the leading order in (15.4.7) holds in either case (and there is even a higher-order result), this distinction isn't a useful one; see Example 15.4.6.

(2) It is sometimes useful to write $\int e^{-sg} f\, dz$ as $\int e^{-sg+\log f}\, dz$ and look for critical points of $sg - \log f$. If there is a point where $g'(z_0) = 0$, it is not hard to see that there is a critical point $z(s)$ for large $s$ so that $z(s) \to z_0$ as $s \to \infty$, but in many cases, the critical point $z(s) \to \infty$ as $s \to \infty$ (see Example 15.4.3).

(3) If the contour can be deformed so that $\min \operatorname{Re} g(z)$ occurs at one or both endpoints, the useful deformation is to the direction of steepest descent, that is, the unique constant phase direction; see Example 15.4.13.

**Example 15.4.3** (Airy Functions). We'll prove (14.5.44) here—as we saw in Example 15.2.12, that allows full asymptotics via Watson's lemma. Begin with (14.7.35)/(14.7.37), that is,

$$Ai(z) = \frac{1}{2\pi i} \int_{\Gamma_1} \exp(-zt + \tfrac{1}{3}t^3)\, dt \tag{15.4.8}$$

where $\Gamma_1$ is given in Figure 14.7.4.

We'll start with $z = x_0 \in (0, \infty)$. Then

$$g(t) = -\tfrac{1}{3}\,t^3 + x_0 t \tag{15.4.9}$$

As $|t| \to \infty$, at the ends $|t^3|$ dominates, and because of the asymptotic phase, $\operatorname{Re} g(t) \to \infty$, so there should be a minimum. Since $g'(t) = -t^2 + x_0$, there is a critical point at $t = t_0 \equiv \sqrt{x_0}$, where $g(t_0) = \frac{2}{3}x_0^{3/2}$. Since $g''(\sqrt{x_0}) = -2\sqrt{x_0}$, the direction of fastest growth is along the direction $\sqrt{x_0} + iy$. This is not the curve of constant phase (which bends) but has the same tangent. So we pick $\widetilde{\Gamma}_1 = \{\sqrt{x_0} + iy \mid -\infty < y < \infty\}$. Since $(\sqrt{x_0} + iy)^3 = x_0^{3/2} + i3x_0 y - 3x_0^{1/2} y^2 - iy^3$, we have

$$\begin{aligned} Ai(x_0) &= \frac{1}{2\pi i} \int_{-\infty}^{\infty} \exp(-\tfrac{2}{3}\,x_0^{3/2} - x_0^{1/2} - \tfrac{1}{3}\,iy^3)\,d(iy) \\ &= \frac{1}{\pi} \int_0^{\infty} \exp(-\tfrac{2}{3}\,x_0^{3/2}) \exp(-x_0^{1/2} y^2) \cos(\tfrac{1}{3}\,y^3)\,dy \end{aligned}$$

Letting $t = y^2$ so $dy = \frac{1}{2}t^{-1/2}\,dt$ yields (14.5.46) for $x_0$ real. Analytic continuation gets it for all $z$ with $|\arg(z)| < \pi$.

In going from $\Gamma_1$ to $\widetilde{\Gamma}_1$, one needs to check the decay of the integrand in the region where one deforms the contour. Since $|\exp(-g(x+iy))| = |\exp(\frac{1}{3}x^3 - xy^2 - x_0 x)|$, this is easy. $\square$

**Example 15.4.4.** (Hankel Functions, $H_\alpha^{(1)}(x)$, for $x \to \infty$ along $\mathbb{R}$, with $\alpha$ fixed). We'll prove the asymptotic formula (14.5.33) for $H_\alpha^{(1)}$ using steepest descent on the integral representation (14.7.14). The initial contour runs from $-\infty + i0$ to $\infty + i\pi$ (see $S_+$ in Figure 14.7.3). The function, $g(t)$ ($t$ is the integration variable), is given by

$$g(t) = x \sinh t \tag{15.4.10}$$

$g'(z) = 0$ at $\cosh t = 0$ or $t = \frac{\pi i}{2}(2n+1)$, $n = 0, \pm 1, \dots$.

It is natural to deform the contour through $t_0 = \frac{i\pi}{2}$.

$$\sinh\left(\frac{i\pi}{2}\right) = \tfrac{1}{2}\,(e^{i\pi/2} - e^{-i\pi/2}) = i$$

$g''(t) = g(t)$, so for $u, v$ real,

$$x^{-1} g\left(\frac{i\pi}{2} + u + iv\right) = i + \frac{i}{2}\,(u^2 - v^2) - 2uv + O(|u|^3 + |v|^3) \tag{15.4.11}$$

$\operatorname{Im} g$ = constant to order $|u|^2 + |v|^2$ occurs if $u = \pm v$. $-g$ has a local minimum at $u = v = 0$ along the line $u = v$.

Thus, we take the deformed curve, $\Gamma_2$, to run from $-\infty$ to $-\frac{\pi}{2}$ along $\mathbb{R}$, then at 45% through $\frac{i\pi}{2}$ to $\frac{\pi}{2} + i\pi$, and finally along $\mathbb{R} + i\pi$ from $\frac{\pi}{2} + i\pi$ to

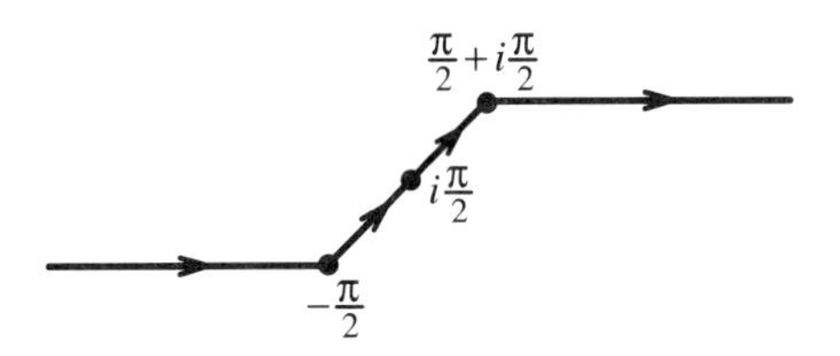


**Figure 15.4.2.** Modified Sommerfeld contour for $H_\alpha^{(1)}$.

$\infty + i\pi$; see Figure 15.4.2. It is easy to see $\operatorname{Re} g(t) > 0$ on this curve except at $t = \frac{i\pi}{2}$, so by (15.4.7)

$$\int_{\Gamma_2} e^{-\alpha t} e^{x \sinh t}\, dt \sim e^{-\pi i \alpha/2} e^{ix} e^{i\pi/4} \sqrt{\frac{2\pi}{x}}\,(1 + O(x^{-1})) \tag{15.4.12}$$

To get $H_\alpha^{(1)}(x)$, we have to multiply by $\frac{1}{\pi i} = \frac{1}{(\sqrt{\pi})^2} e^{-i\pi/2}$ and we obtain (14.5.33), that is,

$$H_\alpha^{(1)}(x) \sim \sqrt{\frac{2}{\pi x}}\, e^{i(x - \frac{\alpha\pi}{2} - \frac{\pi}{4})}(1 + O(x^{-1})) \tag{15.4.13}$$

□

**Example 15.4.5.** ($J_\alpha(x)$ as $\alpha, x \to \infty$ with $\alpha/x$ fixed in $(0,1)$; see Example 15.3.7 for $\alpha/x \in (1,\infty)$ and Example 15.3.8 for $\alpha = x$). This is the problem that Debye invented the method of steepest descent to solve! We will prove that for $\beta \in (0,\infty)$,

$$J_\alpha(\alpha \operatorname{sech} \beta) = \frac{e^{-\alpha(\beta - \tanh\beta)}}{(2\pi\alpha \tanh\beta)^{1/2}} \left(1 + O\Big(\frac{1}{\alpha}\Big)\right) \tag{15.4.14}$$

for $\alpha > 0$, and going to infinity by (14.7.13),

$$J_\alpha(\alpha \operatorname{sech}\beta) = \frac{1}{2\pi i} \int_{S_0} \exp(-\alpha g(t))\, dt \tag{15.4.15}$$

with $g(t) = t - \operatorname{sech}\beta \sinh t$ and $S_0$ the contour in Figure 14.7.2. Since $g'(t) = 1 - \cosh t/\cosh\beta$, the zeros are at $\cosh t = \cosh\beta$, that is, $t = \pm\beta \pm 2\pi i n$. Since $g''(\beta) = -\tanh\beta < 0$, $g(\beta + u + iv) = g(\beta) - \tanh\beta(u^2 - v^2 + 2iuv) + O(|u|^3 + |v|^3)$ the tangents to the curves of constant $\operatorname{Im} g$ are $u = 0$ or $v = 0$ and $\beta$ is a local minimum along the curve $u = 0$.

Thus, it is natural to deform the contour to the one, $\Gamma_2$, in Figure 15.4.3, that runs along $\operatorname{Im} t = -i\pi$ from $-\infty - i\pi$ to $\beta - i\pi$, up through $t = \beta$ to $\beta + i\pi$ and back out to $\infty + i\pi$.

Notice that

$$\operatorname{Re} g(\beta + i\gamma) = \beta - \sinh\beta \cos\gamma > g(\beta) \quad \text{if } \gamma \in \pm[0,\pi] \tag{15.4.16}$$

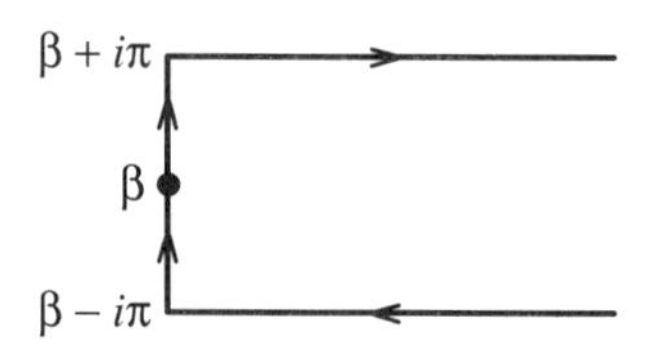


**Figure 15.4.3.** Modified Sommerfeld contour for $J_\alpha(\alpha \operatorname{sech} \beta)$.

and

$$\operatorname{Re} g(\beta \pm i\pi + \eta) = \beta + \eta \frac{\sinh(\beta+\eta)}{\cosh \beta} \geq \beta > g(\beta), \quad \eta \in [0, \infty) \quad (15.4.17)$$

so $\operatorname{Re} g$ is strictly minimized on $\Gamma_2$ at $z = \beta$.

Since $e^{iw} = i$, $g(\beta) = \beta - \tanh \beta$, and $|g''(\beta)| = \tanh \beta$, we get

$$J_\alpha(\alpha \operatorname{sech} \beta) \sim \frac{e^{-\alpha(\beta - \tanh \beta)}}{2\pi i} \, i \sqrt{\frac{2\pi}{\alpha \tanh \beta}} \quad (15.4.18)$$

which is (15.4.14). □

The next two examples describe asymptotics of the counting of partitions of two kinds of sets. The *partition number*, $p(n)$, is the number of partitions of $n$ identical objects into subsets. For example, $p(3) = 3$ corresponding to $\{\{\times,\times,\times\}\}$, $\{\{\times,\times\},\{\times\}\}$, and $\{\{\times\},\{\times\},\{\times\}\}$; the first ten are:

| $n$ | 1 | 2 | 3 | 4 | 5 | 6 | 7 | 8 | 9 | 10 |
|---|---|---|---|---|---|---|---|---|---|---|
| $p(n)$ | 1 | 2 | 3 | 5 | 7 | 11 | 15 | 22 | 30 | 42 |

Equivalently, $p(n)$ is the number of ways of writing $n$ as a sum of positive integers, for example, $3 = 3$, $3 = 2 + 1$, $3 = 1 + 1 + 1$, so $p(n) = \#(m_1, m_2, \dots, m_n \in \{0, 1, \dots, n\} \mid \sum_{j=1}^n j m_j = n)$.

The *Bell numbers*, $d_n$, are the number of ways to partition $n$ distinct objects into subsets. For example, $d_3 = 5$ corresponding to $\{\{1,2,3\}\}$, $\{\{1\},\{2,3\}\}$, $\{\{2\},\{1,3\}\}$, $\{\{3\},\{1,2\}\}$, $\{\{1\},\{2\},\{3\}\}$; the first ten are:

| $n$ | 1 | 2 | 3 | 4 | 5 | 6 | 7 | 8 | 9 | 10 |
|---|---|---|---|---|---|---|---|---|---|---|
| $d_n$ | 1 | 2 | 5 | 15 | 52 | 203 | 877 | 4140 | 21147 | 115975 |

**Example 15.4.6** (Asymptotics of Bell Numbers)**.** Let $d_n$ denote the number of distinct partitions of $\{1, \dots, n\}$. We'll prove here that

$$\begin{aligned} \frac{\log d_n}{n} &= \log n - \log(\log n) - 1 + \frac{\log(\log n)}{\log n} \\ &\quad + \frac{1}{\log n} + \frac{1}{2}\left(\frac{\log(\log n)}{\log n}\right)^2 + O\left(\frac{\log(\log n)}{(\log n)^2}\right) \end{aligned} \quad (15.4.19)$$

We begin by noting that in partitioning $\{1, \dots, n+1\}$, $n+1$ can go in a set with $j = 0, 1, \dots, n$ elements. These $j$ elements can be chosen in $\binom{n}{j}$ ways and we have to then partition the remaining $n-j$ elements. Thus, we get the recursion relation (here $d_0 \equiv 1$)

$$d_{n+1} = \sum_{j=0}^{n} \binom{n}{j} d_{n-j} \tag{15.4.20}$$

Thus, if

$$D(z) = \sum_{n=0}^{\infty} \frac{d_n z^n}{n!} \tag{15.4.21}$$

one has (Problem 1(a)) that

$$D'(z) = e^z D(z) \tag{15.4.22}$$

which is solved by (Problem 1(b))

$$D(z) = \exp(e^z - 1) \tag{15.4.23}$$

Therefore,

$$\frac{d_n}{n!} = \frac{1}{2\pi i e} \int_{C_n} \exp(e^z) z^{-n-1}\, dz \tag{15.4.24}$$

where $C_n$ is any rectifiable Jordan closed contour surrounding 0. Write

$$\exp(e^z) z^{-n-1} = \exp(f_n(z)), \qquad f_n(z) = e^z - (n+1) \log z \tag{15.4.25}$$

On $\mathbb{R}$, we have $f_n'(x) = e^x - (n+1)x^{-1}$. Since $xe^x$ runs monotonically from 0 to $\infty$, as $x$ runs from $x$ to $\infty$, there is a unique solution, $u_n$, in $(0, \infty)$ to

$$u_n e^{u_n} = n+1 \tag{15.4.26}$$

and it is not hard to see (Problem 2) that as $n \to \infty$,

$$u_n = \log n - \log(\log n) + O\biggl(\frac{\log(\log n)}{\log n}\biggr) \tag{15.4.27}$$

The solution of $ye^y = x$ is called the *Lambert* $W$*-function*, $y = W(x)$, so $u_n = W(n+1)$.

On $\mathbb{R}$, $f''(x) = e^x + (n+1)x^{-2} > 0$, so $u_n$ is a local minimum of $\operatorname{Re} f$ along $\mathbb{R}$, and thus, the imaginary direction from $u_n$ is the direction of steepest descent. This suggests that we take $C_n$ to be the line $\{u_n + iy \mid -\pi < y < \pi\}$ together with the counterclockwise parts of the circle

$$|z| = R_n \equiv (u_n^2 + \pi^2)^{1/2} \tag{15.4.28}$$

from $u_n + i\pi$ around to $u_n - i\pi$. Since $dz = i\, dy$, the $i^{-1}$ in (15.4.24) is canceled.

Laplace's method with complex argument (see Theorem 15.2.3) applies to $\frac{1}{2\pi e}\int_{-\pi}^{\pi}\exp(f(u_n+iy))\,dy$ and gives a contribution of

$$C_1 \equiv (2\pi e)^{-1}\sqrt{2\pi}\,e^{f(u_n)}[f''(u_n)]^{-1/2}(1+o(1)) \tag{15.4.29}$$

to $\frac{d_n}{n!}$. Thus,

$$C_1 \geq C\exp(e^{u_n})u_n^{-n-1}e^{-u_n/2} \tag{15.4.30}$$

Let $C_2$ be the contribution of the circular piece. Since $\operatorname{Re} z \leq u_n$ on this contour and $|\exp(e^a)| \leq \exp(|e^a|) \leq \exp(e^{\operatorname{Re} a})$, we see that

$$|C_2| \leq e^{-1}\exp(e^{u_n})(u_n^2+\pi^2)^{-n/2} \tag{15.4.31}$$

Thus,

$$\frac{|C_2|}{C_1} \leq \frac{u_n}{Ce}\,e^{u_n/2}\left(1+\frac{\pi^2}{u_n^2}\right)^{-n/2} \tag{15.4.32}$$

Since $\pi^2/u_n^2$ is small for $n$ large and for all $x$ small, $1+x \geq e^{\frac{1}{2}x}$, we see for $n$ large

$$\left(1+\frac{\pi^2}{u_n^2}\right)^{-n/2} \leq \exp\left(-\frac{\pi^2}{4}\,\frac{n}{u_n^2}\right) \tag{15.4.33}$$

Since $u_n = O(\log n)$, $|C_2|/C_1 \to 0$. Therefore,

$$\frac{d_n}{n!} = (2\pi e)^{-1}\sqrt{2\pi}\exp(e^{u_n})u_n^{-n-1}e^{-u_n/2}(1+o(1)) \tag{15.4.34}$$

This plus Stirling's formula plus (15.4.27) implies (15.4.19) (Problem 3). □

**Example 15.4.7** (Asymptotics of Partition Numbers). Let $p(n)$ denote the number of partitions of $n$ identical objects. We'll prove the *Hardy–Ramanujan–Uspensky formula,*

$$p(n) = \frac{1}{4n\sqrt{3}}\exp\left(\pi\sqrt{\frac{2n}{3}}\right)(1+O(n^{-1/2})) \tag{15.4.35}$$

The proof is involved so we'll formally give it after the example ends; here we'll present the generating function and sketch the idea of the proof.

We begin with the generating function found by Euler:

$$F(z) \equiv \sum_{n=0}^{\infty} p(n)z^n = \prod_{k=1}^{\infty}\frac{1}{1-z^k} \tag{15.4.36}$$

The proof is similar to that of the Euler factorization formula

$$\prod_{k=1}^{\infty}(1-z^k)^{-1} = \prod_{k=1}^{\infty}(1+z^k+z^{2k}+\dots) = \sum_{m_1,m_2,\dots=0}^{\infty} z^{m_1+2m_2+3m_3+\dots}$$

$n = m_1 + 2m_2 + \dots$ in exactly $p(n)$ ways. If $|z| < 1$, there is absolute convergence of the product, which justifies rearrangement. Thus, for any $0 < r < 1$,

$$p(n) = \frac{1}{2\pi i} \oint_{|z|=r} z^{-n-1} F(z)\, dz \tag{15.4.37}$$

The basic idea is that for each $n$, we want to pick $r_n$ so $n \log z + \log F(z)$ has a saddle point on the circle $|z| = r_n$. Since $p(n) > 0$, we expect the most severe singularity of $f(z)$ which controls asymptotics of $p(n)$ should occur at $z = 1$ and we expect a saddle at $z = r_n$.

$F(z)$ has a natural boundary on $\partial\mathbb{D}$ (Problem 7), but one can see that 1 is the worst singularity since $(1-z)^{-1}$ only has a singularity at $z = 1$ and $(1-z^k)^{-1}$ has equal singularities at all $k$-th roots, so, in particular, at $z = 1$. Thus, we want a good approximation near $z = 1$ which is simpler than $F$. We'll show that in a suitable region near $z = 1$, we have

$$F(z) = \sqrt{\frac{1-z}{2\pi}} \exp\biggl(\frac{\pi^2}{6 \log(\frac{1}{z})}\biggr) (1 + O(|1-z|)) \tag{15.4.38}$$

Therefore, in (15.4.37), the approximate quantity in the exponent (if we use $z^{-n} = \exp(n \log(\frac{1}{z}))$) is $g_n(y) = ny + \frac{\pi^2}{6y}$ with $y = \log(\frac{1}{z})$. $g_n(y)$ has a saddle point at

$$y_n = \frac{\pi}{\sqrt{6n}} \tag{15.4.39}$$

where

$$g_n(y_n) = \pi \sqrt{\frac{2n}{3}} \tag{15.4.40}$$

We'll need to establish (15.4.38) near $z = 1$, control the rest of $|z| = 1$ (using the fact that $(1-z)^{-1}$ is only singular near $z = 1$), and use Laplace's method to do the integral near the saddle at $y_n$. The details are below. $\square$

We begin the formal analysis with a formula for the log of the function in (15.4.36):

$$\begin{aligned} \log F(z) &= \sum_{k=1}^{\infty} -\log(1 - z^k) \\ &= \sum_{j,k=1}^{\infty} \frac{z^{kj}}{j} \\ &= \sum_{j=1}^{\infty} \frac{1}{j} \frac{z^j}{1 - z^j} \end{aligned} \tag{15.4.41}$$

Away from $z = 1$, we single out the $j = 1$ term to get a bound that will show the main contribution of the integral comes from the region near $z = 1$:

**Lemma 15.4.8.** *For $z \in \mathbb{D}$, we have*

$$|F(z)| \le \exp\left(\left[\frac{1}{|1-z|} + \left(\frac{\pi^2}{6} - 1\right)\frac{1}{1-|z|}\right]\right) \tag{15.4.42}$$

**Proof.** By the triangle inequality,

$$1 = |1| = |1 - z^j + z^j| \le |1 - z^j| + |z|^j$$

so $|1 - z^j| \ge 1 - |z|^j$ and

$$\frac{|z^j|}{|1 - z^j|} \le \frac{|z|^j}{1 - |z|^j} \tag{15.4.43}$$

If $a \in (0,1)$,

$$1 - a^j = (1-a)(1 + a + \cdots + a^{j-1}) \ge (1-a)ja^j$$

so that

$$\frac{|z|^j}{1 - |z|^j} \le \frac{1}{j(1 - |z|)} \tag{15.4.44}$$

Using (15.4.41), (15.4.43), and (15.4.44), we get

$$\left|\log F(z) - \frac{1}{1-z}\right| \le \frac{1}{1-|z|}\sum_{j=2}^{\infty}\frac{1}{j^2} = \left(\frac{\pi^2}{6} - 1\right)\frac{1}{1-|z|} \tag{15.4.45}$$

Thus,

$$|\log F(z)| \le \frac{1}{|1-z|} + \left(\frac{\pi^2}{6} - 1\right)\frac{1}{1-|z|} \tag{15.4.46}$$

which implies (15.4.42). □

This will allow us to show the contribution to the integral of the region

$$R_n = \{|z| = y_n \mid |1 - z| \ge 3(1 - |z|)\} \tag{15.4.47}$$

is exponentially small compared to the answer we want, (15.4.35). We thus focus on the region

$$M = \{z \in \mathbb{D} \mid |1 - z| \le 3(1 - |z|)\} \tag{15.4.48}$$

It will be helpful to use the change of variables mentioned also above equation (15.4.43), $z = e^{-y}$. Thus, (15.4.41) becomes

$$\log F(e^{-y}) = \sum_{j=1}^{\infty}\frac{1}{j}\,\frac{1}{e^{jy} - 1} \tag{15.4.49}$$

We are interested in what happens as $z \to 1$, that is, $y \to 0$. If we write $\frac{1}{j}$ as $y\frac{1}{jy}$, we see the right side is a Riemann sum approximating $\int_0^\infty \frac{1}{e^x - 1}\frac{dx}{x}$, an integral which diverges at $x = 0$. To handle that (and, after all, we do expect $\log F(e^{-y})$ to diverge!), we make subtractions.

With Binet's formula, (9.7.40) of Part 2A, in mind, we initially aim for $\mu(z=0)$ which has a divergence at $t=\infty(\int_1^\infty \frac{dt}{2t})$, so we'll instead aim for $\int_1^\infty \frac{e^{-t}}{2t}\,dt$, that is, we write

$$\frac{1}{j}\,\frac{1}{e^{jy}-1} = \frac{1}{j^2 y} - \frac{e^{-jy}}{2j} + \frac{1}{j}\left(\frac{1}{e^{jy}-1} - \frac{1}{jy} + \frac{e^{-jy}}{2}\right) \tag{15.4.50}$$

If we define

$$g(x) = \frac{1}{x}\left(\frac{1}{e^x-1} - \frac{1}{x} + \frac{e^{-x}}{2}\right) \tag{15.4.51}$$

then, since $\sum_{j=1}^\infty j^{-2} = \frac{\pi^2}{6}$ and $\sum_{j=1}^\infty j^{-1}(e^{-y})^j = -\log(1-e^{-y})$, we have, by (15.4.49), that

$$\log F(e^{-y}) = \frac{\pi^2}{6y} + \frac{1}{2}\log(1-e^{-y}) + \sum_{j=1}^\infty y g(yj) \tag{15.4.52}$$

The sum is a Riemann sum approximating a Riemann integral, so we need next to evaluate the integral and to discuss estimates on the approximation of sums by integrals.

**Proposition 15.4.9.** (a) *For* $0<\alpha<1$,

$$\int_0^\infty \frac{e^{-x}-e^{-\alpha x}}{x}\,dx = \log\alpha \tag{15.4.53}$$

(b) $$\int_0^\infty \left(\frac{1}{e^x-1} - \frac{1}{x} + \frac{e^{-x}}{2}\right)\frac{dx}{x} = \log\left(\frac{1}{\sqrt{2\pi}}\right) \tag{15.4.54}$$

**Proof.** (a) $\frac{d}{d\beta}e^{-\beta x} = -xe^{-\beta x}$, so

$$\frac{e^{-x}-e^{-\alpha x}}{x} = -\int_\alpha^1 e^{-\beta x}\,d\beta \tag{15.4.55}$$

Plugging into the integral in (15.4.53) and interchanging the integrals (valid by positivity and Fubini's theorem),

$$\text{LHS of (15.4.53)} = -\int_\alpha^1 \left(\int_0^\infty e^{-\beta x}\,dx\right)d\beta = -\int_\alpha^1 \frac{d\beta}{\beta} = \log\alpha$$

(b) We have

$$\begin{aligned}&\text{LHS of (15.4.54)}\\ &= \lim_{\alpha\downarrow 0}\left[\int_0^\infty \left(\frac{1}{e^x-1} - \frac{1}{x} + \frac{1}{2}\right)e^{-\alpha x}\,dx + \frac{1}{2}\int_0^\infty \frac{e^{-x}-e^{-\alpha x}}{x}\,dx\right]\\ &= \lim_{\alpha\downarrow 0}\,[\mu(\alpha) + \tfrac{1}{2}\log\alpha]\end{aligned} \tag{15.4.56}$$

where $\mu(\alpha)$ is given by (9.7.38) of Part 2A, and we've used Binet's formula, (9.7.40) of Part 2A. By the definition (9.7.38) of Part 2A of $\mu(\alpha)$,

$$(15.4.56) = \log\left(\frac{1}{\sqrt{2\pi}}\right) + \lim_{\alpha\downarrow 0}\left[\log(\alpha^{-1}\Gamma(\alpha+1)) + \alpha + (\tfrac{1}{2}-\alpha)\log\alpha + \tfrac{1}{2}\log\alpha\right] \tag{15.4.57}$$

where we used $\Gamma(\alpha) = \alpha^{-1}\Gamma(\alpha+1)$. As $\alpha \downarrow 0$, $\log\Gamma(\alpha+1) \to 0$, $\alpha \to 0$, and $\alpha\log\alpha \to 0$, and the $\log\alpha$ terms cancel, so the limit in (15.4.57) is 0, and we get (15.4.54). □

**Proposition 15.4.10.** *If $g$ is $C^1$ on $(0,\infty)$ with $g, g' \in L^1$, then for any $u \in (0,\infty)$,*

$$\left|\int_0^\infty g(x)\,dx - \sum_{j=1}^\infty ug(uj)\right| \le u\int_0^\infty |g'(x)|\,dx \tag{15.4.58}$$

**Proof.** By the intermediate value theorem,

$$\left|\int_{(j-1)u}^{ju} g(x)\,dx - ug(uj)\right| \le |ug(x_j(u)) - ug(uj)| \tag{15.4.59}$$

for some $x_j \in [(j-1)u, ju]$. Thus,

$$\text{RHS of (15.4.59)} \le u\int_{(j-1)u}^{ju} |g'(x)|\,dx \tag{15.4.60}$$

and (15.4.58) is immediate. □

We can summarize these last few results in

**Proposition 15.4.11.** *Let $\theta_0 \in (0, \frac{\pi}{2})$. Then for $z \in \{z = e^{-y} \mid 0 < |y| < \frac{\pi}{2\sin\theta_0};\ |\arg(y)| < \theta_0\}$,*

$$F(z) = \sqrt{\frac{1-z}{2\pi}}\exp\left(-\frac{\pi^2}{6\log z}\right)[1 + R(z)] \tag{15.4.61}$$

*where, if also $|z| > \frac{1}{2}$, we have*

$$|R(z)| \le C(\theta_0)|1-z| \tag{15.4.62}$$

**Remark.** If $\arg(y) = \theta$, we have $z = e^{-|y|\cos\theta}e^{i|y|\sin\theta}$, so the $z$'s in question have $|z| < 1$, $|\arg(z)| < \frac{\pi}{2}$. Moreover, in this region, $|y| = |\log z| = |\int_z^1 \frac{dw}{w}| \le \frac{1}{|z|}|1-z|$, so

$$|z| > \tfrac{1}{2} \Rightarrow |y| \le 2|1-z| \tag{15.4.63}$$

**Proof.** Write $z = e^{-y}$, $y = \log(\frac{1}{z})$ and exponentiate (15.4.52) to obtain (15.4.61) where

$$1 + R(z) = \exp\left(\left[\sum_{j=1}^\infty yg(yj) - \log\left(\frac{1}{\sqrt{2\pi}}\right)\right]\right) \tag{15.4.64}$$

Write $y = |y|e^{i\theta}$ and let

$$g_\theta(x) = e^{i\theta} g(e^{i\theta} x) \tag{15.4.65}$$

By $\theta_0 < \frac{\pi}{2}$, we see, by a simple estimate,

$$\sup_{|\theta| \le \theta_0} \int [|g_\theta(x)| + |g_\theta'(x)|]\, dx < \infty \tag{15.4.66}$$

so $\theta \mapsto \int_0^\infty g_\theta(x)\, dx$ is analytic in $\{\theta \mid |\operatorname{Re}\theta| < \theta_0\}$ and, by scaling $\theta$, independent, if $\operatorname{Re}\theta = 0$ so, by analyticity, for all $\theta$,

$$\int_0^\infty g_\theta(x)\, dx = \log\biggl(\frac{1}{\sqrt{2\pi}}\biggr) \tag{15.4.67}$$

by (15.4.54).

Noticing that $yg(yj) = |y| g_\theta(|y|j)$, we conclude, by (15.4.66) and Proposition 15.4.10, that uniformly in $|\arg(y)| \le \theta_0$,

$$\biggl| \sum_{j=1}^\infty yg(yj) - \log\biggl(\frac{1}{\sqrt{2\pi}}\biggr) \biggr| \le C_1(\theta_0)|y| \tag{15.4.68}$$

We obtain (15.4.62) from (15.4.63), (15.4.64), and (15.4.68). □

We'll now have all the tools to prove

**Theorem 15.4.12** (Hardy–Ramanujan–Uspensky Asymptotics). *The partition function obeys*

$$p(n) = \frac{1}{4n\sqrt{3}} \exp\biggl(\pi\sqrt{\frac{2n}{3}}\biggr)(1 + O(n^{-1/2})) \tag{15.4.69}$$

*as $n \to \infty$.*

**Remark.** While we don't explicitly mention steepest descent, our choice $|z| = r_n$ comes from (15.4.37), which in turn comes from the fact that, as a function of $y = \log(\frac{1}{z})$, $z^{-n}\exp(\frac{\pi^2}{6\log(1/z)})$ has a saddle point at $y_n = \frac{\pi}{\sqrt{6n}}$. The direction of steepest descent is the imaginary direction in $y$, so a circle in $z$ is tangent to this direction.

**Proof.** On $\mathbb{D} \cap \mathbb{H}_T$, define

$$\widetilde{F}(z) = \sqrt{\frac{1-z}{2\pi}} \exp\biggl(-\frac{\pi^2}{6\log z}\biggr) \tag{15.4.70}$$

We start with

$$p(n) = \frac{1}{2\pi i} \oint_{|z|=r_n} z^{-n-1} F(z)\, dz \tag{15.4.71}$$

where

$$r_n = e^{-y_n}, \qquad y_n = \frac{\pi}{\sqrt{6n}} \tag{15.4.72}$$

With (15.4.47) in mind, we define $\theta_n > 0$ by requiring

$$|1 - r_n e^{i\theta_n}| = 3(1 - r_n) \tag{15.4.73}$$

Thus, if $\rho_n = 1 - r_n$, we see $2r_n(1 - \cos\theta_n) = 8\rho_n^2$, which implies $\theta_n \to 0$ (since $r_n \to 1$, $\rho_n \to 0$). Since $\rho_n = y_n + O(\frac{1}{n})$,

$$\theta_n = 2\sqrt{2}\, y_n \biggl(1 + O\biggl(\frac{1}{\sqrt{n}}\biggr)\biggr) \tag{15.4.74}$$

We write

$$p(n) = p_1(n) + p_2(n) + p_3(n) \tag{15.4.75}$$

$$p_1(n) = \frac{1}{2\pi i} \int_{\substack{|z|=r_n \\ |\arg(z)| \le \theta_n}} z^{-n-1} \widetilde{F}(z)\, dz \tag{15.4.76}$$

$$p_2(n) = \frac{1}{2\pi i} \int_{\substack{|z|=r_n \\ |\arg(z)| \le \theta_n}} z^{-n-1} (F(z) - \widetilde{F}(z))\, dz \tag{15.4.77}$$

$$p_3(n) = \frac{1}{2\pi i} \int_{\substack{|z|=r_n \\ |\arg(z)| \ge \theta_n}} z^{-n-1} F(z)\, dz \tag{15.4.78}$$

$$q(n) = \bigl(4n\sqrt{3}\,\bigr)^{-1} \exp\biggl(\pi \sqrt{\frac{2n}{3}}\biggr) \tag{15.4.79}$$

We'll prove that for some $c < \pi\sqrt{\frac{2}{3}}$,

$$p_1(n) = q(n)(1 + O(n^{-1/2})), \quad p_n(n) = q(n) O(n^{-1/2}), \quad p_3(n) = O(e^{c\sqrt{n}}) \tag{15.4.80}$$

which implies (15.4.69).

We start with $p_3(n)$. Since $|1-z| \ge 3(1-|z|)$ in the region of integration, Lemma 15.4.8 implies that in that region,

$$|F(z)| \le \exp\biggl(\biggl(\frac{\pi^2}{6} - \frac{2}{3}\biggr) \frac{1}{1 - r_n}\biggr) \tag{15.4.81}$$

Of course, $|z|^{-n-1} = \exp((n+1)\log(\frac{1}{r_n}))$ so, since $\log(\frac{1}{r_n})$ and $(1 - r_n)$ are $y_n(1 + \frac{1}{\sqrt{n}})$, we see

$$p_3(n) \le C \exp\biggl(\biggl(\pi\sqrt{\frac{2}{3}} - \frac{2}{3}\frac{\sqrt{6}}{\pi}\biggr)\sqrt{n}\biggr) \tag{15.4.82}$$

proving the $p_3$ bound in (15.4.80).

In $p_1(n)$, we first note that $z^{-1}\, dz = i\, d\theta$. Write $\varphi = y_n^{-1}\theta$, $\varphi_n \equiv \frac{\theta_n}{y_n} = 2\sqrt{2}(1 + O(\frac{1}{\sqrt{n}}))$, and get, since $\log(\frac{1}{z}) = y_n(1 + i\varphi)$,

$$p_1(n) = \frac{y_n}{2\pi} \int_{-\varphi_n}^{\varphi_n} f_n(\varphi) \exp(n y_n g(\varphi))\, d\varphi \tag{15.4.83}$$

where

$$g(\varphi) = (1+i\varphi) + \left(\frac{1}{1+i\varphi}\right), \qquad f_n(\varphi) = \sqrt{\frac{1-z}{2\pi}}\,\Bigg|_{z=r_n e^{iy_n\varphi}} \tag{15.4.84}$$

In (15.4.83), the $y_n$ comes from the change of variables $d\theta = y_n\, d\varphi$.

We notice that $g(\varphi) = 2-\varphi^2+O(\varphi^3)$ and, since $\varphi_n$ is bounded uniformly in $n$, $\operatorname{Re} g(\varphi) \le 2 - c\varphi^2$ for some $c > 0$ and $|\varphi| < \varphi_n$. Moreover,

$$f_n(0) = \sqrt{\frac{y_n}{2\pi}}\,(1+O(n^{-1/2})) \tag{15.4.85}$$

Thus, we can use Laplace's method to see that

$$p_1(n) = \frac{y_n}{2\pi}\sqrt{\frac{\pi}{ny_n}}\sqrt{\frac{y_n}{2\pi}}\exp\biggl(\pi\sqrt{\frac{2}{3n}}\biggr)(1+O(n^{-1/2})) \tag{15.4.86}$$

Here the $\sqrt{\frac{\pi}{ny_n}}$ comes from the Gaussian integral and $\sqrt{\frac{y_n}{2\pi}}$ from (15.4.85). Since $y_n = \pi\sqrt{\frac{\pi}{6n}}$, this proves $p_1(n) = q(n)\bigl(1+O(n^{-1/2})\bigr)$.

Finally, in (15.4.77), we use $|F(z) - \widetilde{F}(z)| = O(\frac{1}{\sqrt{n}})|F(z)|$ in the region in question (by (15.4.62)). The Gaussian integral is different (since we use the upper bound on $\operatorname{Re} g(\varphi)$) but still Gaussian, and we get the $p_1$ bound in (15.4.80). □

**Example 15.4.13.** Let

$$I(\alpha) = \int_0^1 \log t\, e^{i\alpha t}\, dt \tag{15.4.87}$$

This doesn't look like steepest descent and, indeed, $g(t) = t$ has no critical points. The contributions come from the endpoints. We want to deform the contour so that it leaves 0 and enters 1 along a direction of constant phase, that is, $\operatorname{Im} t =$ constant. $e^{i\alpha t} = e^{i\alpha u}e^{-\alpha v}$ if $t = u + iv$ decreases as $v = \operatorname{Im} t$ increases. Thus, we deform to the contour 0 to $iR$, $iR$ to $1+iR$, $1+iR$ to 1, and then take $R \to \infty$. Since $|e^{i(u+iR)}| = e^{-R}$, the top piece goes to zero and we get

$$I(\alpha) = i\int_0^\infty \log(iv)e^{-\alpha v}\, dv - ie^{i\alpha}\int_0^\infty \log(1+iv)e^{-\alpha v}\, dv \tag{15.4.88}$$

By scaling and (see Problem 27(b) of Section 9.6 of Part 2A)

$$\int_0^\infty (\log s)e^{-s}\, dt = -\gamma \tag{15.4.89}$$

where $\gamma$ is Euler's constant, the first integral in (15.4.88) is

$$-\frac{i\log\alpha}{\alpha} - \frac{i\gamma+\frac{\pi}{2}}{\alpha} \tag{15.4.90}$$

We can apply Watson's lemma and $\log(1+iv) = -\sum_{n=1}^\infty (-iv)^n/n$ to get asymptotics of the second integral. We find, as $\alpha \to \infty$,

$$I(\alpha) \sim -\frac{i\log\alpha}{\alpha} - \frac{i\gamma + \frac{\pi}{2}}{\alpha} + ie^{i\alpha}\sum_{n=1}^\infty \frac{(-i)^n(n-1)!}{\alpha^{n+1}} \tag{15.4.91}$$

□

**Notes and Historical Remarks.** The Dutch physicist Peter Debye (1884–1966), who won the Nobel Prize in Chemistry in 1936, developed the method of steepest descent in 1909 [**96**] because he needed the large $\alpha$ behavior of $J_\alpha(\alpha y)$ for some work in theoretical physics. Remarkably, as of November 2010, Debye's Wikipedia biography doesn't even mention this work (although it has a fascinating description of controversies concerning Debye's possible Nazi sympathies—controversies that took place sixty-five years after the events).

Debye mentioned that Riemann, in a work written in Italian in 1863 but only published posthumously [**328**] by Schwarz, had a similar idea in the analysis of hypergeometric integrals of the form

$$\int_0^1 t^{a+n}(1-t)^{b+n}(1-xt)^{c-n}\,dt$$

as $n \to \infty$. For $x \in [0,1]$, this can be analyzed by Laplace's method (see Problem 8 of Section 15.2), but Riemann studied complex $x$ by a version of the method of steepest descent (see Problem 4).

P. A. Nekrasov [**288**] also had a variant of the method of steepest descent, but this work was ignored until its resurrection by Russian historians of mathematics [**314**].

The technique in Examples 15.4.6 and 15.4.13 of studying asymptotics of combinatorial sequences by forming a generating function and using a steepest descent analysis on a contour integral is common. See [**45, 95, 303**] for surveys and [**68, 146, 364**] for specific examples of the technique.

Bell numbers were named by Becker–Riordan [**32**] after E. T. Bell [**36, 37, 39**]. Eric Temple Bell (1883–1960), born in Scotland, was a professor at Caltech known not only for his mathematical contributions but for his popular books on the history of mathematics [**38**] and as the author of thirteen science fiction books written under the pen name John Taine. For more on Bell numbers and their history, see Rota [**335**]. Our approach to their asymptotics follows de Bruijn [**95**]. The *Online Encyclopedia of Integer Sequences* has a lot more about Bell numbers [**358**]. The Lambert $W$-function is named after work of Lambert [**240**].

Partition numbers have long fascinated mathematicians, going back to work of Leibniz and Euler, who found the generating function. For books on the subject, see [**14, 17, 19, 55, 174, 287, 321, 360, 379**]. The asymptotic formula (15.4.35) is due to Hardy–Ramanujan [**177**] in 1918 and, independently, to Uspensky [**384**] in 1920. Both papers use the theory of modular functions and a contour integral/steepest descent approach.

Newman [**298**] tells a story of where the conjectured $\exp(C\sqrt{n})$ behavior of $p(n)$ came from. Percy MacMahon (1854–1929) was so fascinated by $p(n)$, he kept long lists of their values posted on his wall and one day noticed the shape was parabolic. The reader should understand why this is an indication of $\exp(C\sqrt{n})$ behavior.

$p(n)$ has numerous amazing mathematical properties. For example, Ramanujan showed for any $n$, one has $p(5n+4) \equiv 0 \bmod 5$. This and other relations are discussed, for example, in Berndt [**55**].

An elementary proof (i.e., without complex analysis, but more involved) that $\log p(n)/\sqrt{n} \to \pi\sqrt{2/3}$ was found by Erdős [**119**] in 1942; a textbook presentation can be found in Nathanson [**287**]. Newman [**294**] even found the constant by elementary means.

The idea of the proof we give seems to be due independently to Newman [**295**] in 1962 and to G. A. Freiman. The latter appeared first in the book of Postnikov [**321**] (Russian original, 1971), who thanks Freiman for the proof. We follow the strategy of the proof as presented in Newman [**298**], although our tactics are different: for example, we get an $O(n^{-1/2})$ error while Newman gets $O(n^{-1/4})$, Newman needlessly replaces $\log(\frac{1}{z})$ by $\frac{1}{2}\,\frac{1-z}{1+z} + O(|1-z|)$, and our calculation of the needed integral follows Newman's earlier paper [**294**].

The first proofs and most later ones rely on the Dedekind eta function defined for $z \in \mathbb{C}_+$ by

$$\eta(z) = e^{i\pi z/12} \prod_{n=1}^{\infty} (1 - e^{2\pi i n z}) \tag{15.4.92}$$

and use a modular relation $\eta$ obeys closely related to Jacobi inversion for his theta function, (10.5.48) of Part 2A. Notice that if $F$ is the generating function (15.4.36), then

$$F(e^{2\pi i z}) = e^{i\pi z/12}(\eta(z))^{-1} \tag{15.4.93}$$

Stein–Shakarchi [**365**] have a presentation of the modular function approach and use it to obtain the refined Hardy–Ramanujan formula we'll mention below.

Hardy–Ramanujan got more than the leading asymptotics. They analyzed the successively most important remaining singularities: after $z = 1$, we have $z = -1$, then the two primitive third roots of $1, \dots$. In fact, the order the roots arise is given by the Farey sequence discussed in the Notes to Section 8.3 of Part 2A! This was extended by Rademacher [**322, 323**] who got a remarkable convergent series for $p(n)$ in terms successively important asymptotically. He used the Ford circles mentioned in the same notes. Andrews [**14**] presents this expansion.

This idea of pushing a contour to the boundary of the circle of convergence and picking the successively most singular remaining points was used by Hardy–Littlewood [**176**], subsequent to the Hardy–Ramanujan work, in their work on Waring's problem. It appeared in a nascent form already in Ramanujan's notebooks before his coming to England. Despite this, it has come to be called the Hardy–Littlewood circle method and is discussed in [**276, 394, 416**].

The *Online Encyclopedia of Integer Sequences* has more on $p(n)$ [**359**].

**Problems**

1. (a) Show that (15.4.20) leads to the differential equation (15.4.22).
   (b) Solve (15.4.22) with initial condition $D(0) = 1$ to get (15.4.23).
2. Show that the solution to (15.4.26) obeys (15.4.27).
3. Show that Stirling's formula, (15.4.27), and (15.4.22) lead to (15.4.19).
4. Following Riemann [**328**], analyze the integral $I_n(a,b,c)$ for complex $x \notin [1,\infty)$.
5. Do an analysis of asymptotics of
$$F(\alpha) = \int_1^\infty \exp(-\alpha(\tfrac{1}{2}\,x^4 - ix))\,dx$$
   as $\alpha \to \infty$ in $\mathbb{R}$.
6. Let $p_d(n)$ be the number of partitions of $n$ identical objects into sets with a distinct set of points. For example, $p_d(5)$ is 3 from $\{\{\times\times\times\times\times\}\}$, $\{\{\times\times\times\},\{\times\times\}\}$, $\{\{\times\times\times\times\},\{\times\}\}$. Let $p_o(d)$ be the partitions into sets with an odd number of elements, for example $p_o(5)$ is 3 from $\{\{\times\times\times\times\times\}\}$, $\{\{\times\times\times\},\{\times\},\{\times\}\}$ and $\{\{\times\},\{\times\},\{\times\},\{\times\},\{\times\}\}$. This problem will prove a remarkable relation of Euler that $p_d(n) = p_o(n)$ for all $n$.
   (a) Prove that
$$\sum_{n=1}^\infty p_d(n)z^n = \prod_{k=1}^\infty (1+z^k), \qquad \sum_{n=1}^\infty p_o(n)z^n = \prod_{k=0}^\infty \frac{1}{1-z^{2k+1}}$$
   (b) Prove that $\prod_{k=1}^\infty (1+z^k)\prod_{k=1}^\infty (1-z^k) = \prod_{k=1}^\infty (1-z^{2k})$.
   (c) Prove that $\prod_{k=1}^\infty (1-z^k)/(1-z^{2k}) = \prod_{k=0}^\infty (1-z^{2k+1})$.
   (d) Prove that $p_d(n) = p_o(n)$.

7. This problem will use the ideas of Problem 18 of Section 2.3 of Part 2A to prove that the function, $F(z)$, of (15.4.36) has a natural boundary on $\partial\mathbb{D}$.

   (a) If $g$ is analytic and nonvanishing on $\partial\mathbb{D}$ and $\log g$ has a natural boundary on $\partial\mathbb{D}$, prove that $g$ has a natural boundary on $\partial\mathbb{D}$.

   (b) If $z = re^{2\pi ip/q}$, prove that

$$\sum_{j\equiv 0 \bmod q} \frac{1}{j}\,\frac{z^j}{1-z^j} \geq q^{-1}(1-r)^{-1}$$

   (c) If $z = re^{2\pi ip/q}$ with $p$ and $q$ relatively prime and $r < 1$ and $n \not\equiv 0 \bmod q$, prove that $|1-z^n|^2 \geq 4r^n \sin^2(\frac{\pi}{q})$ and conclude that

$$\left| \sum_{j\not\equiv 0 \bmod q} \frac{1}{j}\,\frac{z^j}{1-z^j} \right| \leq \frac{1}{2\sin(\pi/q)}\,[\log 2 - \log(1-r)]$$

   (d) Prove that $F(z)$ has a natural boundary on $\partial\mathbb{D}$.

## 15.5. The WKB Approximation

> I regard as quite useless the reading of large treatises of pure analysis: too large a number of methods pass at once before the eyes. It is in the works of applications that one must study them; one judges their ability there and one apprises the manner of making use of them.
>
> —*Joseph Louis Lagrange* (1736-1813)

Thus far, we've only discussed asymptotics of integrals and deduced asymptotics of solutions of ODEs by finding integral representations for their solutions. Here we'll look at ODEs whose solutions are not known to have integral representations, especially

$$-u''(x) + W(x)u(x) = 0 \tag{15.5.1}$$

in particular, if $W$ is a polynomial. If $\deg(W) = 1$, we have a translate of Airy's equation and we've found asymptotics, and if $\deg(W) = 2$, the solutions are parabolic cylinder functions, including the special case of harmonic oscillator wave functions where we have explicit formulae. But for $\deg(W) \geq 3$, no integral representation method is known from which to get the asymptotics as we will below. (15.5.1) will be studied when $W \in L^1$ locally where it can be interpreted as an integral equation. Alternatively, if $u$ is continuous, (15.5.1) makes sense as a statement about distributions.

The methods will determine asymptotics at infinity but not relate them directly to initial values. The ideas are also useful for other purposes, especially the method of variation of parameters (Theorem 15.5.4), which is a perturbation theory result.

Let us look for solutions that are nonvanishing near infinity. Even if the real-valued solutions are oscillatory (e.g., $W \equiv -1$, $u(x) = \sin x$ or $\cos x$), there are complex solutions (e.g., $e^{\pm ix}$ if $W \equiv -1$) that are everywhere nonvanishing. Suppose that $u(x) = \exp(Q(x))$. Logs can be unwieldy, but their derivatives are simple, so we'll focus on $Q'$; equivalently, we define $v(x)$ so that

$$u(x) = \exp\biggl(\int_{x_0}^{x} v(y)\, dy\biggr) \tag{15.5.2}$$

solves (15.5.1). Thus,

$$v(x) = \frac{u'(x)}{u(x)} \tag{15.5.3}$$

**Proposition 15.5.1.** *If $u$ is nonvanishing near $x = x_0$ and $v$ is given by* (15.5.3), *then $u$ solves* (15.5.1) *if and only if $v$ solves*

$$v'(x) = W(x) - v(x)^2 \tag{15.5.4}$$

**Remark.** (15.5.4) is called the *Riccati equation.*

**Proof.** One computes

$$v'(x) + v(x)^2 = \frac{u''(x)}{u(x)} \tag{15.5.5}$$

to see that (15.5.4) is equivalent to (15.5.1). □

Rewrite (15.5.4) as

$$v(x)^2 = W(x) - v'(x) \tag{15.5.6}$$

We'll see there are many cases where $v'$ is small compared to $v^2$, for example, if $v(x) \to c \neq 0$ and $v' \to 0$ or if $v(x) \sim x^\ell$ ($\ell > 0$) with $v'(x) \sim x^{\ell-1}$, so $v'/v^2 \sim x^{\ell-1}/x^{2\ell} = O(x^{-\ell-1})$. Thus,

$$\begin{aligned} \sqrt{W(x) - v'(x)} &= \sqrt{W(x)}\left(1 - \frac{v'(x)}{W(x)}\right)^{1/2} \\ &\sim \sqrt{W(x)} - \frac{1}{2}\,\frac{v'(x)}{\sqrt{W(x)}} \end{aligned} \tag{15.5.7}$$

To first-order, $v$ is $\sqrt{W(x)}$, so

$$\frac{v'(x)}{\sqrt{W(x)}} \sim \left(\frac{W'(x)}{2W(x)^{1/2}}\right)\frac{1}{W(x)^{1/2}} = \frac{1}{2}\,\frac{d}{dx}\,(\log W(x))$$

and thus, (15.5.6)/(15.5.7) become

$$v(x) \sim \sqrt{W(x)} - \frac{1}{4}\frac{d}{dx}\log|W(x)| \tag{15.5.8}$$

We replaced $(\log(W(x))$ by $\log(|W(x)|)$ since $\log(\operatorname{sgn} W)$ is constant in a region where $W$ doesn't change sign and has zero derivative. Using $\pm$ for the two branches of square root (and dropping a constant $\sqrt{W(x_0)}$), we thus find "approximate" solutions to (15.5.1)

$$b_\pm(x) = |W(x)|^{-1/4}\exp\biggl(\int_{x_0}^x \mp\sqrt{W(y)}\,dy\biggr) \tag{15.5.9}$$

This is only expected to be "valid" away from zeros of $W$ which are called *turning points*. The approximation (15.5.9) is known as the WKB *approximation*, although it goes back to Carlini, Liouville, and Green a hundred years earlier than the work of the men whose initials are W, K, and B (see the Notes). We will see that in many cases, $u_\pm(x)$ give the asymptotics of the true solutions of (15.5.1) as $x \to \infty$.

For definiteness in regions where $W(x) > 0$, we interpret $\sqrt{W(y)}$ to be positive, and if $W(x) < 0$, $\operatorname{Im}\sqrt{W(y)} > 0$. Remarkably, if $W$ is $C^2$, in regions where $W(x) > 0$ or where $W(x) < 0$, $b_+$ and $b_-$ solve the same differential equation

$$-b_\pm'' + \widetilde{W} b_\pm = 0 \tag{15.5.10}$$

for a fixed $\widetilde{W}$ distinct from $W$.

Here are the calculations. By differentiating $\log u_\pm$,

$$b_\pm'(x) = Q_\pm(x) b_\pm(x), \qquad Q_\pm(x) = \mp W(x)^{1/2} - \frac{1}{4}\frac{W'(x)}{W(x)} \tag{15.5.11}$$

Thus,

$$b_\pm''(x) = \widetilde{W}_\pm(x) b_\pm(x), \qquad \widetilde{W}_\pm(x) = Q_\pm'(x) + Q_\pm^2(x) \tag{15.5.12}$$

A simple calculation (Problem 1) shows that $\widetilde{W}_+ = \widetilde{W}_- \equiv \widetilde{W}$, where

$$\widetilde{W}(x) = W(x) + \frac{5}{16}\biggl(\frac{W'(x)}{W(x)}\biggr)^2 - \frac{1}{4}\frac{W(x)''}{W(x)} \tag{15.5.13}$$

We summarize with

**Theorem 15.5.2.** *Let $W$ be a $C^2$ function nonvanishing in $(a,b)$. Define $b_\pm(x)$ by* (15.5.11) *for some $x_0 \in (a,b)$. Then, with $\widetilde{W}$ given by* (15.5.13), *we have that on $(a,b)$,*

$$-b_\pm''(x) + \widetilde{W}(x) b_\pm(x) = 0 \tag{15.5.14}$$

**Remark.** That $\widetilde{W}_+ = \widetilde{W}_-$ is equivalent to $b_+''/b_+ = b_-''/b_-$ which, in turn, is equivalent to $(b_+'b_- - b_-'b_+)' = 0$ which, in turn, is equivalent to $b_+'b_- - b_-'b_+ = (Q_+ - Q_-)b_+b_-$ is constant, which it is.

We emphasize that $\widetilde{W} - W$ is often quite small as $x \to \infty$. For example, if $W$ is a polynomial of degree $d$, $W'/W$ is $O(x^{-1})$ and $W''/W$ is $O(x^{-2})$, so $\widetilde{W} - W$ is $O(x^{-2})$ and, in particular, is $L^1$ at infinity. We are heading towards a proof that, under some circumstances, $W$ and $\widetilde{W}$ have solutions asymptotic at infinity, that is, (15.5.1) has solutions $u_\pm$ with either $u_\pm/b_\pm \to 1$ at infinity or $|u_\pm - b_\pm| \to 0$ at infinity. We'll think of $W$ as a perturbation of $\widetilde{W}$. Writing a second-order equation as a vector-valued first-order equation, the following will thus be one key idea:

**Theorem 15.5.3.** *Let $A(x)$ be a $k \times k$ complex-valued continuous function with*

$$\int_{x_0}^\infty \|A(x)\|\, dx < \infty \tag{15.5.15}$$

*Then for any $\beta \in \mathbb{C}^k$, there exists a unique $C^1$ function, $\varphi(x)$, on $(x_0, \infty)$ with value in $\mathbb{C}^k$ so that*

$$\varphi'(x) = A(x)\varphi(x), \qquad \varphi(x) \to \beta \quad \text{as } x \to \infty \tag{15.5.16}$$

**Proof.** Since we have solubility on finite intervals (see Theorem 1.4.7 of Part 2A), we only need to establish (15.5.16) near infinity. Pick $x_1$ so

$$a \equiv \int_{x_1}^\infty \|A(x)\|\, dx < 1 \tag{15.5.17}$$

Suppose we find $\varphi$ continuous, $\mathbb{C}^k$-valued, and bounded on $(x_1, \infty)$ so that

$$\varphi = \mathcal{T}\varphi, \qquad \mathcal{T}\psi(x) \equiv \beta - \int_x^\infty A(y)\psi(y)\, dy \tag{15.5.18}$$

$\mathcal{T}$ is a continuous map of all bounded, $\mathbb{C}^k$-valued, continuous functions, $C((x_1, \infty); \mathbb{C}^k)$ to itself. For any such $\psi$, $\mathcal{T}\psi(x) \to \beta$ as $x \to \infty$ and $\mathcal{T}\psi$ is $C^1$ with $(\mathcal{T}\psi)' = A\psi$. Thus, $\varphi = \mathcal{T}\varphi$ is equivalent to (15.5.16).

By (15.5.17), we have

$$\|\mathcal{T}\psi_1 - \mathcal{T}\psi_2\|_\infty \le a\|\psi_1 - \psi_2\|_\infty \tag{15.5.19}$$

so $\mathcal{T}$ is a contraction—implying there is a unique $\varphi$ solving $\varphi = \mathcal{T}\varphi$ (see Theorem 5.12.4 of Part 1). $\square$

**Remark.** In fact, $\varphi$ has a series expansion by iterating $\mathcal{T}$ explicitly, namely,

$$\varphi(x) = \sum_{n=0}^\infty \varphi_n(x), \quad \varphi_0(x) \equiv \beta, \quad \varphi_{n+1}(x) = -\int_x^\infty A(y)\varphi_n(y)\, dy \tag{15.5.20}$$

Because $A$ is continuous, solutions of (15.5.18) are $C^1$ and solve the ODE in the classical sense. Below we'll write (15.5.1) as an integral equation which makes sense even if $W$ is not continuous; indeed, $W$ locally $L^1$ is enough. We'll discuss "solutions" of (15.5.1) to mean solutions in the sense of the integral equation. One can show such a $u$, while not $C^2$, is $C^1$ and so bounded. Thus, $Wu$ is a locally $L^1$ function and $u'$ has a distributional derivative equal to $Wu$ (in the sense of ordinary, not necessarily tempered) distributions. Put differently, $u$ is $C^1$ and obeys (15.5.1) in the sense that if $\varphi \in C_0^\infty$, then $\int [-\varphi''(x)u(x) + \varphi(x)W(x)u(x)]\, dx = 0$.

(15.5.1) is equivalent to

$$\varphi'(x) = A_0(x)\varphi(x), \quad \varphi(x) = \begin{pmatrix} u'(x) \\ u(x) \end{pmatrix}, \quad A_0(x) = \begin{pmatrix} 0 & W(x) \\ 1 & 0 \end{pmatrix} \tag{15.5.21}$$

and because of the 1, this is never $L^1$ at infinity, so at first sight, Theorem 15.5.3 seems irrelevant to second-order equations. That it isn't depends on a method often called *variation of parameters* (a term also sometimes used for something distinct but related; see the Notes).

We start with two comparison functions, $c_\pm(x)$. These are often taken as the WKB solutions, but they are sometimes other putative approximations. In particular, they need not solve the same ODE as each other. What is critical is that at each $x$, $\binom{c_+'(x)}{c_+(x)}$ and $\binom{c_-'(x)}{c_-(x)}$ are linearly independent in $\mathbb{C}^2$, that is, their Wronskian

$$\omega(x) \equiv c_+'(x)c_-(x) - c_-'(x)c_+(x) \tag{15.5.22}$$

is never zero. Thus, we can expand

$$\begin{pmatrix} u'(x) \\ u(x) \end{pmatrix} = \alpha(x)\begin{pmatrix} c_+'(x) \\ c_+(x) \end{pmatrix} + \beta(x)\begin{pmatrix} c_-'(x) \\ c_-(x) \end{pmatrix} \tag{15.5.23}$$

**Theorem 15.5.4** (Variation of Parameters). *If $\alpha, \beta$ are related to $c_\pm$ and $u$ by (15.5.23), then $u''(x) = W(x)u(x)$ is equivalent to*

$$\begin{pmatrix} \alpha \\ \beta \end{pmatrix}'(x) = A(x)\begin{pmatrix} \alpha \\ \beta \end{pmatrix}(x) \tag{15.5.24}$$

*where ($\omega$ given by (15.5.22))*

$$A(x) = \omega(x)^{-1}\begin{pmatrix} -c_-(x)f_+(x) & -c_-(x)f_-(x) \\ c_+(x)f_+(x) & c_+(x)f_-(x) \end{pmatrix} \tag{15.5.25}$$

*with*

$$f_\pm(x) = -c_\pm''(x) + W(x)c_\pm(x) \tag{15.5.26}$$

*If $c_\pm$ obey*

$$c_\pm''(x) = Z_\pm(x)c_\pm(x) \tag{15.5.27}$$

*then*

$$A(x) = \omega(x)^{-1} \begin{pmatrix} \delta_+(x)c_+(x)c_-(x) & \delta_-(x)c_-(x)^2 \\ -\delta_+(x)c_+(x)^2 & -\delta_-(x)c_+(x)c_-(x) \end{pmatrix} \tag{15.5.28}$$

*where*

$$\delta_\pm(x) = Z_\pm(x) - W(x) \tag{15.5.29}$$

**Proof.** We provide the proof modulo straightforward but tedious calculations left to the reader (Problem 2). (15.5.28) is immediate from (15.5.25) and the definition (15.5.27) of $Z_\pm$, so we need only prove (15.5.25).

Let $M$ be given by

$$M(x) = \begin{pmatrix} c'_+(x) & c'_-(x) \\ c_+(x) & c_-(x) \end{pmatrix} \tag{15.5.30}$$

$\Gamma(x) = \binom{\alpha}{\beta}(x)$, $\Phi(x) = \binom{u'}{u}(x)$, and $A_0$ by the last equation in (15.5.21). Then

$$M(x)^{-1} = \omega(x)^{-1} \begin{pmatrix} c_-(x) & -c'_-(x) \\ -c_+(x) & c'_+(x) \end{pmatrix} \tag{15.5.31}$$

We have, by (15.5.24),

$$\Phi = M\Gamma, \qquad \Gamma = M^{-1}\Phi \tag{15.5.32}$$

and $\Phi' = M'\Gamma + M\Gamma'$ or

$$\Gamma' = -M^{-1}M'\Gamma + M^{-1}\Phi' \tag{15.5.33}$$

Thus,

$$\Phi' = A_0\Phi \Leftrightarrow \Gamma' = A\Gamma \tag{15.5.34}$$

where

$$A = M^{-1}A_0M - M^{-1}M' \tag{15.5.35}$$

(15.5.25) is then the explicit calculation of (15.5.35). □

**Example 15.5.5** (Weyl's Limit Point/Limit Circle Classification). Let $(a, b) \subset \mathbb{R}$ where $a$ can be $-\infty$ and/or $b$ $+\infty$. Let $u_1, u_2$ be two distinct solutions (i.e., not multiples of one another) of

$$-u'' + Vu = E_0u \tag{15.5.36}$$

where $V$ is locally $L^1$ and $E_0 \in \mathbb{C}$. Suppose $u_j$ is "$L^2$ at $b$," that is, for $c \in (a, b)$,

$$\int_c^b (|u_j(y)|^2 + (|u'_j(y)|^2)\, dy < \infty \tag{15.5.37}$$

Now suppose $E_1 \in \mathbb{C}$ and we seek solutions of

$$-w'' + Vw = E_1w \tag{15.5.38}$$

We can apply Theorem 15.5.4, where now $\delta_\pm(x) = E_0 - E_1$, $\omega(x)$ is a nonzero constant, and by (15.5.37),

$$\int_c^b \|A(y)\| \, dy < \infty \tag{15.5.39}$$

Thus, we can use Theorem 15.5.3 to construct solutions $w_j$ of (15.5.38) where $w_j(x) = \alpha_j(x)u_1(x) + \beta_j(x)u_2(x)$ and $\alpha_j, \beta_j$ bounded near $b$. Thus, we have proven

$$\text{All solutions of (15.5.36) } L^2 \text{ at } b \text{ for one } E_0 \Rightarrow \text{the same for all } E_0 \tag{15.5.40}$$

□

This result is important in the self-adjointness theory of $-\frac{d^2}{dx^2}+V(x)$ and is discussed further in Section 7.4 of Part 4 where the name limit point/limit circle will be explained.

**Example 15.5.6** ($V \in L^1$ solved at positive energy). We are going to look at

$$-u''(x) + V(x)u(x) = Eu(x) \tag{15.5.41}$$

where $E$ is a real constant ("energy"). Suppose $V$ is $L^1$ and $E > 0$. When $V \equiv 0$, the solutions are (here $k = \sqrt{E} > 0$)

$$c_\pm(x) = e^{\pm ikx} \tag{15.5.42}$$

Then $W(x) = V(x) - k^2$, $Z_\pm(x) \equiv k^2$, $\omega(x) \equiv 2ik$, and by (15.5.28),

$$A(x) = (2ik)^{-1}V(x)\begin{pmatrix} -1 & -e^{-2ikx} \\ e^{2ikx} & 1 \end{pmatrix} \tag{15.5.43}$$

is in $L^1$ if $V$ is. We conclude for $E > 0$ and $V$ in $L^1$ that (15.5.41) has solutions $u_\pm$ with $|u_\pm(x) - e^{\pm ikx}| + |u'_\pm(x) \mp ik\, e^{\pm ikx}| \to 0$ as $x \to \infty$. □

**Example 15.5.7.** ($V = V_1 + V_2$, $V_1 \in L^1$, $V_2' \in L^1$, $V_2 \to 0$ at infinity solved at positive energy). We look at (15.5.41) again, but now $V$ is not $L^1$—rather $V = V_1 + V_2$, where $V_1 \in L^1$, $V_2$ is $C^1$, $V_2(x) \to 0$ at infinity, and $V_2' \in L^1$ (e.g., $V(x) = (1+x)^{-\alpha}$, any $\alpha > 0$ and if $0 < \alpha < 1$, $V$ is not $L^1$). Let $W_2(x) = V_2(x) - k^2$. Then $k^{1/2}|W_2(x)|^{-1/4} \to 1$, so in WKB, the $|W_2(x)|^{-1/4}$ shouldn't be important. We thus pick $x_1$ so

$$V_2(x) < \tfrac{1}{2}k^2 \quad \text{if } x > x_1 \tag{15.5.44}$$

and, on $(x_1, \infty)$, define

$$c_\pm(x) = \exp\biggl(\int_{x_1}^x \pm i\sqrt{k^2 - V_2(y)}\, dy\biggr) \tag{15.5.45}$$

By dropping the $W_2^{-1/4}$, we'll only need $C^1$ conditions, not $C^2$. (However, by moving some of $V_2$ into $V_1$, one can always find a new decomposition with $V_2$ $C^2$ and push through $b_\pm$ (see Problem 4). Indeed, one can handle $V_2$ of bounded variation rather than $C^1$; see Problems 3 and 4.)

Straightforward calculations show that

$$Z_\pm(x) = \mp \frac{iV_2'(y)}{2\sqrt{k^2 - V_2(y)}} + V_2(y) - k^2 \tag{15.5.46}$$

Thus, with $W(x) = V(x) - k^2$,

$$W(x) - Z_\pm(x) = V_1(x) \pm \frac{iV_2'(y)}{2\sqrt{k^2 - V_2(y)}} \tag{15.5.47}$$

lies in $L^1$ and $A(x)$, given by (15.5.28), is in $L^1$.

It follows by Theorem 15.5.3 that there are solutions $u_\pm$ of (15.5.1) obeying

$$|u_\pm(x) - c_\pm(x)| + |u_\pm'(x) - c_\pm'(x)| \to 0 \tag{15.5.48}$$

Notice that $|u_\pm(x)| \to 1$, but the phase is not asymptotically $\pm ikx$ if $V_2 \notin L^1$ but has a "WKB correction." □

If $E < 0$, say $E = -\kappa^2$, it is natural (in the pure $V \in L^1$ case) to take $c_\pm(x) = e^{\mp\kappa x}$. Now it appears the method breaks down since $c_-(x)^2 = e^{2\kappa x}$ is huge and $V \in L^1$ doesn't make $A \in L^1$.

Before giving up hope though, note that $A$ now has the form

$$A(x) = \begin{pmatrix} b_{11}(x) & b_{12}(x)g(x)^{-1} \\ b_{21}(x)g(x) & b_{22}(x) \end{pmatrix} \tag{15.5.49}$$

where $B(x)$ is $L^1$ at infinity and $g(x)$ (which is $e^{-2\kappa x}$ in this case) goes to zero (so $g^{-1}(x)$ goes to infinity). $A\binom{\alpha_0}{\beta_0}$ has top component $b_{11}(x)\alpha_0 + b_{12}(x)g^{-1}(x)\beta_0$ which is not, in general, $L^1$ if $\beta_0 \neq 0$, but if $\beta_0 = 0$, it is. In second-order, we'll have $A(x)A(y)\binom{1}{0}$ for $x < y$. There are now some $g^{-1}(x)$'s but only in the top component, and they are actually multiplied by $g(y)$ and so are harmless if $g(x) \geq g(y)$, that is, $g$ is monotone. Thus, we might hope to get solutions asymptotic to $c_+(x)$ this way but not (at least by this method) $c_-$. The key is the following:

**Theorem 15.5.8.** *Suppose $A$ has the form* (15.5.49), *where $B$ is a matrix with $\int_{x_0}^\infty \|B(y)\|\, dy < \infty$, and $g(x)$ is monotone decreasing to 0 on $(x_0, \infty)$. Then there exists a solution of*

$$\varphi'(x) = A(x)\varphi(x), \qquad \varphi(x) \to \begin{pmatrix} 1 \\ 0 \end{pmatrix} \tag{15.5.50}$$

*and which also obeys*

$$g^{-1}(x)\varphi_2(x) \to 0 \quad \text{as } x \to \infty \tag{15.5.51}$$

**Proof.** Pick $x_1$ so that

$$\int_{x_1}^{\infty} [|b_{11}(x)| + |b_{12}(x)| + |b_{21}(x)| + |b_{22}(x)|]\, dx \le \tfrac{1}{2} \tag{15.5.52}$$

Let $X$ be the Banach space of functions $\varphi\colon (x_1, \infty) \to \mathbb{C}^2$ that obey (15.5.51) and are bounded. Put the norm on $X$:

$$\|\varphi\| = \sup_{x \ge x_1} [|\varphi_1(x)| + g^{-1}(x)|\varphi_2(x)|] \tag{15.5.53}$$

It is easy to see that $X$ is complete in this norm.

Define $\mathcal{T}\colon X \to X$ by the second equation in (15.5.18) with $\beta = \binom{1}{0}$. It is easy to see that the integral in (15.5.18) exists since $g^{-1}\varphi_2$ is bounded, and that (with $\|\cdot\|$ given by (15.5.53))

$$\left\| \mathcal{T}\varphi - \begin{pmatrix} 1 \\ 0 \end{pmatrix} \right\| \le \tfrac{1}{2}\, \|\varphi\| \tag{15.5.54}$$

by (15.5.52) and $g(x)^{-1} \le g(y)^{-1}$ if $y \ge x$. Also, by the integral expressions and $g(x)^{-1} \le g(y)^{-1}$, we have $g^{-1}(x)(\mathcal{T}\varphi)_2(x) \to 0$ so that $\mathcal{T}$ is a map onto $X$.

Since $\mathcal{T}\varphi - \mathcal{T}\psi = [\mathcal{T}\varphi - \binom{1}{0}] - [\mathcal{T}\psi - \binom{1}{0}] = \mathcal{T}(\varphi - \psi) - \binom{1}{0}$, (15.5.54) implies that $\mathcal{T}$ is a contraction, so there is a $\varphi \in X$ that obeys $\varphi = \mathcal{T}\varphi$. This has all the required properties once extended from $(x_1, \infty)$ to $(x_0, \infty)$ by local solubility (see Theorem 1.4.7 of Part 2A). □

The above will let us construct one solution of certain second-order ODEs. One can get a second from

**Theorem 15.5.9.** *Let $u_1$ solve (15.5.1) in $(a, b)$ with $u_1$ positive there. Then, for $x_0 \in (a, b)$,*

$$u_2(x) = u_1(x) \int_{x_0}^{x} \frac{1}{u_1(y)^2}\, dy \tag{15.5.55}$$

*is a second linearly independent solution.*

**Proof.** $u_2 = u_1 f$ obeys

$$\begin{aligned} u_2'' &= W u_1 f + 2u_1' f' + u_1 f'' \\ &= W u_2 + u_1 f' \left[ 2\,\frac{u_1'}{u_1} + \frac{(f')'}{f'} \right] \end{aligned} \tag{15.5.56}$$

so $u_2$ obeys (15.5.1) if and only if

$$2\,\frac{u_1'}{u_1} + \frac{(f')'}{f'} = \frac{d}{dx}\,[\log f' + \log u_1^2] = 0 \tag{15.5.57}$$

This happens if and only if $f' = c_1 u_1^{-2}$, that is, $f = [c_1 \int_{x_0}^{x} 1/u_1(y)^2] + c_2$. This shows $u_2$ given by (15.5.55) solves (15.5.1) (and explains where the choice (15.5.55) comes from).

$u_2$ is independent of $u_1$, since $(u_2/u_1)' = u_1^{-2}$ is not zero. $\square$

Before turning to examples, we want to note something about the general scheme of applying these last two theorems in the context of (15.5.28). We'll apply this method when $c_\pm > 0$ near infinity. We'll need to have

$$\omega(x)^{-1}(Z_\pm(x) - W(x))c_+(x)c_-(x) \in L^1 \quad \text{at infinity} \tag{15.5.58}$$

We take

$$g(x) = \frac{c_+(x)}{c_-(x)} \tag{15.5.59}$$

so we'll need $g \to 0$ monotonically.

Theorems 15.5.4 and 15.5.8 will then give a solution of the form (15.5.23) with $\alpha(x) \to 1$ and $\beta(x)c_1(x)/c_+(c) \to 0$. Thus,

$$\frac{u(x)}{c_+(x)} = \alpha(x) + \beta(x)\,\frac{c_-(x)}{c_+(x)} \to 1 \tag{15.5.60}$$

We'll also have that

$$\frac{u_+'(x)}{c_+'(x)} = \alpha(x) + \beta(x)\,\frac{c_-'(x)}{c_+'(x)} \to 1 \tag{15.5.61}$$

so long as have a constant $K$ with

$$\left|\frac{c_-'(x)}{c_-(x)}\right| \le K \left|\frac{c_+'(x)}{c_+(x)}\right| \tag{15.5.62}$$

near infinity. Theorem 15.5.9 will then give a second solution which one can sometimes show behaves like $c_-$.

**Example 15.5.10** ($V \in L^1$ solved at negative energy)**.** We have (15.5.41) with $V$ in $L^1$, but now $E = -\kappa^2 < 0$, where $\kappa > 0$. It is natural to take

$$c_\pm(x) = e^{\mp\kappa x} \tag{15.5.63}$$

Then $\omega(x) = -2\kappa$, $c_+c_- \equiv 1$, $|Z_\pm - W| = |V|$, so (15.5.58) is just $V \in L^1$. $g(x) = e^{-2\kappa x}$ is indeed monotone and (15.5.62) holds with $K = 1$. We thus have a solution $u_+$ with

$$\left|\frac{u_+(x)}{c_+(x)} - 1\right| + \left|\frac{u_+'(x)}{c_+'(x)} - 1\right| \to 0 \tag{15.5.64}$$

Define $u_-(x) = 2\kappa u_+(x)\,f(x)$, where $f(x) = \int_{x_0}^{x} u_+(y)^{-2}\,dy$. Then by (15.5.64) and $\int_{x_0}^{x} e^{2\kappa y}\,dy = (2\kappa)^{-1}(e^{2\kappa y} - 1)$, we have

$$\frac{u_-(x)}{c_-(x)} \to 1 \tag{15.5.65}$$

Moreover, by the formula for $u_-$, we have that

$$\frac{u_-'}{u_-} = \frac{u_+'}{u_+} + \frac{2\kappa}{u_+ u_-} \to -\kappa + 2\kappa = \kappa$$

that is,

$$\frac{u_-'(x)}{c_-'(x)} \to 1 \tag{15.5.66}$$

□

**Example 15.5.11** ($xV \in L^1$ solved at zero energy)**.** If $W(x) \to 0$, that is, $W(x) = V(x) - E$ with $E = 0$, then it is natural to take

$$c_+(x) = 1, \qquad c_-(x) = x \tag{15.5.67}$$

$Z_\pm = 0$ and $\omega(x) = -1$. Since $c_+ c_- = x$, we need more than $V \in L^1$. Instead, we need

$$\int_{x_0}^\infty x|V(x)|\, dx < \infty \tag{15.5.68}$$

(15.5.62) doesn't hold, but we don't expect that $u_+'/c_+' \to 0$ since $c_+' \equiv 0$. However, $u_+'(x) = \beta(x) = o(1/x)$, so we have a solution $u_+$ with

$$|u_+(x) - 1| + |xu_+'(x) - 1| \to 0 \tag{15.5.69}$$

As for the second solution, $f(x) \equiv \int_{x_0}^x 1/u_+(y)\, dy$ has $f(x)/x \to 1$, so we get a solution $u_-(x)$ with

$$\left|\frac{u_-(x)}{x} - 1\right| + |u_-'(x) - 1| \to 0 \tag{15.5.70}$$

□

We summarize the $V \in L^1$ case in

**Theorem 15.5.12.** *Let $V(x) \in L^1(x_0, \infty)$. Then*

(a) *For $E = k^2 > 0$, there are solutions, $u_\pm$, of* (15.5.41) *obeying*

$$|u_\pm(x) - e^{\pm ikx}| + |u_\pm'(x) \mp ik\, e^{\pm ikx}| \to 0 \tag{15.5.71}$$

(b) *For $E = -\kappa^2 < 0$, there are solutions, $u_\pm$, of* (15.5.41) *obeying*

$$|u_\pm(x)e^{\pm\kappa x} - 1| + |u_\pm'(x)e^{\pm\kappa x} \mp \kappa| \to 0 \tag{15.5.72}$$

(c) *If, in addition,* (15.5.68) *holds, then for $E = 0$, there are solutions, $u_\pm$, obeying*

$$|u_+(x) - 1| + |xu_+'(x) - 1| + \left|\frac{u_-(x)}{x} - 1\right| + |u_-'(x) - 1| \to 0 \tag{15.5.73}$$

We'll leave the $V = V_1 + V_2$ with $V_1 \in L^1$, $V_2' \in L^1$, $V_2 \to 0$ at infinity and $E < 0$ to the Problems (see Problem 5).

**Example 15.5.13** (Bessel's Equation). We are interested in solutions of

$$-f'' - \frac{1}{x} f' \pm \frac{\alpha^2}{x^2} f = \pm f \tag{15.5.74}$$

The $+f$ equation is solved by $J_\alpha, Y_\alpha, H_\alpha^{(1)}, H_\alpha^{(2)}$ and the $-f$ equation by $K_\alpha$ and $I_\alpha$. This doesn't have the form of (15.5.42) because of the $x^{-1}f$ term. Since $(hf)'' = hf'' + 2h'f' + h''f$, if we pick $h$ so $2h'/h = x^{-1}$, that is, $h = x^{1/2}$, then $u = hf$ will not have a first derivative term. We thus pick

$$u(x) = x^{1/2} f(x) \tag{15.5.75}$$

and find (15.5.74) is equivalent to

$$-u'' + V_\pm u = \pm u, \qquad V_\pm(x) = (-\tfrac{1}{4} \pm \alpha^2) x^{-2} \tag{15.5.76}$$

Theorem 15.5.12 is applicable (since $E = \pm 1$, $V \in L^1(1, \infty)$ is required; we don't care about $E = 0$).

For the $+$ equation, $u$ has solutions asymptotic to $e^{\pm ix}$; for the $-$ equation, to $e^{\pm x}$. We see that (15.5.74) has solutions asymptotic to $x^{-1/2} e^{\pm ix}$ for the $H_\alpha^{(1)}$ equation and to $x^{-1/2} e^{-x}$ for the $K_\alpha$ equation, consistent with (14.5.26), (15.2.57), and (15.2.60). □

We turn finally to the case $W \to \pm\infty$ at infinity, including the promised polynomial case. The key condition will be

$$\int_{x_0}^{\infty} \left[ \frac{|W'(x)|^2}{|W(x)|^{5/2}} + \frac{|W''(x)|}{|W(x)|^{3/2}} \right] dx < \infty \tag{15.5.77}$$

which, in terms of the $\widetilde{W}$ of (15.5.13), implies

$$\int_{x_0}^{\infty} |W(x)|^{-1/2} |W(x) - \widetilde{W}(x)| \, dx < \infty \tag{15.5.78}$$

As already noted, (15.5.77) holds if $W$ is a polynomial, but it also holds in some cases with some strong oscillations, for example, $W(x) = -x^4(2 + \cos x)$ or even $W(x) = -x^8(2 + \cos x^2)$. We need to separately consider the cases $W(x) \to -\infty$ or $W(x) \to \infty$. We construct the decaying solution in the latter case and leave the growing solution to the Problems (see Problem 6).

**Theorem 15.5.14.** *Let $W(x)$ be a $C^2$ function with $W(x) \to -\infty$ as $x \to \infty$ and so that* (15.5.77) *holds. Let $b_\pm$ be the* WKB *approximations* (15.5.9). *Then* (15.5.1) *has solutions, $u_\pm(x)$, so that*

$$|b_\pm(x)|^{-1} |u_\pm(x) - b_\pm(x)| \to 0 \tag{15.5.79}$$

*If, moreover,*

$$\frac{|W'(x)|}{|W(x)|^{3/2}} \to 0 \tag{15.5.80}$$

*then*

$$|b'_\pm(x)|^{-1}|u'_\pm(x) - b'_\pm(x)| \to 0 \tag{15.5.81}$$

**Remarks.** 1. If $W$ is a polynomial with $W(x) \to -\infty$ and (15.5.77) and (15.5.80) hold, we have WKB solutions.

2. $|c_\pm(x)| = |W(x)|^{-1/4}$. If (15.5.80) holds, $|c_\pm(x)'|$ can be replaced by $|W(x)|^{1/4}$ in (15.5.59).

**Proof.** Let $c_\pm = b_\pm$. By (15.5.77), we have (15.5.78) and so, by Theorems 15.5.4 and 15.5.3, there are solutions of the form (15.5.23) with $\alpha(x) \to 1$, $\beta(x) \to 0$. Thus, since $|c_+|/|c_-| = 1$, we have (15.5.57).

If (15.5.80) holds and $Q_\pm$ are given by (15.5.13), then $|Q_\pm(x)|/|W_\pm(x)|^{1/2} \to 1$, so $|c'_+/c'_-| \to 1$. Thus, (15.5.62) and so (15.5.81) holds. □

Given the constancy of the Wronskians of two solutions, one might have guessed that (15.5.1) cannot have independent solutions both going to zero at infinity, but if $W$ goes to $-\infty$ and is so regular that (15.5.77) holds, then it does. This can happen because the constancy of the Wronskian is saved by derivatives going to infinity. It can even happen (e.g., $W(x) = -x^\alpha$ ($\alpha > 2$)) that there are multiple solutions $L^2$ at infinity. This will have dramatic consequences (see Section 7.4 in Part 4).

**Theorem 15.5.15.** *Let $W(x)$ be a $C^2$ function with $W(x) \to \infty$ as $x \to \infty$ and so that* (15.5.77) *holds. Let $b_\pm$ be the* WKB *solutions ($b_+$ is the decaying one). Then* (15.5.1) *has a solution, $u_+$, with*

$$\frac{u_+(x)}{b_+(x)} \to 1 \tag{15.5.82}$$

*as $x \to \infty$. If* (15.5.80) *holds, then also*

$$\frac{u'_+(x)}{b'_+(x)} \to 1 \tag{15.5.83}$$

**Remark.** In most cases, for example, for some $\alpha > 0$, $W(x)/x^\alpha$ is bounded above and away from zero, then $u_+(x) \to 0$ and $u'_+(x) \to 0$ and, by Wronskian arguments, there is a unique such solution.

**Proof.** Let $g(x) \equiv c_+(x)/c_-(x) = \exp(-2\int_{x_0}^x \sqrt{W(y)}\,dy)$, where $x_0$ is chosen so $W(y) > 0$ for $y > x_0$. Then $g \to 0$ since $W(y) \to \infty$ and is monotone.

By (15.5.77) and Theorems 15.5.4 and 15.5.8, there is a solution, $u_+$, of the form (15.5.23) where $\alpha(x) \to 1$ and $\beta(x)c_-(x)/c_+(x) \to 0$. Thus, (15.5.61) holds.

By (15.5.80) with $Q_\pm$ given by (15.5.13), we have $Q_+/Q_- \to 1$, so by (15.5.13), we have (15.5.62). Thus, (15.5.83) holds. □

We end this section with some examples: two connected with special functions and one remark about the more general cases.

**Example 15.5.16** (Airy Functions)**.** $W(x) = x$. We have both (15.5.77) and (15.5.80). $|W(x)|^{1/4}$ is $|x|^{1/4}$ "explaining" the prefactor and $\int_0^x \sqrt{|y|}\,dy = \frac{2}{3}x^{3/2}$ "explaining" the exponential. As $x \to \infty$, the decaying solution has $u(x) \sim x^{-1/4}\exp(-\frac{2}{3}x^{3/2})$ and $u'(x) \sim x^{1/4}\exp(-\frac{2}{3}x^{3/2})$. As $x \to -\infty$, all real-valued solutions are asymptotic to $c_1|x|^{-1/4}\cos(\frac{2}{3}|x|^{3/2} + c_2)$ for suitable $c_1$ and $c_2$ □

**Example 15.5.17** (Harmonic Oscillator Basis)**.** Recall (see Section 6.4 of Part 1) that the harmonic oscillator basis solved

$$-u''(x) + x^2 u(x) = Eu(x) \tag{15.5.84}$$

for $E = 2n+1$, $n = 0, 1, 2, \dots$. Thus, $W(x) = x^2 - E$ which obeys (15.5.77) and (15.5.80). $|W(x)|^{1/4} = x^{1/2}$ and $W(x)^{1/2} = x(1 - E/x^2)^{1/2} = x - E/2x + O(x^{-3/2})$. Since the $O(x^{-3/2})$ is integrable, it contributes an additive constant asymptotically to the integral, so

$$\int_{x_0}^{x} \sqrt{W(y)}\,dy = \frac{1}{2}\,x^2 - \frac{E}{2}\,\log x + C + o(1) \tag{15.5.85}$$

Thus, the decaying solution of (15.5.83) is asymptotic to

$$\widetilde{C}|x|^{(E-1)/2}\exp(-\tfrac{1}{2}\,x^2) \tag{15.5.86}$$

For $E = 2n+1$, this is consistent with solutions of the form $q_n(x)e^{-\frac{1}{2}x^2}$ with $q_n$ a polynomial of degree $n$. □

**Example 15.5.18** (Degree $P > 2$)**.** Now $V(x)$ is a polynomial of degree $d > 2$. If $W(x) = V(x) - E$, the main difference is that $\sqrt{W(x)} = \sqrt{V(x)}\,(1 - E/V(x))^{1/2} = \sqrt{V(x)} + O(|V(x)|)^{-1/2}$. Since $d > 2$, $O(|V(x)|^{-1/2})$ is in $L^1$. Thus, if $u(x, E)$ is the decaying solution, we have

$$\frac{u(x,E)}{u(x,E')} \to \text{constant} \tag{15.5.87}$$

Thus, in the case of degree 2, the leading asymptotics is $E$-dependent, but it is $E$-independent if $d > 2$. □

**Notes and Historical Remarks.** The approximation (15.5.9), at least in one special case, goes back at least to Carlini [**71**] in 1817, but the general form is associated mainly with 1837 work of Green [**162**] and Liouville [**263**], so much so that it is sometimes called the Liouville–Green approximation (see Schlissel [**340**] for an historical review).

In the aftermath of the discovery of quantum mechanics, three physicists, Brillouin [**65**], Kramers [**231**], and Wentzel [**407**], independently developed the connection formulae we'll discuss shortly. While the basic formulae (15.5.8) predated this work by a hundred years, the name WKB or BWK or BKW approximation stuck. In fact, shortly before this work, the connection formulae were found by the mathematician Jeffreys [**219**], so the method is sometimes called the JWKB or WKBJ approximation.

The WKB approximation in the complex plane sheds light on Stokes' phenomenon. $\exp(\int_{z_0}^{z} \sqrt{W(y)}\,dy)$ can be defined along a contour integral. If the contour is in a region where $\operatorname{Re}\sqrt{W(y)} > 0$, then $b_+$ is the decaying solution. If it is in the region where $\operatorname{Re}\sqrt{W(y)} < 0$, $b_-$ is (and if, say, $\operatorname{Re}\sqrt{W(z_0)} = 0$, this is possible). Stokes lines are asymptotics of the curve where $\operatorname{Re}\sqrt{W(y)} = 0$. More generally, the set $\{y \mid \operatorname{Re}\sqrt{W(y)} = 0\}$ are the curves, called *Stokes curves*, where there is a shift between asymptotically dominant approximate solutions.

On $\mathbb{R}$, the analogs are the turning points, $y_0$, where $W(y_0) = 0$, and one shifts between $\sqrt{W(y)}$ real and $\sqrt{W(y)}$ imaginary, that is, regions where the solutions are growing/decaying and where they are oscillatory. *Connection formulae* try to combine the WKB solutions on the two sides. (Obviously, the decaying solution is real so it has to combine to have the form $|W(x)|^{-1/4}\cos(\theta_0 + \int_{x_0}^{x} \sqrt{-W(y)}\,dy)$ and the issue is what to take for $\theta_0$.) The key is to assume at points where $W(x_0) \neq 0$, we have $W'(x_0) \neq 0$, so $W$ is linear near $x_0$, and one uses the asymptotics of an Airy function across the gap. Of course, these are only approximate solutions and so are only exact in some limit. It cannot be a limit of $x \to \infty$, so there have to be other parameters, for example, studying solutions of $-\hbar^2 u''(x) + V(x)u(x) = Eu(x)$ in the limit as $\hbar \to 0$.

A particular use of the WKB formula is to justify the Bohr–Sommerfeld quantization rules in one-dimensional quantum mechanics in the small $\hbar$ limit. They say if $V(x) < E$ between $x_-(E)$ and $x_+(E)$, then the $L^2$ eigenvalues of $-\hbar^2 u'' + Vu = Eu$ are given by (in the $\hbar \to 0$ limit)

$$\int_{x_-}^{x_+} \sqrt{E - V(x)}\,dx = (n + \tfrac{1}{2})\hbar, \qquad n = 0, 1, 2 \ldots$$

The extra $\frac{1}{2}$ comes from connection formulae. For further discussion of this in textbooks, see [**283**], or for more mathematical approaches, see [**141, 142, 378**].

Variation of parameters was originally introduced by Lagrange to solve $Du(x) = f(x)$, where $D$ is a linear differential operator. If $\{u_j\}_{j=1}^n$ are a basis for the solutions of $Du = 0$, then one tries to solve $Du = f$ by $u(x) = \sum_{j=1}^n a_j(x)u_j(x)$ with $\{a_j\}_{j=1}^n$ restricted by $u^{(k)} = \sum_{j=1}^n a_j u_j^{(k)}$, $k = 0, 1, \ldots, n-1$. For example, in case $n = 2$, $a_1$ and $a_2$ obey first-order rather than second-order equations. In one sense, our use of "variation of parameters" is related to $Du + (\delta V)u = 0$, which is equivalent to $Du = -(\delta V)u$. It has become a standard perturbation technique.

Our approach in this section follows papers of Harrell [**179**] and Harrell–Simon [**180**], which apply WKB techniques to analyze some subtle problems in quantum theory. Example 15.5.7 follows ideas in Simon [**353**]. See Problem 7 for an example with $r|V(r)|$ bounded at infinity, so barely not $L^1$, where $-u'' + V(r)u = u$ (i.e., $E > 0$) has a solution which is $L^2$ at infinity (the first examples of this type go back to Wigner–von Neumann [**399**]).

## Problems

1. Given $Q_\pm$ defined by (15.5.11) and $\widetilde{W}_\pm$ defined by (15.5.12), verify (15.5.13).

2. Verify that if $A$ is given by (15.5.35), then $A$ is given by (15.5.25)/(15.5.26).

3. Let $V\colon [0,\infty)$ be of bounded variation with $\lim_{x\to\infty} V(x) = 0$. The purpose of this problem is to show that $V$ can be written as $V_1 + V_2$, where $V_1 \in L^1(0,\infty)$ and $V_2$ is $C^\infty$ with $V_2^{(k)} \in L^1$ for $k = 1, 2, \dots$ and $V_2(x) \to 0$ as $x \to \infty$. Pick $f \in C_0^\infty(-1,1)$ with $f \geq 0$ and $\int_{-1}^{1} f(z)\, dz = 1$. Define $V_2 = f * V$, where $V$ is extended to $(-\infty, 0)$ by setting it equal to 0 there. Let $V_1 = V - V_2$.

   (a) Prove there is a signed measure $\mu$ on $\mathbb{R}$ with $|\mu|(\mathbb{R}) < \infty$ so that $V(x) = \mu([x,\infty))$, except perhaps at the countable set of discontinuities of $V$.

   (b) For $x$ not a discontinuity of $V$, prove that $|V_1(x)| \leq |\mu|([x-1, x+1])$.

   (c) Prove that $V_1 \in L^1$.

   (d) Prove that $V_2$ is $C^\infty$.

   (e) For $k \equiv 1$, prove that $V_2^{(k)} = f^{(k-1)} * \mu$ and use this to prove that $V_2^{(k)} \in L^1$.

   (f) Prove that $V_2 \to 0$ at infinity.

4. (a) Using Problem 3, extend Example 15.5.7 to the case $V_2' \in L^1$ is replaced by $V_2$ of bounded variation.

   (b) If $V = V_1 + V_2$ with $V_1 \in L^1$, $V_2 \to 0$ at infinity, and $V_2$ in $C^2$ with $V_2', V_2'' \in L^1$, show that one can use the WKB solutions instead of $c_\pm$ of (15.5.45).

   (c) Show how, in the generality of Example 15.5.7, one can modify $V_1, V_2$ to use the WKB solution.

5. In the context of the assumptions of Example 15.5.7, let $E = -\kappa^2$ and

$$c_\pm(x) = \exp\biggl(\mp \int_{x_1}^{x} \sqrt{\kappa^2 + V_2(y)}\, dy\biggr) \tag{15.5.88}$$

where $x_1$ is such that $-V_2(y) < \kappa^2$ if $y > x_1$.

(a) By following Example 15.5.10, prove there is a solution $u_+$ of (15.5.1) asymptotic to $c_+(x)$.

(b) Use Theorem 15.5.9 to find a second solution and show it is asymptotic to $c_-(x)$.

6. Let $W(x)$ be a polynomial of degree $d$, positive at infinity.

(a) Prove that $[b_-(x)]^{-1} \int_{x_1}^{x} b_+(y)^{-2}\, dy \to 1$ as $x \to \infty$. (*Hint*: If $f$ is $\log b_+(x)$, prove that $\int_0^{x_2 - x_1} \exp(f(x-y) - f(x))\, dx \sim 1/f'(x)$.)

(b) Construct a solution asymptotic to $u_-$ for such a $W$.

7. This problem will study $-u'' + Vu = Eu$, where

$$V(x) = \sum_{j=1}^{m} \gamma_j x^{-1} \sin(\alpha_j x) + Q(x)$$

with $|Q(x)| \le C(1+x)^{-1-\varepsilon}$ with each $\alpha_j > 0$ and all unequal. The point will be to prove that if $E = k^2$ with $k = \frac{1}{2}\alpha_j$, then there is a solution $u_+$ so that as $x \to \infty$,

$$u_+(x) = \begin{cases} x^{-\gamma_j/2\alpha_j}(\cos(\frac{1}{2}\,\alpha_j x) + o(1)) & \text{if } \gamma_j/\alpha_j > 0 \\ x^{\gamma_j/2\alpha_j}(\sin(\frac{1}{2}\,\alpha_j x) + o(1)) & \text{if } \gamma_j/\alpha_j < 0 \end{cases}$$

In particular, if $|\gamma_j| > \alpha_j$, there is a solution $L^2$ at infinity at a positive energy $E = (\frac{1}{2}\alpha_j)^2$.

Let $\gamma_j/\alpha_j > 0$. Let

$$c_+(x) = x^{-\gamma_j/2\alpha_j} \cos\biggl(\frac{\alpha_j x}{2}\biggr), \qquad c_-(x) = x^{\gamma_j/2\alpha_j} \sin\biggl(\frac{\alpha_j x}{2}\biggr)$$

Follow the methods of Example 15.5.10 to prove there is a solution $u_+$ with $x^{\gamma_j/2\alpha_j}|u_+(x) - c_+(x)| \to 0$.

Chapter 16

# Univalent Functions and Loewner Evolution

> But the old problem was not dead. We had left behind a crucial issue. How can one characterize the shapes of spin clusters that form in critical phenomena? We did not even realize that we had left something behind.
>
> —*Leo Kadanoff* [**223**][1]

**Big Notions and Theorems:** Univalent Function, Schlicht Function, Gronwall Area Theorem, Koebe Function, Bieberbach Inequality, Distortion Estimates, Slit Domain, Loewner Equation, Schramm-Loewner Evolution (SLE), Domain Markov Property, Percolation Exploration Process

The central theme of this chapter is the study of one-one analytic maps on $\mathbb{D}$ called *univalent* (or sometimes by the German name *schlicht*, meaning simple—the symbol $S$ below is from "schlicht"). In one sense, this is a subset of the theory of conformal maps. If $f$ is univalent on $\mathbb{D}$ with $f'(0) > 0$, $\Omega \equiv f[\mathbb{D}]$ is a simply connected region, and if $z_0 = f(0)$, $f$ is the inverse of the unique Riemann mapping, $g$ (guaranteed by Theorem 8.1.1 of Part 2A), with $g(z_0) = 0$, $g'(z_0) > 0$. Thus, the family of univalent maps on $\mathbb{D}$ can be viewed as a parametrization of pairs of a simply connected region, $\Omega$, and a distinguished point $z_0 \in \Omega$.

[1] talking about the impact of SLE on statistical physics.

By tradition, one restricts to $f(0) \equiv z_0 = 0$ and $f'(0) = 1$ removing translation and scaling covariance, that is, we will study

$$S = \{f\colon \mathbb{D} \to \mathbb{C} \mid f \text{ is one-one; } f(0) = 0, f'(0) = 1\} \tag{16.0.1}$$

Section 16.1 will present some basic properties of $S$, including that it is compact in the topology of uniform convergence on compact subsets of $\mathbb{D}$.

An especially important example of a function in $S$ is $f_{\text{Koebe}}$, the function of Example 8.4.2 of Part 2A,

$$f_{\text{Koebe}}(z) = z(1-z)^{-2} \tag{16.0.2}$$

$$= \sum_{n=1}^{\infty} nz^n \tag{16.0.3}$$

Perhaps the central result of this theory is de Branges' theorem, aka the Bieberbach conjecture, that if

$$f(z) = \sum_{n=1}^{\infty} a_n z^n \tag{16.0.4}$$

lies in $S$, then

$$|a_n| \le n \tag{16.0.5}$$

Notice $f_{\text{Koebe}}(z)$ has equality in (16.0.5) and so do "rotations" of $f_{\text{Koebe}}$, $\omega^{-1}f(\omega z)$, for $\omega \in \partial\mathbb{D}$. A refined version of de Branges' theorem says that equality holds in (16.0.5) even for one $n \ge 2$ if and only if $f$ is the Koebe function or one of its rotations.

We will not prove de Branges' theorem here, but only Bieberbach's original result that $|a_2| \le 2$ (in Section 16.1) and Loewner's result that $|a_3| \le 3$ (in Section 16.2). Loewner's result depended on showing that the slit maps, that is, ones where $\Omega = \mathbb{C} \setminus \text{Ran}(\gamma)$ with $\gamma$ a continuous Jordan curve from a point of $\mathbb{C}$ to $\infty$, are dense in $S$ and the analysis of how the map $f_\gamma\colon \mathbb{D} \to \mathbb{C}\setminus\text{Ran}(\gamma)$ changes as $\gamma$ is wound up, that is, replaced by the curve run from a point on the curve out to infinity. This produces a differential equation for $f$ called Loewner evolution. This equation will be discussed initially in Section 16.2.

In fact, a main object of this chapter is to discuss Loewner evolution—indeed, a family of stochastic processes where the driving term in the differential equation is a Brownian curve—these are the Schramm–Loewner evolutions (SLE), one of the most heavily studied analytic subjects of the past fifteen years. If it were not for SLE, this chapter might not even be here. The Bieberbach conjecture was a dominant theme in complex analysis for sixty years, but as much because of its remaining unsolved as because of its intrinsic importance. Especially since we will not prove the full de Branges theorem, this chapter is here because, in Section 16.3, we wish to give a brief introduction to SLE, which has become a central part of mathematical analysis.

## 16.1. Fundamentals of Univalent Function Theory

**Definition.** Let $\Omega$ be a region in $\mathbb{C}$. A function $f\colon \Omega \to \mathbb{C}$ is called *univalent* (aka *schlicht*) if $f$ is analytic on $\Omega$ and $f(z) = f(w) \Rightarrow z = w$. The set, $S$, is defined by

$$S = \{f \text{ is univalent on } \mathbb{D} \mid f(0) = 0, f'(0) = 1\} \tag{16.1.1}$$

$$\Sigma = \{f \text{ is univalent on } \mathbb{C} \setminus \overline{\mathbb{D}} \mid f \text{ has a simple pole at } \infty \text{ with residue } 1\} \tag{16.1.2}$$

Thus, $f \in S$ and $g \in \Sigma$ have the form

$$f(z) = z + \sum_{n=2}^{\infty} a_n z^n, \qquad g(w) = w + \sum_{n=0}^{\infty} b_n w^{-n} \tag{16.1.3}$$

$\Sigma$ is natural, given our interest in $S$, because

$$g(z) = [f(1/z)]^{-1} = z - a_2 + (a_2^2 - a_3)z^{-1} + \dots \tag{16.1.4}$$

sets up to a one-one correspondence between $S$ and

$$\Sigma' = \{g \in \Sigma \mid 0 \notin \operatorname{Ran}(g) = g[\mathbb{C} \setminus \overline{\mathbb{D}}]\} \tag{16.1.5}$$

$\widehat{\mathbb{C}} \setminus \overline{\mathbb{D}}$ is simply connected in $\widehat{\mathbb{C}} = \mathbb{C} \cup \{\infty\}$, so for $g \in \Sigma$,

$$E_g \equiv \mathbb{C} \setminus g[\mathbb{C} \setminus \overline{\mathbb{D}}] \tag{16.1.6}$$

is always a connected compact subset of $\mathbb{C}$ (by Theorem 8.1.2 of Part 2A), and if $g \in \Sigma'$, $0 \in E_g$. We are heading towards showing $|a_2| \le 2$ for $f \in S$, which is not only of intrinsic interest but will imply $\mathbb{D}_{1/4}(0) \subset \operatorname{Ran}(f)$ for any $f \in S$ and also that functions in $S$ have uniform bounds (in $S$) on each compact subset of $\mathbb{D}$. The key will be:

**Theorem 16.1.1** (Gronwall's Area Theorem). *Let*

$$g(w) = w + b_0 + \sum_{n=1}^{\infty} b_n w^{-n}, \qquad |w| > 1 \tag{16.1.7}$$

*be in $\Sigma$. Then*

$$\sum_{n=1}^{\infty} n|b_n|^2 \le 1 \tag{16.1.8}$$

*with equality if and only if $|E_g| = 0$, where $|\cdot|$ is the two-dimensional Lebesgue measure.*

**Proof.** We use the Lusin area formula, (2.2.22) of Part 2A. Let $\mathbb{A}_{r,R}$ be $\{z \mid r < |z| < R\}$. Then, by that formula, for $1 < r < R < \infty$,

$$\text{Area } g[\mathbb{A}_{r,R}] = \int_r^R \biggl[\int_0^{2\pi} |g'(\rho e^{i\theta})|^2 \, d\theta\biggr] \rho \, d\rho \tag{16.1.9}$$

Since $1 < r \leq \rho$, uniformly in $\rho$, we have that

$$g'(\rho e^{i\theta}) = 1 - \sum_{n=1}^\infty n b_n e^{-i(n+1)\theta} \rho^{-n-1} \tag{16.1.10}$$

Since $\{e^{in\theta}\}_{n=0}^\infty$ are orthogonal in $L^2(\partial\mathbb{D}, d\theta)$, we see

$$\int_0^{2\pi} |g'(\rho e^{i\theta})|^2 \, d\theta = 2\pi \biggl[1 + \sum_{n=1}^\infty |nb_n|^2 \rho^{-2n-2}\biggr] \tag{16.1.11}$$

Plugging into (16.1.9), we find

$$\text{Area } g[\mathbb{A}_{r,R}] = \pi \biggl[R^2 - r^2 + \sum_{n=1}^\infty n|b_n|^2 [r^{-2n} - R^{-2n}]\biggr]$$

Let $h(w) = w + b_0 + b_1 w^{-1}$. The same calculation shows

$$\text{Area } h[\mathbb{D}_R(0)] = \pi R^2 + O(R^{-2}) \tag{16.1.12}$$

Since $|g(Re^{i\theta}) - h(Re^{i\theta})| = O(R^{-2})$, we see

$$\text{Area inside } g[\partial\mathbb{D}_R] - \text{Area inside } h[\partial\mathbb{D}_R] = O(R^{-1}) \tag{16.1.13}$$

so, subtracting the area inside $g[\partial\mathbb{D}_R]$ and Area $g[\mathbb{A}_{r,R}]$, we find

$$\text{Area inside } g[\partial\mathbb{D}_r] = \pi \biggl[r^2 - \sum_{n=1}^\infty n|b_n|^2 r^{-2n}\biggr] \tag{16.1.14}$$

If $\widetilde{E}^{(r)}$ is the region inside $g[\partial\mathbb{D}_r]$, the $\widetilde{E}^{(r)}$ are decreasing as $r$ decreases with $r > 1$ and their intersection is $E$. Thus,

$$|E| = \lim_{r \downarrow 1} |\widetilde{E}^{(r)}| = \pi \biggl[1 - \sum_{n=1}^\infty n|b_n|^2\biggr] \tag{16.1.15}$$

Since $|E| \geq 0$, we find (16.1.8) with equality only if $|E| = 0$. $\square$

**Lemma 16.1.2.** *Let $f \in S$ have the form* (16.1.3). *Then*

$$h(z) = [f(z^2)]^{1/2} = z + \frac{a_2}{2} z^3 + \dots \tag{16.1.16}$$

*is in $S$ also and*

$$h(-z) = -h(z) \tag{16.1.17}$$

**Remark.** Equivalently, any $f \in S$ is the square of an odd function in $S$ evaluated at $\sqrt{z}$.

**Proof.** $f(z^2) = z^2[1 + \sum_{n=2}^{\infty} a_n z^{2n-2}]$. Since $f(z) \neq 0$ for $z \neq 0$ (since $f$ is one-one), the function in [ ] is analytic on $\mathbb{D}$ and nonvanishing, so it has a square root analytic in $\mathbb{D}$. Thus, we can define $h(z) = z[1 + \sum_{n=2}^{\infty} a_n z^{2n-2}]^{1/2} = z + \frac{a_2}{2} z^3 + O(z^5)$ is odd.

Since $h(z)^2 = f(z^2)$, we see if $h(z) = h(w)$, then $f(z^2) = f(w^2) \Rightarrow z^2 = w^2 \Rightarrow z = \pm w \Rightarrow h(z) = \pm h(w) \Rightarrow z = w$ (if $h(z) = h(w)$ since $h(-z) = -h(z)$). Thus, $h \in S$. □

**Example 16.1.3** (Koebe Function). Recall the Koebe function (16.0.2) has

$$f_{\text{K}}(z) = z(1-z)^{-2} \tag{16.1.18}$$

so

$$\begin{aligned} h_{\text{K}}(z) &= (f_{\text{K}}(z^2))^{1/2} = (z^2(1-z^2)^{-2})^{1/2} \\ &= z(1-z^2)^{-1} && (16.1.19) \\ &= z + z^3 + z^5 + \dots && (16.1.20) \\ &= (z^{-1} - z)^{-1} && (16.1.21) \end{aligned}$$

The $g$ associated to $h$ is

$$(h_{\text{K}}(1/z))^{-1} = z - z^{-1} \tag{16.1.22}$$

Since $f_{\text{K}}$ is

$$f_{\text{K}}(z) = \frac{1}{z + z^{-1} - 2} \tag{16.1.23}$$

the associated $g$ is

$$g_{\text{K}}(z) = z + z^{-1} - 2 \tag{16.1.24}$$

is, except for a translation, the Joukowski map of Example 8.4.3 of Part 2A. Notice $b_1 = 1$, up to $b_0$, the unique element in $\Sigma$ with $b_1 = 1$ (because of (16.1.8), $b_1 = 1 \Rightarrow b_n = 0$ for $n \geq 2$). □

**Theorem 16.1.4** (Bieberbach's Inequality). *If $f \in S$, then $|a_2| \leq 2$ with equality if and only if $f$ is the Koebe function or a rotation of the Koebe function.*

**Proof.** By the lemma, $f(z^2) = h(z)^2$ for $h$ odd, with $h(z) = z + \frac{1}{2} a_2 z^3 + \dots$. The $g(z)$ associated to $h(z)$ has (by (16.1.4) the "$a_2$" here is 0 and "$a_3$" is $\frac{1}{2} a_2$)

$$g(z) = z - \frac{a_2}{2} z^{-1} + O(z^{-3}) \tag{16.1.25}$$

Since (16.1.8) implies that $|b_1| \leq 1$, we see $|a_2/2| \leq 1$ with equality only if

$$g(z) = z - e^{i\theta} z^{-1} \tag{16.1.26}$$

which leads (Problem 1) to $f(z) = z(1 - e^{i\theta} z)^{-2} = e^{-i\theta} f_{\text{K}}(e^{i\theta} z)$. □

**Corollary 16.1.5** (Koebe One-Quarter Theorem). *For any $f \in S$, $f[\mathbb{D}]$ contains $\mathbb{D}_{1/4} = \{z \mid |z| < \frac{1}{4}\}$. If $f$ is not the Koebe function or a rotation of it, then $f[\mathbb{D}]$ contains some $\mathbb{D}_{1/4+\delta}$ with $\delta > 0$.*

**Remark.** $f_K[\mathbb{D}] = \mathbb{C} \setminus (-\infty, -\frac{1}{4}]$, so $-\frac{1}{4} \notin f[\mathbb{D}]$ and one cannot do better than $\mathbb{D}_{1/4}$.

**Proof.** Suppose $\omega \notin f[\mathbb{D}]$. Then

$$q(z) = \frac{\omega f(z)}{\omega - f(z)} = z + \left(a_2 + \frac{1}{\omega}\right) z^2 + \dots \tag{16.1.27}$$

lies in $S$ since $w \to \frac{\omega w}{\omega - w}$ is a bijection of $\mathbb{C} \setminus \{\omega\}$ and $\mathbb{C} \setminus \{-1\}$. Thus, by Bieberbach's inequality,

$$\left| a_2 + \frac{1}{\omega} \right| \le 2 \tag{16.1.28}$$

Therefore,

$$\left| \frac{1}{\omega} \right| \le \left| a_2 + \frac{1}{\omega} \right| + |a_2| \le 4 \tag{16.1.29}$$

that is, $|\omega| \ge \frac{1}{4}$. Thus, $\mathbb{C} \setminus f[\mathbb{D}] \subset \mathbb{C} \setminus \mathbb{D}_{1/4} \Rightarrow \mathbb{D}_{1/4} \subset f[\mathbb{D}]$. In order for $|\omega| = \frac{1}{4}$ to be possible, we must have $|a_2| = 2$ in (16.1.29), which implies the final assertion. $\square$

**Corollary 16.1.6.** *For all $f \in S$ and $z \in \mathbb{D}$,*

(a) $$\left| \frac{zf''(z)}{f'(z)} - \frac{2|z|^2}{1-|z|^2} \right| \le \frac{4|z|}{1-|z|^2} \tag{16.1.30}$$

(b) $$\frac{1-|z|}{(1+|z|)^3} \le |f'(z)| \le \frac{1+|z|}{(1-|z|)^3} \tag{16.1.31}$$

(c) $$\frac{|z|}{(1+|z|)^2} \le |f(z)| \le \frac{|z|}{(1-|z|)^2} \tag{16.1.32}$$

(d) *$S$ is a compact subset of $\mathfrak{A}(\mathbb{D})$, the Fréchet space of functions analytic in $\mathbb{D}$ in the topology of uniform convergence on each $\mathbb{D}_r$, $r < 1$.*

**Remark.** (16.1.31) are called *distortion estimates* and (16.1.32) are called *growth estimates.*

**Proof.** (a) Fix $\zeta \in \mathbb{D}$. Taking a fractional linear transformation $T_\zeta$ that maps $\mathbb{D}$ to $\mathbb{D}$ and $\zeta$ to 0, for $\alpha, \beta$ suitable $F(z) = \alpha f(T_\zeta(z)) + \beta$ will map $\zeta$ to 0 and have derivative 1, so we'll get information on $F''(\zeta)$ at 0 from the Bieberbach inequality. Explicitly,

$$F(z) = \frac{f(\frac{z+\zeta}{1+\bar{\zeta}z}) - f(\zeta)}{(1-|\zeta^2|)f'(\zeta)} = z + \alpha(\zeta)z^2 + O(z^3) \tag{16.1.33}$$

$$\alpha(\zeta) = \frac{1}{2}\left[(1-|\zeta|^2)\,\frac{f''(\zeta)}{f'(\zeta)} - 2\bar{\zeta}\right] \tag{16.1.34}$$

$F$, called the *Koebe transform* of $f$, lies in $S$ so $|\alpha(\zeta)| \le 2$, which (Problem 2) simplifies to (16.1.30).

(b) If $|z| = r$, $-|\alpha| \le \operatorname{Re}(\alpha) \le |\alpha|$ yields

$$\frac{2r^2 - 4r}{1-r^2} \le \operatorname{Re}\left[\frac{zf''(z)}{f'(z)}\right] \le \frac{2r^2+4r}{1-r^2} \tag{16.1.35}$$

Noting that

$$\operatorname{Re}\left[\frac{zf''(z)}{f'(z)}\right]\Bigg|_{z=re^{i\theta}} = r\,\frac{\partial}{\partial r}\log|f'(re^{i\theta})| \tag{16.1.36}$$

and that $\log|f'(re^{i\theta})|\big|_{r=0} = 0$, we can integrate (16.1.35) from 0 to $r$ and then exponentiate. (16.1.31) results.

(c) Since $\int_0^r \frac{1+\rho}{(1-\rho)^3}\,d\rho = \frac{r}{(1-r)^2}$ (check by differentiation), we immediately get the upper bound. We leave the lower bound to Problem 3.

(d) Immediate from the upper bound in (c) and Montel's theorem (Theorem 6.2.1 of Part 2A). $\square$

We saw $|a_2| \le 2$ is quite powerful! The compactness result in the last corollary implies $\sup_{f\in S}|a_n(f)| < \infty$. Looking at the extreme function for $a_2$ and the power series for the Koebe function, Bieberbach made his famous conjecture in 1916,

$$f \in S \Rightarrow |a_n| \le n \tag{16.1.37}$$

de Branges finally proved it almost seventy years later in 1985!

**Notes and Historical Remarks.** The area theorem (Theorem 16.1.1) is due to Gronwall [**168**] in 1914. The more usual proof of the area theorem (Problem 4) uses Green's theorem to compute the area inside $g[\partial\mathbb{D}_r]$ rather than the Lusin area theorem—we prefer the latter since it is more directly related to area, although the proof is slightly more lengthy.

Bieberbach's estimate $|a_2| \le 2$ and his conjecture is from a 1916 paper [**57**]. This paper also proved distortion and growth estimates and the Koebe one-quarter theorem (Corollary 16.1.5) which Koebe [**228**] conjectured in 1907. Koebe proved there was some $\rho > 0$ so all $f \in S$ had $\mathbb{D}_\rho \subset f[\mathbb{D}]$ and conjectured $\rho = \frac{1}{4}$ based on a fact that because of his function, it could not be any larger. Hurwitz [**204**] earlier had this result implicitly. In 1910, Koebe [**229**] had the first form of distortion and growth estimates.

That every $f \in S$ can be uniquely written, $f(z^2) = h(z)^2$ for an odd $h$ in $S$, makes odd univalent functions of special interest. In 1932, Littlewood–Paley [**265**] proved that there is a constant, $A$, so $h$ odd and univalent on $\mathbb{D}$ and $h'(0) = 1$ implies $|a_n(h)| \le An$. They conjectured that this was true for $A = 1$, something that, given Lemma 17.1.2, would imply the Bieberbach conjecture. But their conjecture was promptly disproved by Fekete–Szegő [**133**] who used Loewner's method (see the next section) to prove the optimal bound for $a_5$ for odd functions was $\frac{1}{2} + e^{-2/3} = 1.013\dots$, not 1. In 1936, Robertson [**330**] conjectured for odd $h \in S$ that

$$\sum_{k=1}^{n} |a_{2k+1}|^2 \le n \tag{16.1.38}$$

a result which, Robinson noted, implied the full Bieberbach conjecture. It was proven by de Branges when he settled the issues.

We will not attempt to summarize the partial results ($|a_n| \le n$ for $n = 4, 5, 6$) prior to de Branges' solution of the Bieberach conjecture—the book of Duren [**111**], published the year before de Branges announced his result, has an admirable summary of the state of the art in 1983; see also the book of Pommerenke [**320**].

de Branges proved the Bieberbach conjecture in 1985 [**94**] using ideas of Loewner (see the next section), Robertson [**330**], and Milin [**274**], plus a fair amount of specialized information about Jacobi polynomials (especially some inequalities of Askey–Gasper [**23**]). Monographs that include streamlined versions of de Branges' proof are Conway [**89**], Gong [**154**], Hayman [**186**], and Segal [**344**].

## Problems

1. (a) If $g(z) = \left[h(1/z)\right]^{-1}$ and $g(z) = z - e^{i\theta} z^{-1}$, find $h(z)$.

   (b) If $f(z^2) = h(z)^2$ for the $h$ of (a), prove that $f(z) = z(1 - e^{i\theta z})^{-2}$.

2. If $\alpha(\zeta)$ is given by (16.1.34), prove that $|\alpha(\zeta)| \le 2$ is equivalent to (16.1.30).

3. (a) Prove that the lower bound in (16.1.32) holds trivially if $|f(z_0)| \ge \frac{1}{4}$.

   (b) If $|f(z_0)| < \frac{1}{4}$, find a curve $\gamma\colon [0,1] \to \mathbb{D}$ so $f(\gamma(t)) = tf(z_0)$.

   (c) Obtain the lower bound in (16.1.32) when $|f(z_0)| < \frac{1}{4}$ by using $f(z_0) = \int_\gamma f'(\omega)\, d\omega$.

4. (a) If $\gamma$ is a Jordan curve in $\mathbb{C}$ which contains 0 in its interior, prove that the area surrounded by $\gamma$ is $\frac{1}{2\pi} \int \bar{z}\, dz$. (*Hint*: Green's theorem.)

(b) If $g$ has the form (16.1.7), prove that for $r > 1$,

$$\frac{1}{2\pi}\int_{|\omega|=r} \overline{g(z)}\, g'(z)\, dz = \pi\bigg\{r^2 - \sum_{n=1}^{\infty} n|b_n|^2 r^{-2n}\bigg\}$$

providing another proof of (16.1.14).

The next three problems involve $H^p(\mathbb{D})$ properties of univalent functions on $\mathbb{D}$, i.e., for which $p \in (0,\infty)$, one has

$$\sup_{0<r<1} \int_0^{2\pi} |f(re^{i\theta})|^p \,\frac{d\theta}{2\pi} < \infty \tag{16.1.39}$$

It is easy to see there is no loss in restricting $f$ to lie in $S$. $H^p$ spaces are the subject of Chapter 5 of Part 3 and in the third problem, the reader will need some facts about such functions proven there. The first problem proves a preliminary needed in the second which will prove that $S \subset H^p(\mathbb{D})$ for any $p < \frac{1}{2}$. The third will show the Koebe function is not in $H^{1/2}(\mathbb{D})$ showing the result in Problem 7 is optimal.

5. Let $\gamma_1$, $\gamma_2$ be two smooth Jordan curves where the bounded region surrounded by $\gamma_2$ contains $\gamma_1$ and the bounded region surrounded by $\gamma_1$ contains 0. Let $r$, $\theta$ be polar coordinates on $\mathbb{C}\setminus 0$, i.e.,

$$z = x + iy; \quad x = r\cos\theta,\ y = r\sin\theta \tag{16.1.40}$$

$\theta$ is not a global function but the differential form $d\theta$ is. The purpose of this problem is to prove that for $0 < p$,

$$\int_{\gamma_1} r^p\, d\theta \le \int_{\gamma_2} r^p\, d\theta \tag{16.1.41}$$

(a) From (16.1.40) compute $dx$ and $dy$ in terms of $dr$ and $d\theta$ and prove that

$$dr \wedge d\theta = r^{-1}\, dx \wedge dy = r^{-1}\, d^2 z \tag{16.1.42}$$

(b) Prove that

$$d(r^p\, d\theta) = p\, r^{p-1}\, dr \wedge d\theta \tag{16.1.43}$$

(c) Using the abstract Stokes' theorem, prove that

$$\int_{\gamma_2} r^p\, d\theta - \int_{\gamma_1} r^p\, d\theta = \int_{\Omega} p\, r^{p-2}\, d^2 z \ge 0 \tag{16.1.44}$$

where $\Omega$ is the region inside $\gamma_2$ intersected with the region outside $\gamma_1$.

(d) Prove that

$$\int_{\gamma_1} r^p\, d\theta \le 2\pi \max\{|z|^p \mid z \in \gamma_1\} \tag{16.1.45}$$

6. This problem will prove $S \subset H^p(\mathbb{D})$ for any $0 < p < \frac{1}{2}$. So pick such a $p$ and $f \in S$. Define

$$\begin{aligned} f(re^{i\theta}) &= R(r,\theta)\exp\bigl(i\Phi(r,\theta)\bigr) \\ A(r) &= \int |f(re^{i\theta})|^p \,\frac{d\theta}{2\pi} \\ M(r) &= \sup_\theta |f(re^{i\theta})| \end{aligned} \tag{16.1.46}$$

(a) Using that $f$ is univalent show that $\{R(r,\theta) \mid 0 \le \theta \le 2\pi\}$ is a Jordan curve so that the associated Jordan regions increase as $r$ increases. Let $\gamma_r$ be this Jordan curve.

(b) Using the radial Cauchy–Riemann equations $\bigl($(2.1.24) of Part 2A$\bigr)$, prove that

$$\frac{dA}{dr} = \frac{p}{2\pi r}\int_{\gamma_r} R^p \, d\Phi \tag{16.1.47}$$

(c) Using (16.1.45) prove that

$$\frac{dA}{dr} \le \frac{p\bigl[M(r)\bigr]^p}{r} \tag{16.1.48}$$

(d) Using (16.1.32), prove that if $p < \frac{1}{2}$

$$\sup_{0\le r\le 1} A(r) \le p\int_0^1 r^{p-1}(1-r)^{-2p}\,dr \tag{16.1.49}$$

$$= p\,\Gamma(p)\,\Gamma(1-2p)\bigl[\Gamma(1-p)\bigr]^{-1} < \infty \tag{16.1.50}$$

(e) Conclude that for $0 < p < \frac{1}{2}$, $S \subset H^p(\mathbb{D})$.

7. This problem will rely on the fact (Theorem 5.3.5 of Part 3) that if $f \in H^p(\mathbb{D})$, then $f^*(e^{i\theta}) = \lim_{r\uparrow 1} f(re^{i\theta})$ exists for Lebesgue a.e. $\theta$ and that $\int |f^*(e^{i\theta})|^p \frac{d\theta}{2\pi} < \infty$.

(a) For $f$ the Koebe function given by (16.1.18) compute $|f^*(e^{i\theta})|$.

(b) Prove for the Koebe function that $\int |f^*(e^{i\theta})|^{1/2}\frac{d\theta}{2\pi} = \infty$ and conclude that the condition $p < \frac{1}{2}$ in Problem 6 is optimal.

8. Prove that if $\Omega \subset \mathbb{C}$ is convex and $\operatorname{Re} f'(z) > 0$ for all $z \in \Omega$, then $f$ is univalent on $\Omega$. (*Hint*: Fundamental theorem of calculus.)

9. If $f(z)$ has the form (16.1.3) converging for $|z| < 1$ and if $\sum_{n=z}^\infty n\,|a_n| < 1$, prove that $f$ is univalent.

## 16.2. Slit Domains and Loewner Evolution

One goal in this section will be to prove $|a_3| \le 3$ for $f \in S$ but even more, the method we use will rely on Loewner evolution which we'll introduce here.

A *single slit domain* or *slit domain* is a region of the form

$$\Omega = \mathbb{C} \setminus \operatorname{Ran}(\gamma) \tag{16.2.1}$$

where $\gamma$ is a Jordan curve from $z_0 \in \mathbb{C}$ to $\infty$, that is, $\gamma\colon [0,1] \to \widehat{\mathbb{C}}$, $\gamma(0) = z_1$, $\gamma(1) = \infty$, $\gamma(s) \neq \gamma(t)$ if $s \neq t$.

The two main results that we'll need to prove $|a_3| \le 3$ are:

**Theorem 16.2.1.** *The conformal maps of $\mathbb{D}$ onto slit domains are dense in $S$ (in the topology of $\mathfrak{A}(\mathbb{D})$); indeed, the ones where $\gamma$ is piecewise analytic with analytic corners are dense.*

**Theorem 16.2.2.** *Let $\Omega$ be a slit domain where $\gamma$ is analytic with analytic corners and let $f\colon \mathbb{D} \to \Omega$ a map in $S$ onto $\Omega$. Then there exist a continuous function, $\kappa(s)\colon [0,\infty) \to \partial\mathbb{D}$, and a family, $\{f_s\}_{s=0}^\infty$, of one-one maps of $\mathbb{D}$ with*

$$f_s(0) = 0, \qquad f_s'(0) = e^{-s}, \qquad f_{s=0}(z) = z \tag{16.2.2}$$

$$\frac{\partial f_s}{\partial s} = -f_s \left[\frac{1+\kappa(s) f_s}{1-\kappa(s) f_s}\right] \tag{16.2.3}$$

*for all $z \in \mathbb{D}$ and so that*

$$\lim_{s\to\infty} e^s f_s(z) = f(z) \tag{16.2.4}$$

**Remarks.** 1. (16.2.3) is called the *Loewner equation* or sometimes the *radial Loewner equation.*

2. Theorem 16.2.2 is valid when $\gamma$ is merely a continuous function. The proof relies on the fact that we have continuity up to the boundary for Riemann maps of $\mathbb{D}$ to $\mathbb{D}$ with a slit removed. This is true (see the remark after Theorem 8.2.4 of Part 2A) for any continuous slit, but we have only proved it for piecewise analytic slits. For this reason, we've only stated it for this case. Given the continuity up to the boundary result for more general slit disks yields the result for more general slit domains.

3. We'll construct $f_s(z)$ directly, so the issue of existence of solutions of (16.2.3) won't come up here, but it will be relevant in the next section; see also the Notes.

Given this pair of results, we can prove $|a_3| \le 3$.

**Theorem 16.2.3** (Loewner's $a_3$ Estimate). *For any $f \in S$, we have*

$$|a_3| \leq 3 \tag{16.2.5}$$

**Remark.** Our proof will show that among maps to slit domains, the only ones with $|a_3| = 3$ are the Koebe maps. Since we get general $f$ by density, it does not show that the only maps in $S$ with $|a_3| = 3$ are Koebe maps. But this has been proven by other means.

**Proof.** By Theorem 16.2.1, if we prove (16.2.5) for mappings to slit domains with slits that are piecewise analytic with analytic corners, then we have it in general. In addition, if $f_\omega(z) = \omega^{-1} f(\omega z)$ for $\omega \in \partial\mathbb{D}$, then

$$a_3(f_\omega) = \omega^2 a_3(f) \tag{16.2.6}$$

Moreover, $f_\omega$ is a suitable slit map if $f$ is, so it suffices to prove

$$\operatorname{Re} a_3 \leq 3 \tag{16.2.7}$$

to get (16.2.5).

Given $f$, let $f_t$ be the interpolations between $f_{t=0}(z) = z$ and $f$ given in Theorem 16.2.2, so if

$$f(z) = z + \sum_{n=2}^\infty a_n z^n, \qquad f_t(z) = e^{-t}\biggl(z + \sum_{n=2}^\infty a_n(t) z^n\biggr) \tag{16.2.8}$$

we have, by (16.2.4),

$$a_n = \lim_{t\to\infty} a_n(t) \tag{16.2.9}$$

(16.2.3) can be rewritten (converging for $|z| < 1$)

$$e^{-t}\biggl(\sum_{n=2}^\infty a_n'(t) z^n\biggr) = -2\kappa f_t^2 - 2\kappa^2 f_t^3 + O(z^4) \tag{16.2.10}$$

(the $-f$ term on the right is canceled by $\frac{d}{dt}(e^{-t})$ term on the left). Thus,

$$a_2'(t) = -2\kappa(t)e^{-t} \tag{16.2.11}$$

$$a_3'(t) = -4\kappa(t)e^{-t}a_2(t) - 2\kappa(t)^2 e^{-2t} \tag{16.2.12}$$

$$= \frac{d}{dt}\, a_2(t)^2 - 2\kappa(t)^2 e^{-2t} \tag{16.2.13}$$

The first equation is solved by

$$a_2(t) = -2\int_0^t \kappa(s) e^{-s}\, ds \tag{16.2.14}$$

which implies, since $|\kappa(s)| = 1$ and we have (16.2.9),

$$|a_2| \leq 2\int_0^\infty e^{-s}\, ds \leq 2 \tag{16.2.15}$$

yielding a second proof of Bieberbach's inequality $|a_2| \leq 2$.

(16.2.13) and (16.2.9) imply that

$$a_3 = 4\left(\int_0^\infty \kappa(s)e^{-s}\,ds\right)^2 - 2\int_0^\infty \kappa(s)^2 e^{-2s}\,ds \tag{16.2.16}$$

Write $\kappa(s) = e^{i\theta(s)}$ and use the fact that $\operatorname{Re}\omega^2 \le (\operatorname{Re}\omega)^2$ to see that

$$\operatorname{Re} a_3 \le 4\left(\int_0^\infty \cos(\theta(s))e^{-s}\,ds\right)^2 - 2\int_0^\infty \cos(2\theta(s))e^{-2s}\,ds \tag{16.2.17}$$

Next, note $e^{-s}\,ds$ is a probability measure and the Schwarz inequality plus $\cos(2y) = -1+2\cos^2 y$ and $\int_0^\infty e^{-2s}\,ds = \frac{1}{2}$ to see that (since $0 \le \cos^2\theta \le 1$)

$$\operatorname{Re} a_3 \le 1 + 4\int_0^\infty \cos^2\theta(s)[e^{-s} - e^{-2s}]\,ds \tag{16.2.18}$$

$$\le 1 + 4\int_0^\infty [e^{-s} - e^{-2s}]\,ds = 3 \tag{16.2.19}$$

which is (16.2.7). □

**Example 16.2.4.** At first sight, Theorem 16.2.1 looks unlikely! After all, $f(z) = z$ lies in $S$ with $\operatorname{Ran}(f) = \mathbb{D}$ and it might seem impossible that $f$ can be approximated by conformal maps onto unbounded domains. But define for $s \in [0, 2\pi)$, $\gamma_s$ to be the curve that runs from $\infty$ to 1 along $[1,\infty)$, and then runs along $\partial\mathbb{D}$ from $e^{i\theta} = 1$ to $e^{is}$ (see Figure 16.2.1). Let $f_s$ be the corresponding conformal map from $\mathbb{D}$ to $\mathbb{C} \setminus \operatorname{Ran}(\gamma_s)$ with $f_s(0) = 0$, $f_s'(0) > 0$. It is not hard to see (and we'll show below) that $f_s'(0) \in [1,4]$, so by applying (16.1.32) to $f_s/f_s'(0)$, we see that $|f_s(z)|$ is bounded on each $\mathbb{D}_{1-\varepsilon}$, $\varepsilon > 0$. Thus, by Montel's theorem, it has limit points and one might expect the limit exists. A little thought suggests that the only possible limit should be $f_{2\pi}(z) = z$. This is what we'll show below—one reason slit maps are dense is this phenomenon of pinching domains. □

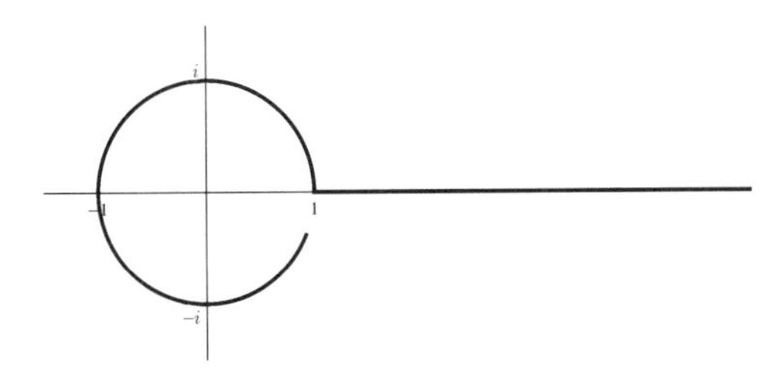

**Figure 16.2.1.** A slit domain approaches a pinch.

Let $\widetilde{S} = \cup_{\lambda>0}\lambda S$, that is, the univalent maps on $\mathbb{D}$ with $f(0) = 0$ and $f'(0) > 0$. If $\Omega = f[\mathbb{D}]$, $\Omega$ is an open, simply connected region with $0 \in \Omega$ and there is a one-one correspondence between such $\Omega$'s and $\widetilde{S}$. We'll often think of elements of $\widetilde{S}$ as pairs $(f, \Omega)$. The following theorem is useful:

**Theorem 16.2.5.** *Let $(f,\Omega)\in\widetilde{S}$. Then*

(a) *If $z_0\notin\Omega$, then*

$$f'(0)\le 4|z_0| \tag{16.2.20}$$

(b) *If $\mathbb{D}_r(0)\subset\Omega$, then*

$$f'(0)\ge r \tag{16.2.21}$$

**Proof.** (a) By the Koebe one-quarter theorem (Corollary 16.1.5) for $f(z)/f'(0)$, $\mathbb{D}_{\frac{1}{4}f'(0)}\subset\Omega$. Thus, $|z_0|\ge\frac{1}{4}f'(0)$, which is (16.2.20).

(b) Let $g=f^{-1}\restriction\mathbb{D}_r(0)$. $h(z)=g(rz)$ maps $\mathbb{D}$ to $\mathbb{D}$. By the Schwarz lemma (Theorem 3.6.7 of Part 2A), since $h(0)=g(0)=0$, $|h'(0)|\le 1$. Thus, $|g'(0)|\le r^{-1}$. Since $g'(0)f'(0)=1$ by the chain rule, this implies (16.2.21). □

We continue the proof of Theorem 16.2.1 by considering limits of $(f_n,\Omega_n)\in\widetilde{S}$. While the method below handles more general $\{\Omega_n\}_{n=1}^\infty$'s, we only require the case of increasing or of decreasing $\Omega_n$'s, so we'll restrict to those cases. If $\Omega_n$ are an increasing family of open, simply connected domains, it is not hard to see that

$$\Omega_\infty=\bigcup_{n=1}^\infty\Omega_n \tag{16.2.22}$$

is also simply connected. If $\Omega_n$ is decreasing, $\bigcup_{n=1}^\infty(\widehat{\mathbb{C}}\setminus\Omega_n)$ may not be closed, but it is connected and so its closure is connected and contains $\{\infty\}$. The complement

$$\widetilde{\Omega}_\infty=\biggl(\bigcap_{n=1}^\infty\Omega_n\biggr)^{\text{int}} \tag{16.2.23}$$

may not be connected and may not contain 0, but if $0\in\widetilde{\Omega}_\infty$, we take

$$\Omega_\infty=\text{connected component of }\widetilde{\Omega}_\infty\text{ containing }0 \tag{16.2.24}$$

It is easy to see that $\Omega_\infty$ is simply connected.

**Proposition 16.2.6** (Carathéodory Convergence Theorem). *Let $(f_n,\Omega_n)$ be a sequence of elements of $\widetilde{S}$ so that either*

(i) *$\Omega_n\subset\Omega_{n+1}$ and $\Omega_\infty$ given by (16.2.22) is not all of $\mathbb{C}$, or*

(ii) *$\Omega_n\supset\Omega_{n+1}$ and for some $r>0$ and all $n$, $\mathbb{D}_r(0)\subset\Omega_n$, in which case $\Omega_\infty$ is given by (16.2.23)/(16.2.24).*

*Let $f_\infty$ be the element of $\widetilde{S}$ associated to $\Omega_\infty$. Then $f_n\to f_\infty$ uniformly on compact subsets of $\mathbb{D}$ and $f_n^{-1}\to f_\infty^{-1}$ in the sense that for any compact $K\subset\Omega_\infty$, $K\subset\operatorname{Ran}(f_n)=\Omega_n$ for all large $n$ and $f_n^{-1}\restriction K\to f_\infty^{-1}\restriction K$ uniformly.*

**Proof.** In either case, there is $z_0 \in \bigcap_{n=1}^\infty (\mathbb{C} \setminus \Omega_n)$. If $\Omega_n$ is increasing, since $\Omega_\infty \neq \mathbb{C}$, pick $z_\infty \notin \Omega_\infty$. If $\Omega_n$ is decreasing, pick any $z_\infty \in \mathbb{C} \setminus \Omega_1$. Moreover, for some $r > 0$, $\mathbb{D}_r(0) \subset \Omega_n$ for all $n$. If $\Omega_n$ is increasing, pick $\mathbb{D}_r(0) \subset \Omega_1$, and if $\Omega_n$ is decreasing, this is part of the hypothesis. Thus, by Theorem 16.2.5, there is $c > 0$ so that for all $n$,

$$c \leq f_n'(0) \leq c^{-1} \tag{16.2.25}$$

By Corollary 16.1.6 applied to $f_n'(0)^{-1} f_n$, we see that

$$|f_n(z)| \leq c^{-1}(1 - |z|)^{-2} \tag{16.2.26}$$

so, by Montel's theorem, $\overline{\{f_n\}_{n=1}^\infty}$ is compact in $\mathfrak{A}(\mathbb{D})$. If $f_\infty$ is a limit point of $\{f_n\}_{n=1}^\infty$, pick $n(j)$ so $f_{n(j)} \to f_\infty$ in $\mathfrak{A}(\mathbb{D})$. By (16.2.25), $f_\infty'(0) \geq c > 0$, so $f_\infty$ is not constant. Thus, by Corollary 6.4.3 of Part 2A, $f_\infty$ is univalent. Since $f_\infty(0) = 0$, $\mathrm{Ran}(f_\infty) \subset \Omega_\infty$ (using Corollary 6.4.2 of Part 2A).

Define $g_n^{-1} = f_n$ as a map of $\Omega_n$ to $\mathbb{D}$. In case $\Omega_n$ is decreasing, view $g_n$ as a map from $\Omega_\infty$ to $\mathbb{D}$. In case $\Omega_n$ is increasing, $g_n$ is defined on $\Omega_m$ once $n > m$. In either case, since $\|g_n\|_\infty \leq 1$, we can pass to a subsequence $\tilde{n}(k) = n(j(k))$ of $\{n(j)\}_{j=1}^\infty$ so that $g_{\tilde{n}(k)} \to g_\infty$ uniformly on compacts of $\Omega_\infty$. Again, by (16.2.25), $g_\infty$ is not constant, so $g_\infty$ is univalent and its range lies in $\mathbb{D}$.

For any $z \in \mathbb{D}$, $f_{\tilde{n}(k)}(z) \to f_\infty(z) \in \Omega_\infty$ as we noted, so $\{f_{\tilde{n}(k)}(z)\}_{k=1}^\infty$ lies in a compact subset of $\Omega_\infty$. By the uniform convergence and uniform equicontinuity, $z = g_{\tilde{n}(k)}(f_{\tilde{n}(k)}(z)) \to g_\infty(f_\infty(z))$. It follows that $\mathrm{Ran}(g_\infty) = \mathbb{D}$ and thus, $g_\infty$ is the unique conformal map of $\Omega_\infty$ to $\mathbb{D}$ with $g_\infty(0) = 0$, $g_\infty'(0) > 0$. It follows from $g_\infty(f_\infty(z)) = z$ that $f_\infty$ is the inverse map of $g_\infty$, that is, $\mathrm{Ran}(f_\infty) = \Omega_\infty$ and $f_\infty$ is the unique map of $\mathbb{D}$ to $\Omega_\infty$ with $f_\infty(0) = 0$, $f_\infty'(0) > 0$. These uniqueness results imply there is a unique limit point of $f_n$ and then of $g_n$, so as claimed, $f_n \to f_\infty$, $g_n \to g_\infty = f_\infty^{-1}$. □

**Proof of Theorem 16.2.1.** It suffices to prove density in $\widetilde{S}$ since if $\{f_n\}_{n=1}^\infty \in \widetilde{S}$ and $f_n \to f$ (with $f \in S$), then $f_n(0)^{-1} f_n \in S$ and $f_n(0)^{-1} f_n \to f$.

Let $\widetilde{A}$ be the set of univalent maps on $\mathbb{D}$, $g$ with $g(0) = 0$, $g'(0) > 0$ so that $g[\mathbb{D}]$ is the interior of an analytic Jordan curve, $\gamma$. Then $\widetilde{A}$ is dense in $\widetilde{S}$ since if $f \in \widetilde{S}$, $g_r(z) = f(rz)$ for $r < 1$ lies in $\widetilde{A}$ and $g_r \to f$ uniformly on compacts.

If $g \in \widetilde{A}$, let $\gamma$ be the curve $g(\partial\mathbb{D})$. Pick $e^{i\theta_0}$ so $|g(e^{i\theta_0})| = \sup_\theta |g(e^{i\theta})|$. Let $h_n$ be the slit mapping corresponding to the curve $\Gamma_n$, which follows $\gamma$ from $g(e^{i\theta_0 + \frac{1}{n}})$ counterclockwise around to $g(e^{i\theta_0})$ and then out to infinity along $[\rho g(e^{i\theta_0}) \mid \rho \geq 1]$. By Proposition 16.2.6, $h_n$ converges to $g$. Thus, the slit maps are dense. □

We now turn to the proof of Theorem 16.2.2. As a preliminary, we need:

**Theorem 16.2.7** (Subordination Principle). *Let $f, g \in \widetilde{S}$ so that $f[\mathbb{D}] \subset g[\mathbb{D}]$ with $f \not\equiv g$. Then*

(a) $f'(0) < g'(0)$

(b) *For every $r \in (0,1)$, $f[\mathbb{D}_r] \subset g[\mathbb{D}_r]$.*

**Proof.** Let $h = g^{-1} \circ f$ which is a well-defined map of $\mathbb{D} \to \mathbb{D}$ since $f[\mathbb{D}] \subset g[\mathbb{D}]$. By the Schwarz lemma, $|h'(0)| < 1$ with $|h'(0)| = 1$ eliminated since $f \not\equiv g \Rightarrow |h'(0)| < 1$. Since $h'(0) = g'(0)^{-1} f'(0)$, we have (a).

Moreover, by the proof of the Schwarz lemma, $|h(z)| \le |z|$. Applying this to $z \in \mathbb{D}_r$, we see $g^{-1}(f[\mathbb{D}_r]) \subset \mathbb{D}_r$. Since $g$ is a bijection, $f[\mathbb{D}_r] \subset g[\mathbb{D}_r]$. □

Given a slit domain, $\Omega = \widehat{\mathbb{C}} \setminus \gamma([0,t_0])$ with $\gamma\colon [0,t_0] \to \widehat{\mathbb{C}}$ continuous, one-one, with $\gamma(t_0) = \infty$, we define for $t \in [0,t_0)$, $\Omega_t = \widehat{\mathbb{C}} \setminus \gamma([t,t_0])$. Let $g_t$ be the unique map in $\widetilde{S}$ with $g_t(0) = 0$, $g_t'(0) > 0$, and $\text{Ran}(g_t) = \Omega_t$. By (16.2.21),

$$g_t'(0) \ge \inf_{s \ge t} |\gamma(s)| \to \infty \quad \text{as } t \uparrow t_0 \tag{16.2.27}$$

By Theorem 16.2.7, $g_t'(0)$ is strictly increasing. If $f = g_{t=0}$ is in $S$, we conclude $t \mapsto g_t'(0)$ is a strictly monotone map of $[0,t_0)$ to $[1,\infty)$. Thus, we can reparametrize $t$ so $t_0 = \infty$ and

$$g_t'(0) = e^t \tag{16.2.28}$$

Thus,

$$g_t(z) = e^t \left[ z + \sum_{n=2}^{\infty} b_n(t) z^n \right] \tag{16.2.29}$$

**Lemma 16.2.8.** *Uniformly on compact subsets of $\mathbb{C}$,*

$$e^t g_t^{-1}(w) \to w \tag{16.2.30}$$

**Proof.** $e^{-t} g_t \in S$, so the growth estimates, (16.1.32), imply for $|z| < 1$ that

$$\frac{|z|}{(1+|z|)^2} \le e^{-t} g_t(z) \le \frac{|z|}{(1-|z|)^2} \tag{16.2.31}$$

Let $w = g_t(z)$. Then (16.2.31) becomes

$$[1 - |g_t^{-1}(w)|]^2 \le e^t \left| \frac{g_t^{-1}(w)}{w} \right| \le [1 + |g_t^{-1}(w)|]^2 \tag{16.2.32}$$

for any $w \in \Omega_t$.

The right inequality plus $g_t^{-1}(w) \in \mathbb{D}$ implies

$$|g_t^{-1}(w)| \le 4|w|e^{-t} \tag{16.2.33}$$

So since $\Omega_t$ increases to $\mathbb{C}$, we conclude uniformly on compact sets of $w \in \mathbb{C}$,

$$g_t^{-1}(w) \to 0 \tag{16.2.34}$$

Therefore, by (16.2.32), uniformly on compact sets of $w$,

$$e^t \left| \frac{g_t^{-1}(w)}{w} \right| \to 1 \tag{16.2.35}$$

Let $h_t(w) = e^t g_t^{-1}(w)/w$. Since $g_t'(0) = e^t$, $(g_t^{-1})'(0) = e^{-t}$, so $h_t(0) = 1$. By (16.2.33), $|h_t(w)| \le 4$, so by Montel's theorem, $h_t(w)$ has analytic limits. By (16.2.35), any such limit has $|h_\infty(w)| = 1$, so by the maximum principle, $h_\infty$ is constant, and by $h_t(0) = 1$, $h_\infty(w) \equiv 1$. Thus, by Montel and uniqueness of the limit, $h_t(w) \to 1$ uniformly on compact subsets of $\mathbb{C}$, which is (16.2.30). $\square$

Now, we define on $\mathbb{D}$,

$$f_s(z) = g_s^{-1}(f(z)) \tag{16.2.36}$$

By $g_s'(0) = e^s$ and $g_{s=0} = f$, we have (16.2.2) and (16.2.30) is (16.2.4). Notice also that $f_s(z) \in \mathbb{D}$. Thus, we need to define $\kappa(s)$ and prove (16.2.3).

By Theorem 8.2.4 of Part 2A, each $g_s$ extends to a continuous map of $\overline{\mathbb{D}}$ to $\widehat{\mathbb{C}}$. There is a unique point, $\lambda(s) \in \partial\mathbb{D}$, with

$$g_s(\lambda(s)) = \gamma(s) \tag{16.2.37}$$

a point $\tilde{\lambda}(s)$ with $g_s(\tilde{\lambda}(s)) = \infty$, and $g_s$ maps the rest of $\partial\mathbb{D}$ two-one to $\gamma[(s,\infty)]$, each interval between $\lambda(s)$ and $\tilde{\lambda}(s)$ maps one-one to $\gamma[(s,\infty)]$; see Figure 16.2.2.

The first key will be to prove $\lambda(s)$ is continuous in $s$. Eventually we'll set $\kappa(s) = \overline{\lambda(s)}$. Fix

$$0 \le s < t < \infty \tag{16.2.38}$$

and define

$$h_{s,t}(z) = g_t^{-1}(g_s(z)) \tag{16.2.39}$$

Initially, we define $h_{s,t}$ for $z \in \mathbb{D}$. Since $\Omega_s \subset \Omega_t$, $h_{s,t}$ is well-defined there. $g_s(z)$ has an extension to $\overline{\mathbb{D}}$, as we've seen. There is an interval $B_{st}$ containing $\lambda(s)$ in its interior that gets mapped by $g_s$ two-one onto $\gamma([s,t])$ (except

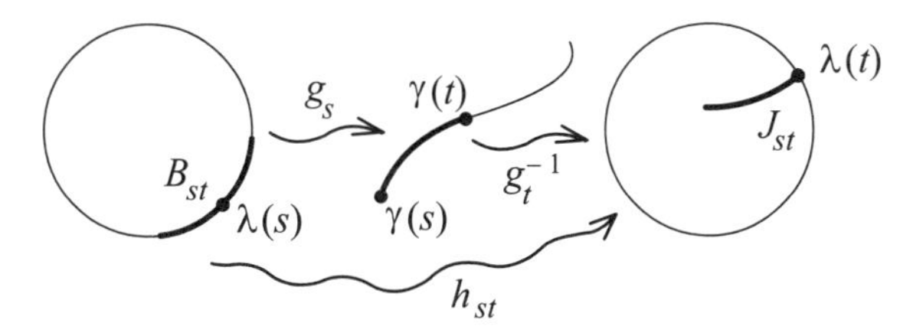


**Figure 16.2.2.** Definition of $B_{st}$ and $J_{st}$.

for $\lambda(s)$ which is the unique solution of (16.2.32)). $g_t^{-1}$ is well-defined on $\Omega_t \cup \gamma(t)$, so $h_{s,t}$ can be extended to a map on $B_{st}$ with image $J_{st}$, a curve in $\mathbb{D}$ ending on $\partial\mathbb{D}$ at $\lambda(t)$ (see Figure 16.2.2). In fact, $h_{s,t}$ has a continuous extension to all of $\overline{\mathbb{D}}$ since as $z \to z_0 \in \partial\mathbb{D} \setminus B_{st}$, $g_s(z)$ approaches $\gamma$ on one side and $g_t^{-1}$ has a well-defined one-sided limit.

**Lemma 16.2.9.** (a) *For $s$ fixed, $B_{st}$ shrinks to $\lambda(s)$ as $t \downarrow s$.*

(b) *$\lambda(s)$ is continuous in $s$.*

(c) *For $t$ fixed, $B_{st}$ shrinks to the point $\lambda(t)$ as $s \uparrow t$ in the sense that $\{z \mid \exists z_n \in B_{s_n t}$ so that $z_n \to z$ and $s_n \uparrow t\} = \{\lambda(t)\}$.*

**Proof.** (a) On $\mathbb{C} \setminus \gamma([s,\infty))$ with the two sides considered as distinct, $g_s^{-1}$ is continuous. Since $\gamma([s,t])$ shrinks to a point as $t \downarrow s$, its image under $g_s^{-1}$ shrinks to zero.

(b) $h_{s,t}$ maps $\mathbb{D}$ to $\mathbb{D}$ and $\partial\mathbb{D} \setminus B_{st}$ to $\partial\mathbb{D}$. Thus, by the reflection principle (using that $h_{s,t}(z) = 0 \Rightarrow z = 0$), $h_{s,t}$ extends analytically to a map of $\mathbb{C} \setminus B_{st}$ to $\mathbb{C}$. Our first goal is to prove that as $t \downarrow s$, $h_{s,t}(z) \to z$ uniformly on compact subsets of $\mathbb{C} \setminus \{\lambda(s)\}$.

To this end, we first prove that

$$\left| \frac{h_{s,t}(z)}{z} \right| \le 4e^{t-s} \tag{16.2.40}$$

By the maximum principle, it suffices to prove this at $0, \infty$ and on the two sides of $B_{st}$. At zero,

$$\lim_{z \to 0} \frac{h_{s,t}(z)}{z} = e^{t-s} \tag{16.2.41}$$

so by reflection,

$$\lim_{z \to \infty} \frac{z}{h_{s,t}(z)} = e^{t-s} \tag{16.2.42}$$

so (16.2.40) holds at 0 and $\infty$. As we approach $B_{st}$ from inside $\mathbb{D}$, $h_{s,t}(z)$ approach $J_{st}$ and $|z| \to 1$, so $|h_{s,t}(z)|/|z|$ is bounded by 1 so (16.2.40) holds on that side.

Since $e^{t-s} h_{s,t} \in S$, the Koebe one-quarter theorem implies $h_{s,t}[\mathbb{D}] \supset \mathbb{D}_{\frac{1}{4}e^{s-t}}$, so $\inf\{|z| \mid z \in J_{st}\} \ge \frac{1}{4}e^{s-t}$. If $J_{st}^*$ is the reflection of $J_{st}$ in $\partial\mathbb{D}$, we conclude

$$\sup\{|z| \mid z \in J_{st}^*\} \le 4e^{t-s} \tag{16.2.43}$$

But the values of the limit of $h_{s,t}$ on the outside of $B_{st}$ lie in $J_{st}^*$, so we have proven (16.2.40).

Let $q_\infty(z)$ be a uniform limit on compacts of $h_{s,t}(z)/z$, which we know exists by (16.2.40). $q_\infty(0) = 1$ and $q_\infty$ is bounded on $\mathbb{C} \setminus \{\lambda(s)\}$. But then $\lambda(s)$ is a removable singularity and, by Liouville's theorem, $q_\infty(z) \equiv 1$.

Thus, every limit point is 1, so $h_{s,t}(z) \to z$ uniformly on compact subsets of $\mathbb{C} \setminus \{\lambda(s)\}$.

Fix $\varepsilon > 0$. Let $C$ be the circle $\{z \mid |z - \lambda(s)| < \varepsilon\}$. By (a), for $t$ close to $s$, $B_{st}$ lies inside $C$. Thus, for such $t$, $J_{st} = h_{s,t}[B_{st}]$ lies in $h_{s,t}[C]$. Since $h_{s,t} \to z$ on $C$, $\operatorname{diam}(h_{s,t})[C] \to \varepsilon$, so for $t$ near $s$, and any $z_0 \in C$, $|h_{s,t}(z_0) - \lambda(t)| \leq 2\varepsilon$. Thus, $|\lambda(t) - \lambda(s)| \leq |\lambda(s) - z_0| + |h_{s,t}(z_0) - z_0| + |h_{s,t}(z_0) - \lambda(t)| \leq \varepsilon + 2\varepsilon + |h_{s,t}(z_0) - z_0| \to 3\varepsilon$. Since $\varepsilon$ is arbitrary, $\lambda$ is continuous as $t \downarrow s$.

Now fix $t$ and consider $s \uparrow t$. By continuity of $g_t$, $J_{st}$ shrinks to $\lambda(t)$. By (16.2.40), $|w/h_{s,t}^{-1}(w)|$ is bounded by $4e^{t-s}$ on $\mathbb{C} \setminus (J_{st} \cup J_{st}^*)$ and goes to 1. Thus, repeating the above with $h_{s,t}, B_{st}$ replaced by $h_{s,t}^{-1}, J_{st} \cup J_{st}^*$, we see that $\lambda(s) - \lambda(t) \to 0$, that is, $\lambda$ is continuous also from the right.

(c) $B_{st}$ is the image of $J_{st} \cup J_{st}^*$, so as above, as $s \uparrow t$, the size of the interval $B_{st}$ shrinks to zero. It contains $\lambda(s)$ and $\lambda(s) \to \lambda(t)$, so $B_{st}$ shrinks to $\{\lambda(t)\}$. □

**Proof of Theorem 16.2.2.** $h_{s,t}(z)/z$ is nonvanishing on $\mathbb{D}$. Let

$$\Phi_{st}(z) = \log\left[\frac{h_{s,t}(z)}{z}\right] \tag{16.2.44}$$

be the branch of log with $\Phi_{st}(0) = s - t$. By the properties of $h_{s,t}$, $\Phi_{st}$ has a continuous extension to $\overline{\mathbb{D}}$ and $\operatorname{Re}[\Phi_{st}(e^{i\theta})]$ is 0 on $\partial\mathbb{D} \setminus B_{st}$ and $< 0$ on $B_{st}$ (since $J_{st} \subset \mathbb{D}$). Let $\alpha_{st}, \beta_{st}$ be the endpoints of $B_{st}$. Then, by the complex Poisson formula (see Theorem 5.3.2 of Part 2A), for $z \in \mathbb{D}$,

$$\Phi_{st}(z) = \frac{1}{2\pi} \int_{\alpha_{st}}^{\beta_{st}} \operatorname{Re}[\Phi_{st}(e^{i\theta})] \, \frac{e^{i\theta} + z}{e^{i\theta} - z} \, d\theta \tag{16.2.45}$$

Taking $z = 0$,

$$s - t = \frac{1}{2\pi} \int_{\alpha_{st}}^{\beta_{st}} \operatorname{Re}[\Phi_{st}(e^{i\theta})] \, d\theta \tag{16.2.46}$$

Since $h_{s,t}(f_s(z)) = f_t(z)$ and $f_s(z) \in \mathbb{D}$, (16.2.45) implies

$$\begin{aligned} \log \frac{f_t(z)}{f_s(z)} &= \frac{1}{2\pi} \int_{\alpha_{st}}^{\beta_{st}} \operatorname{Re}[\Phi_{st}(e^{i\theta})] \left[\frac{e^{i\theta} + f_s(z)}{e^{i\theta} - f_s(z)}\right] d\theta \\ &= (s-t)\left[\operatorname{Re}\left[\frac{e^{i\sigma} + f_s(z)}{e^{i\sigma} - f_s(z)}\right] + i \operatorname{Im}\left[\frac{e^{i\tau} + f_s(z)}{e^{i\tau} - f_s(z)}\right]\right] \end{aligned} \tag{16.2.47}$$

where we use (16.2.46) and the mean value theorem separately on Re[ ] and Im[ ] (this theorem says if $f \geq 0$ on $(a,b)$, $g$ is real-valued and continuous, then for some $c \in (a,b)$, $\int_a^b f(s)g(s)\,ds = g(c)\int_a^b f(s)\,ds$). Here $\sigma, \tau \in (\alpha_{st}, \beta_{st})$.

By (a), (c) of Lemma 16.2.9 as $t \downarrow s$ or $s \uparrow t$, $|\alpha_{st} - \beta_{st}| \to 0$ and any point in between goes to $\lambda$. Thus,

$$\frac{\partial}{\partial s} \log(f_s(z)) = -\frac{\lambda(s) + f_s(z)}{\lambda(s) - f_s(z)} \tag{16.2.48}$$

With $\kappa(s) = 1/\lambda(s) = \overline{\lambda(s)}$, this is (16.2.3). □

**Notes and Historical Remarks.** The Carathéodory convergence theorem is from his 1912 paper [**69**]. The result is much more general than the version we need and so state. Let $\Omega_n$ be a sequence of simply connected regions with $0 \in \Omega_n$. Their *kernel* is set to be $\{0\}$ if $\inf_n[\inf\{z \in \mathbb{C} \setminus \Omega_n\}] = 0$ and otherwise is the largest region, $\Omega_\infty$, containing 0 so that for any $K \subset \Omega_\infty$ compact, eventually $K \subset \Omega_n$. We say $\Omega_n$ converges to its kernel, $\Omega_\infty$, if and only if $\Omega_\infty$ is the kernel of every subsequence $\Omega_{n(j)}$. The full Carathéodory convergence theorem says that if $f_n\colon \mathbb{D} \to \Omega_n$ is the conformal map with $f_n(0) = 0$, $f_n'(0) > 0$, then $f_n \to f_\infty$, some function on $\mathbb{D}$, uniformly on compacts, if and only if $\Omega_n$ converges to $\Omega_\infty$, and in that case, either $\Omega_\infty = \{0\}$, $f \equiv 0$, or $\Omega_\infty$ is simply connected, $f_\infty$ is a conformal map of $\mathbb{D}$ to $\Omega_\infty$, and $f_n^{-1} \to f_\infty^{-1}$ uniformly on compact subsets of $\Omega_\infty$. For a proof, see Duren [**111**].

Loewner's theorem and differential equation are from his 1923 paper [**267**]. The title has a "I" at the end. The story is told that Loewner submitted the paper to Bieberbach who was an editor of *Mathematische Annalen.* Bieberbach was excited by the paper, asserted to many that he was certain Loewner's method would yield his full conjecture, and given his certainty of this, Bieberbach added the "I." Some versions of the story assert that Loewner was so annoyed by this that he stopped working on the problem.

Charles Loewner (1893–1968) was born Karel Löwner in Prague, started using Karl when he attended the German (aka Charles) University of Prague where he got a Ph.D. in 1917 under Pick's supervision. In turn, his students included Lipman Bers, Roger Horn, and Adriano Garsia. Besides his work on univalent functions, Loewner proved a beautiful theorem on monotone matrix functions (discussed briefly in Problem 2 of Section 7.5 of Part 4 and, in more detail in Simon[**355**]).

In 1929, he returned to Prague. He became uneasy about the situation in Germany and in 1936 began learning English. When the Germans occupied Prague in 1939, he was immediately arrested, probably more for his left-wing political views than because he was Jewish. His wife spent a week negotiating with the Gestapo until they agreed to release him if he paid twice the exit tax and emigrated. (In a commentary of life's changes, I talked to one of Loewner's colleagues at Stanford who remembers Mrs. Loewner as a real estate broker!)

Arriving in America, he changed his name to Charles Loewner, and this is the usual contemporary spelling instead of the Löwner he published his greatest paper under. Loewner spent his first years in America at various universities until Szegő, in 1951, brought him to Stanford. Loewner's work was given greater acclaim after his death than before due to the work of de Branges and Schramm and a growing appreciation of his work on monotone matrix functions. As with Bergman, we have respected Loewner's decision to change his name to the extent of changing how it appears in the bibliography so that it agrees with his choice rather than what you'll find on the original paper.

## 16.3. SLE: A First Glimpse

In this section, we'll introduce the reader to the theory of Schramm–Loewner evolution. The section will be descriptive without explicit theorems or proofs (and with some technicalities suppressed) but we hope the reader will get a glimpse of this beautiful subject as it links to the earlier themes of this chapter.

To understand the context, it helps to first understand why Brownian motion occurs so often in models of nature. Of course, one reason is its connection to $e^{t\Delta}$ as seen in (4.16.37) of Part 1. But another is the following. Let $\{x(t)\}_{t\geq 0}$ be a family of random variables, $L^2$ continuous in $t$ with $x(0)=0$, $\mathbb{E}(x(t))=0$, and so that $x$ has the strong Markov-type condition of independent increments: explicitly, for any $t_0\geq 0$ and $\Delta t>0$ and any $n$, $\{x(t_0+(j+1)\Delta t)-x(t_0+j\Delta t)\}_{j=0}^{n-1}$ are independent and identically distributed with distribution only depending on $\Delta t$ and not on $t_0$. This $x(t)$ is "essentially" Brownian motion. For $x(t)$ is a limit of sums of iids $\{x(\frac{j+1}{n})-x(\frac{j}{n})\}_{j=0}^{n-1}$, so Gaussian by the central limit theorem (see Section 7.3 of Part 1). By the independence and identical distribution $\mathbb{E}(x(t)^2)=n\,\mathbb{E}(x(\frac{t}{n})^2)$ so by continuity $\mathbb{E}(x(t)^2)=Ct$. Moreover, by the independence, for $t>s$, $\mathbb{E}(x(t)x(s))=\mathbb{E}([(x(t)-x(s))+x(s)]x(s))=\mathbb{E}(x(s)^2)$, i.e.,

$$\mathbb{E}(x(t)x(s))=C\min(t,s)$$

so $x(s)=\sqrt{C}\,b(s)$. If one prefers, one can absorb the constant into time and get $x(s)=b(Cs)$.

We thus expect any reasonable limit of discrete time processes independent increments (e.g., random walks) should be Brownian motion. There are limitations to this connected with Levy processes and the fact that the limits have got to have finite variance, but it is a wonderful guideline as seen by Donsker's theorem as discussed in the Notes to Section 4.16 of Part 1.

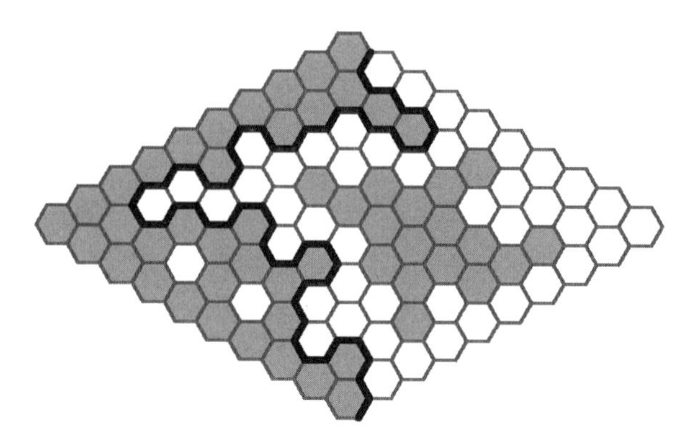

**Figure 16.3.1.** A $10 \times 10$ grid percolation.

We want to consider some random curves which typically arise as scaling limits of boundary interfaces. We'll look at the percolation exploration process.

Figure 16.3.1 shows a $10 \times 10$ grid of a two-dimensional hexagonal lattice. We color the 19 hexagons along the left edge grey and the 17 hexagons along the right edge white. For all the remaining hexagons, we pick to color them grey or white randomly as independent choices with probability $1/2$. The reader can convince themselves that if they only consider the hexagon boundary edges between hexagons of opposite colors, they'll get several closed curves (three in the above figure) and one curve, shown here in black that extends to the boundary of the array.

Next, we consider the $40 \times 40$ array shown in Figure 16.3.2. Again, we fix the boundary regions and choose the other hexagons randomly of each other (and of the $10 \times 10$ choices) and look at the unique color boundary curve stretching to the edges of the array.

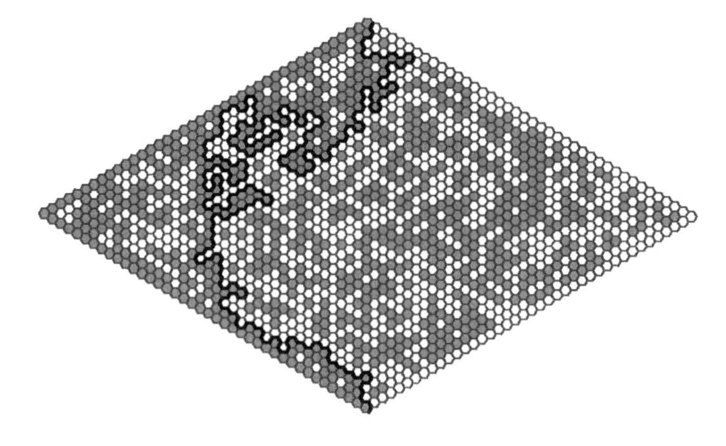

**Figure 16.3.2.** A $40 \times 40$ grid percolation.

We are interested in the scaling limit—taking the number of hexagons to infinity but shrinking hexagon size so the area is constant. The hope is that, as with Brownian motion and random walks, the limit might be a mathematically accessible object—a set of random curves with a simple probability law.

The limiting interface curve should have a description as a function of $t$. It is certainly not Markovian—indeed, the curve is not self-intersecting (as is two-dimensional Brownian motion by recurrence, i.e., return to the origin) which means the future depends not just on the present but on the entire past.

Two concepts are critical to the understanding of the scaling limit—one is that it is *conformally invariant*, that is, it was observed, originally numerically, that suitable probabilities are invariant under conformal transformation.

The other is that the limit has the *domain Markov property*: namely, if, as with the Loewner parameterization, we describe the curve by a set of suitably normalized conformal maps $f_s$ taking $\mathbb{D}$ to our region with the curve to time $s$ removed, then $f_s$ has independent increments in the sense that $f_s^{-1} \circ f_{s+\Delta s}$ and $f_{s+\Delta s}^{-1} \circ f_{s+2\Delta s}$ are independent random univalent functions which are identically distributed.

This leads to a Loewner evolution where the forcing function, $K(s)$, is a multiple of Brownian motion, the celebrated stochastic Loewner evolution.

For many problems, the relevant Loewner evolution is not (16.2.3) which describes what is called *radial Loewner evolution* which has conformal maps from $\mathbb{D}$ to its exterior. Instead, one has a family of conformal maps of $\mathbb{C}_+$ minus a curve to $\mathbb{C}_+$ called *chordal Loewner evolution.* These maps obey

$$\dot{g}_t(z) = \frac{2}{g_t(z) - \sqrt{\kappa}B(t)}, \qquad g_{t=0}(t) = z \tag{16.3.1}$$

Here, $B(t)$ is a one-dimensional Brownian path thought of as a real number subtracted from the complex number $g_t(z)$, so for each such path there is a $g$ and the probability measure on Brownian paths induces one on the $g$'s. $\kappa > 0$ is a real parameter and there are families of SLE, denoted by $\text{SLE}_\kappa$.

One initial subtlety in the theory concerns the solvability of (16.3.1). In the last section, one started with a curve and derived the Loewner ODE and forcing function, $K(t)$. In the SLE definition, one starts with $K(t) = \sqrt{\kappa}B(t)$ and wants the curve. Prior to the work on SLE, the best existence results required $K$ to be Hölder continuous of order $1/2$. Brownian motion is, with probability 1, Hölder continuous of order $\alpha < \frac{1}{2}$ but not order $1/2$ (see Theorems 4.16.6 and 4.16.7 and (4.16.40) of Part 1) so the construction of SLE required a new existence result (see the Notes).

It has been proven that if $\kappa \leq 8$, $\mathrm{SLE}_\kappa$ curves are of Hausdorff dimension $1 + \frac{\kappa}{8}$ with probability 1 and they are space filling if $\kappa \geq 8$. There is also a transition at $\kappa = 4$. For $\kappa \leq 4$, the curve is simple almost surely and does not touch itself or the boundary. For $\kappa > 4$, it has touching points of itself and the boundary. For scaling limits of certain models, it has been rigorously proven that $\mathrm{SLE}_\kappa$ describes the limit for suitable values of $\kappa$. In particular, the percolation exploration process converges to $\mathrm{SLE}_6$.

The reader should have a glimpse into why a synthesis of the material earlier in this chapter and probability theory has become important in pure mathematics and also in statistical physics.

**Notes and Historical Remarks.** SLE was invented by Oded Schramm (1961–2008) in 2000 [**342**] whose partial motivation came from the discovery (non-rigorously) that many scaling limits in two-dimensional statistical physics were conformally invariant. He conjectured that $\mathrm{SLE}_\kappa$ precisely described the scaling limit of several models of probability theory and/or statistical physics and among other things would prove conformal invariance. Schramm named them SLE for "stochastic Loewner evolution" but especially since his death in a hiking accident, it is used for Schramm–Loewner evolution.

Two subsequent developments caused an explosion of interest in the subject. In 2001, Stanislav Smirnov [**361**] proved a crossing probability for critical bond percolation on a triangular lattice was given by a formula the theoretical physicist John Cardy found [**70**] assuming conformal invariance and his work indicated a close connection to $\mathrm{SLE}_6$. Also, in 2001, Lawler, Schramm, and Werner [**255**] used $\mathrm{SLE}_6$ to prove a conjecture of Mandelbrot [**270**] that the dimension of the boundary of planar Brownian motion was 4/3. Werner and Smirnov received the Fields Medal for this and related work.

In a series of papers [**256, 257, 258, 259**], Lawler, Schramm, and Werner analyzed $\mathrm{SLE}_\kappa$ further and, in particular, proved that loop-erased random walk converged to $\mathrm{SLE}_2$. That the percolation exploration model converges to $\mathrm{SLE}_6$ can be found in Camia–Newman [**67**]; see also Werner [**409**], that loop-erased random walks converge to $\mathrm{SLE}_2$ is due to Lawler–Schramm–Werner [**260**], and that interfaces in the critical Ising model converge to $\mathrm{SLE}_3$ is due to Chelkak-Smirnov [**84, 85**].

The result on the dimension of $\mathrm{SLE}_\kappa$ mentioned in the text is from Beffara [**34**] and the existence result for the SLE solvability mentioned is due to Rohde–Schramm [**334**]. The transition at $\kappa = 4$ is from Alberts–Sheffield [**11**].

Two general introductions to SLE are a book of Lawler [**254**] and the St. Fleur lecture notes of Werner [**408**]. Some additional references are [**33, 35, 46, 54, 166, 224, 239, 301**].

*Chapter 17*

# Nevanlinna Theory

> When he [Ahlfors] received his Fields Medal the following year, Carathéodory remarked that it was hard to say which was more surprising: that Nevanlinna could develop his entire theory without the geometric picture to go with it or that Ahlfors could condense the whole theory into fourteen pages.
>
> —*Robert Osserman* [**64**]

**Big Notions and Theorems:** Counting Function, Proximity Function, Nevanlinna Characteristic, First Main Theorem (FMT), Second Main Theorem (SMT), Relation of Nevanlinna Characteristic to Order, Newton's Lemma, Cartan's Identity, Critical Characteristic, Deficiency, Critical Deficiency, SMT Implication for Deficiencies, Picard's Theorem, Borel's Theorem, Four Totally Ramified Values Theorem, Two Totally Ramified Values Theorem, Four Common Values Theorem, Ahlfors–Shimizu Characteristic, Ahlfors' Proof of the SMT

This chapter will look into refinements of Picard's theorem. One result we have in mind is intermediate between the great and little Picard theorem: namely, that an entire meromorphic function[1] that is not a rational function takes all but two values infinitely often. (The little Picard theorem says all values but two are taken, while the great Picard theorem says infinitely often but only requires the function to be defined and meromorphic in a neighborhood of infinity.)

Here are the kinds of questions we have in mind. Suppose $a$ and $b$ are two distinct values that $f$ takes. Is there a sense that the number of times

[1] Recall we use this to indicate a function meromorphic in all of $\mathbb{C}$.

$f$ takes the value $a$ in $\mathbb{D}_R(0)$ is asymptotically the same as the number of times the value $b$ is taken in $\mathbb{D}_R(0)$?

It will be natural to look at certain weighted averages of the number of times a value is taken.

**Definition.** For any $a \in \widehat{\mathbb{C}}$ and entire meromorphic function, $f$, $n_f(r, a)$ is the number of times $f$ takes the value $a$ in $\overline{\mathbb{D}_r(0)}$ counting multiplicity, that is, if the points in $\overline{\mathbb{D}_r(0)}$ where $f(\zeta) = a$ are $\{\zeta_j\}_{j=1}^{V_r}$ with $m_j$ the multiplicity of the zero of $f(\zeta) - a$ at $\zeta_j$, then

$$n_f(r, a) = \sum_{j=1}^{V_r} m_j \tag{17.0.1}$$

The *counting function*, $N_f(r, a)$, is given by

$$N_f(r, a) = \int_0^r \frac{n_f(s, a) - n_f(0, a)}{s}\, ds + n_f(0, a) \log r \tag{17.0.2}$$

$$= n_f(0, a) \log r + \sum_{\zeta_j \neq 0} m_j \log\left(\frac{r}{|\zeta_j|}\right) \tag{17.0.3}$$

It is a theorem of Valiron [**385**] and Littlewood [**264**] that for all $a, b \notin S$, a subset of $\mathbb{C}$ of Lebesgue measure zero (depending on $f$), one has that

$$\lim_{r \to \infty} \frac{N_f(r, a)}{N_f(r, b)} = 1 \tag{17.0.4}$$

and Ahlfors [**6**] proved that $S$ is a polar set (i.e., has logarithmic capacity zero). One of our later results (Theorem 17.3.5) will prove a weaker result.

Besides these results, we'll see:

(1) There are at most four values, $a_j$, for which every $z_0$ with $f(z_0) = a_j$, we have $f'(z_0) = 0$ (see Theorem 17.3.8).
(2) For entire functions (i.e., no poles), there are at most two values with the property in (1) (see Theorem 17.3.10).
(3) If $f, g$ are distinct, nonconstant entire meromorphic functions, there are at most four values, $a_j$, with $\{z \mid f(z) = a_j\} = \{z \mid g(z) = a_j\}$ (see Theorem 17.3.12).

Moreover, we'll see by simple examples that the numbers four, two, four are optimal.

To understand the ideas behind this remarkable set of results, we begin with a simple example.

**Example 17.0.1.** Let $f(z) = \exp(z)$, the simplest example of a function missing two values, namely, zero and infinity. Fix $R$ large. Then

$$|f(Re^{i\theta})| = \exp(R \cos \theta) \tag{17.0.5}$$

If $z_0 \neq 0, \infty$, $f(Re^{i\theta})$ can be near $z_0$ on a set of $\theta$ of size $O(R^{-1})$ but on each interval $\pm(-\frac{\pi}{2}+\varepsilon, \frac{\pi}{2}-\varepsilon)$, $|f(Re^{i\theta})|$ goes to $+\infty$ (respectively, 0) for the $+(\dots)$ (respectively, $-(\dots)$) case. This says that the missing values of $f$ occur as the most common approximate value for $|z|$ near $\infty$. $\square$

The first key notion of Nevanlinna theory is that while $N_f(r,a)$ can be highly dependent on $a$, the sum of $N_f$, and a measure of the closeness of $f$ to $a$ is asymptotically independent of $a$. One defines the *proximity function*, $m_f(r,a)$, by

$$m_f(r,\infty) = \frac{1}{2\pi}\int_0^{2\pi} \log_+|f(re^{i\theta})|\, d\theta \tag{17.0.6}$$

where, as usual, $\log_+(x) = \max(0, \log x)$ for $x \geq 0$. For $a \neq \infty$,

$$m_f(r,a) = \frac{1}{2\pi}\int_0^{2\pi} \log_+[|f(re^{i\theta}) - a|^{-1}] \tag{17.0.7}$$

The *Nevanlinna characteristic* is

$$T_f(r) = m_f(r,\infty) + N(r,\infty) \tag{17.0.8}$$

In Section 17.1, we'll prove Nevanlinna's first main theorem (FMT): For each $a \in \mathbb{C}$,

$$m_f(r,a) + N_f(r,a) = T_f(r) + O(1) \tag{17.0.9}$$

as $r \to \infty$. This will essentially be a clever rewriting of Jensen's formula. We'll also show for rational functions, $f$, of degree $d$, $T_f(r) = d\log r + O(1)$ and for all other entire meromorphic functions,

$$\lim_{r\to\infty} \frac{T_f(r)}{\log r} = \infty \tag{17.0.10}$$

We'll also discuss the relation of $T_f(r)$ to the order for entire functions of finite order. Notice that since $m_f \geq 0$, (17.0.9) implies that

$$\limsup_{r\to\infty} \frac{N_f(r,a)}{T_f(r)} \leq 1 \tag{17.0.11}$$

In Section 17.2, we'll prove a remarkable identity of H. Cartan (when $f(0) \neq \infty$):

$$T_f(r) = \int_0^{2\pi} N_f(r, e^{i\theta})\, \frac{d\theta}{2\pi} + \log_+|f(0)| \tag{17.0.12}$$

This will, first of all, prove that $T_f(r)$ is a convex function of $\log r$. (If $f$ is entire, $T_f(r) = m_f(r,\infty)$, which is a convex function of $\log r$, by a subharmonicity argument; see Theorem 5.1.2 of Part 3. But if $f$ takes all values, this convexity argument doesn't follow from subharmonicity.) More importantly, (17.0.12) is a first hint that for most $a$, $N_f$ dominates $m_f$. Indeed, (17.0.12) implies the lim sup in (17.0.11) is 1 for $d^2z$-Lebesgue a.e. $a$.

Sections 17.3 and 17.4 turn to the second main theorem (SMT) of Nevanlinna theory that demonstrates that for most $a$, $m_f(r,a)$ is small compared to $T_f(r)$ for a sequence of $a$'s going to infinity. Section 17.3 states and applies the SMT and Section 17.4 provides a proof.

$N_{1,f}(r)$ is the counting function associated to critical points of $f$. If $f$ is entire, $N_{1,f}(r) = N_{f'}(r,0)$, but if $f$ has poles, we want $N_{1,f}$ to count higher-order poles (just as $N_{1,f}$ counts points where, for some $a$, $(f(z)-a)$ vanishes to order $D>1$, giving that point with weight $D-1$, so for poles of order $D>1$, we count the points with weight $D-1$). The SMT says that for any distinct $\{a_1,\dots,a_q\}\in\widehat{\mathbb{C}}$, we have for $r$ not in some set $E$ of finite Lebesgue measure

$$\sum_{j=1}^{q} m_f(r,a_j) + N_{1,f}(r) \le 2T_f(r) + O(\log(rT_f(r))) \tag{17.0.13}$$

If $f$ is not a rational function, we have, by (17.0.10), that $\log r = o(T_f(r))$, so the error in (17.0.13) is $o(T_f(r))$. In particular, (17.0.13) immediately implies Picard's theorem, since if $N_f(r,a_j)=0$, then by the FMT, $m_f(r,a_j) = T_f(r)+O(1)$. Thus, the left side of (17.0.13) for $q=3$ and $\{a_1,a_2,a_2\}$ three excluded values $\ge 3T_f(r)+O(1)$, violating (17.0.13).

**Notes and Historical Remarks.** R. Nevanlinna developed his theory in an attempt to understand Borel's refinement of Picard's theorem that for finite-order, entire analytic functions, the exponent of convergence of the zeros of $f(z)-a$ is the same for all values of $a$ with at most one exception. Nevanlinna did this in a 1925 paper [**291**] and the theory codified in a 1929 book [**292**] (and in a later 1936 book [**293**]).

Rolf Nevanlinna (1895–1980) was a Finnish mathematician with a German mother. Both he and his older brother Frithiof (1894–1977) were students of their cousin Ernst Lindelöf (and, in turn, Rolf was Ahlfors' teacher). Both brothers developed their celebrated theory (as we'll see, Frithiof was central to the geometric approach to Section 17.4) while working in nonacademic jobs: Frithiof in an insurance company and Rolf as a teacher in a high school, which only allowed him to do research in the evenings. After his great 1925 paper, he was appointed a professor in the University of Helsinki. He was elected Rector in 1941.

His actions during the war have caused some controversy. He was attracted to right-wing politics throughout his life and was a strong German sympathizer during the war, being in charge of a committee whose task was to recruit young Finns to the Finnish branch of the SS. After the war, because of these activities, he was dismissed as Rector and spent some time in Zurich. He was rehabilitated enough to become president of the ICM in 1959.

His defenders claim he was only motivated by patriotism and anti-Russian sentiments—Finland had broken from the Russian empire at the time of the Russian revolution (Nevanlinna had fought in the following Finnish civil war on the anti-Communist side) and, in 1939, the Russians had attacked Finland. It is also true that there is no evidence of anti-Semitic actions and he was half-German. That said, given this history, I am among those who question the appropriateness of an ICM prize named in his honor. Books on the subject of Nevanlinna theory include Sario et al. [**338**], Stoll [**373**], Hayman [**183**], Goldberg–Ostrovskii [**151**], and Cherry–Ye [**86**].

We should also mention the unpublished note of Eremenko [**121**], from which we borrowed heavily in our presentation in this chapter.

In his paper, Nevanlinna called his two theorems "Hauptsatz," which can be translated "main theorem" or "fundamental theorem," and both terms have been used in the English language literature—about equally with, perhaps, "main theorem" being more frequent in the recent literature.

In this chapter, we'll only discuss Nevanlinna theory for entire meromorphic functions, but there have been many other arenas where it has been developed. Nevanlinna himself discussed functions on the disk. If $f$ has no poles, the Nevanlinna characteristic is finite if $\sup_{0<r<1} \int_0^{2\pi} \log_+|f(re^{i\theta})| \frac{d\theta}{2\pi} < \infty$. Nevanlinna called such functions of finite characteristic—the more common name is the Nevanlinna class—they will play a major role in Chapter 5 of Part 3. Other names are "of bounded characteristic" or "bounded type."

Going back to Cartan [**73, 75**], with later contributions by Ahlfors [**9**], Weyl [**410**], Shiffman [**347**], and Eremenko–Sodin [**122, 123**], are extensions to curves in $\mathbb{CP}[n]$ (an entire meromorphic function being $n = 1$). A recent summary of this area is Lang [**247**]. Not only have there been extensions that allow the range to have higher dimension, there is work, initiated by Griffiths [**163, 164**], allowing the domain to be a higher dimension. See Noguchi–Winkelmann [**302**].

A different theme, initiated by Milloux [**277**], involves relations between values of $f$ and its derivatives. A high point is the Hayman alternative [**182**]: If $f$ is an entire nonpolynomial function and $f$ fails to take any value infinitely often, then its $k$-th derivative, $f^{(k)}$, takes every value, except perhaps 0, infinitely often. For a recent paper in this area, see Bergweiler–Langley [**53**]. For this and applications in the theory of complex ODEs, see Laine [**238**] and Eremenko [**120**].

Finally, we mention a deep analogy found by Osgood [**309**] and Vojta [**396**] between value distribution theory and Diophantine approximation in number theory. From this point of view, the 2 in the SMT (instead of coming from the Euler characteristic of $\widehat{\mathbb{C}}$) is the analog of the 2 in Roth's theorem on approximation by rationals. See the book of Ru [**336**].

There is an enormous literature that has evolved, especially on refined error estimates—it is reviewed in the books of Lang–Cherry [**248**] and Cherry–Ye [**86**].

## 17.1. The First Main Theorem of Nevanlinna Theory

Our main goal in this section is to prove the FMT, (17.0.9). We begin by extending Jensen's formula, Theorem 9.8.1 of Part 2A, that if $f$ is entire with $f(0) \neq 0$, then for any $r$,

$$\int_0^{2\pi} \log|f(re^{i\theta})| \frac{d\theta}{2\pi} = \log|f(0)| + \sum_{j=1}^{n_f(r,0)} \log\left(\frac{r}{|z_j|}\right) \tag{17.1.1}$$

where $\{z_j\}_{j=1}^{n_f(r,0)}$ are the zeros of $f$ in $\mathbb{D}_r(0)$, counting multiplicity.

First, we wish to allow entire meromorphic functions. $f$ can be written $g/h$ with $g, h$ entire (see Corollary 9.4.4 of Part 2A; alternatively, one has (17.1.1) for functions analytic in a neighborhood of $\overline{\mathbb{D}_r(0)}$, in which case we can take $h$ as a polynomial). If $\{w_j\}_{j=1}^{n_f(r,0)}$ are the poles subtracting (17.1.1) for $g$ from that for $h$, we get

$$\int_0^{2\pi} \log|f(re^{i\theta})| \frac{d\theta}{2\pi} = \log|f(0)| + \sum_{j=1}^{n_f(r,0)} \log\left(\frac{r}{|z_j|}\right) - \sum_{j=1}^{n_f(r,\infty)} \log\left(\frac{r}{|w_j|}\right) \tag{17.1.2}$$

Second, we want to allow zeros or poles at zero. Write

$$f(z) = \eta(f) z^{m_0(f)}(1 + O(z)) \tag{17.1.3}$$

where $\eta(f) \neq 0$. Then (17.1.2) for $f/z^{m_0(f)}$, we get

$$\begin{aligned}\int_0^{2\pi} \log|f(re^{i\theta})| \frac{d\theta}{2\pi} &= \log|\eta(f)| + \sum_{j=1}^{n_f(r,0)} \log\left(\frac{r}{|z_j|}\right) \\ &\quad - \sum_{j=1}^{n_f(r,\infty)} \log\left(\frac{r}{|w_j|}\right) + m_0(f)\log r\end{aligned} \tag{17.1.4}$$

Recognizing (17.0.3) for $N_f(r,0)$ and $N_f(r,\infty)$ and noting that if $m_0(f) > 0$, then $m_0 = n(0, a = 0)$, and if $m_0(f) < 0$, then $m_0 = -n(0, a = \infty)$, (17.1.4) can be rewritten.

**Theorem 17.1.1** (Jensen's Formula). *For an entire meromorphic function, $f$, and all $r$,*

$$\int_0^{2\pi} \log|f(re^{i\theta})| \frac{d\theta}{2\pi} = \log|\eta(f)| + N_f(r,0) - N_f(r,\infty) \tag{17.1.5}$$

**Lemma 17.1.2.** *For $0 < x, y$, we have*

$$\log_+(x+y) \le \log_+(x) + \log_+(y) + \log 2 \tag{17.1.6}$$

$$\log_+(|x-y|) \le \log_+(x) + \log_+(y) \tag{17.1.7}$$

$$\log_+(xy) \le \log_+(x) + \log_+(y) \tag{17.1.8}$$

**Proof.** Without loss, suppose $0 < y \le x$. Then since $\log_+(\cdot)$ is monotone,

$$\log_+(x+y) \le \log_+(2x) \le \log_+(x) + \log 2 \tag{17.1.9}$$

implying (17.1.6). (17.1.7) follows immediately from monotonicity and positivity of $\log_+(\cdot)$. (17.1.8) is immediate from $(u+v)_+ \le u_+ + v_+$ for all $u, v \in \mathbb{R}$ with $(u)_+ = \max(u, 0)$. □

**Theorem 17.1.3** (FMT). *Define $T_f(r)$ by*

$$T_f(r) = N_f(r,\infty) + m_f(r,\infty) \tag{17.1.10}$$

*where $N_f$ is given by* (17.0.2) *and $m_f(r,\infty)$ by* (17.0.6). *Then for any $a \in \widehat{\mathbb{C}}$,*

$$N_f(r,a) + m_f(r,a) = T_f(r) + O(1) \tag{17.1.11}$$

**Remark.** The proof shows the $O(1)$ term is bounded by

$$|\log|\eta(f-a)|| + \log_+(|a|) + \log 2$$

**Proof.** The result holds by definition if $a = \infty$, so we suppose $a \in \mathbb{C}$. Noting that

$$\int_0^{2\pi} \log|f(re^{i\theta})| \frac{d\theta}{2\pi} = m_f(r,\infty) - m_f(r,0) \tag{17.1.12}$$

we see (17.1.5) says that

$$N_f(r,0) + m_f(r,0) = T_f(r) - \log|\eta(f)| \tag{17.1.13}$$

This is (17.1.11) for $a = 0$.

We note next that, by (17.1.6)/(17.1.7),

$$\big|\log_+(|f(z)-a|) - \log_+(|f(z)|)\big| \le \log_+(|a|) + \log 2 \tag{17.1.14}$$

so

$$|m_{f-a}(r,\infty) - m_f(r,\infty)| \le \log_+(|a|) + \log 2 \tag{17.1.15}$$

Since $N_f(r,\infty) = N_{f-a}(r,\infty)$, we have

$$|T_{f-a}(r) - T_f(r)| \le \log_+(|a|) + \log 2 \tag{17.1.16}$$

Therefore, (17.1.13) for $f-a$ implies

$$\begin{aligned} N_f(r,a)+m_f(r,a) &\equiv N_{f-a}(r,0)+m_{f-a}(r,0) \\ &= T_{f-a}(r)-\log|\eta(f-a)| \\ &= T_f(r)-\log|\eta(f-a)|+\log_+(|a|)+\log 2 \end{aligned} \tag{17.1.17}$$

□

**Example 17.1.4** (Example 16.0.1, cont.). Let $f(z)=\exp(z)$. $\log|f(re^{i\theta})|$ $=r\cos\theta$, so

$$T_f(r)\equiv m_f(r,\infty)=\int_{-\pi/2}^{\pi/2} r\cos\theta\,\frac{d\theta}{2\pi}=\frac{r}{\pi} \tag{17.1.18}$$

On the other hand, if $a=1$, $f(z)=1$ for $z=2\pi i n$; $n\in\mathbb{Z}$. If $r=2\pi m$, we have

$$\begin{aligned} N_f(2\pi m,1) &= \log m+2\sum_{j=1}^{m-1}\log\left(\frac{m}{j}\right) \\ &= (2m+1)\log m-2\log(m!) \end{aligned} \tag{17.1.19}$$

By (9.7.1) of Part 2A,

$$\log(m!)=(m+\tfrac{1}{2})\log m-m+\log\sqrt{2\pi}+o(1) \tag{17.1.20}$$

so

$$N_f(2\pi m,1)=\frac{2\pi m}{\pi}+O(1)$$

By monotonicity of $N_f(r,1)$ in $r$,

$$N_f(r,1)=\frac{r}{\pi}+O(1) \tag{17.1.21}$$

It is easy to see (Problem 1) that $m_f(r,1)\to 0$. Thus, we see the tradeoff between $N_f$ and $m_f$ illustrated quantitatively. We also see that a naive idea that $N_f(r,a)\sim n_f(r,a)\log r$ is wrong. In this case,

$$\frac{n_f(r,1)}{N_f(r,1)}\to 1 \tag{17.1.22}$$

In fact, this is true for all $a\neq 0,\infty$.

In the above, we used Stirling's formula to verify the FMT but the argument works backwards—the FMT for $f(z)=e^z$ implies Stirling's formula! □

**Example 17.1.5.** Let $f(z)=2\cosh z=e^z+e^{-z}$. It is easy to see that $\log|f(re^{i\theta})|=r|\cos\theta|+O(e^{-2r|\cos\theta|})$ so

$$T_f(r)=\frac{2r}{\pi}+O(r^{-1}) \tag{17.1.23}$$

The extra factor of 2 vis-à-vis $e^z$ is accommodated in $N_f$ by values being taken twice per period (so there are twice as many solutions of $f(z) = a$). □

**Proposition 17.1.6.** *For two entire meromorphic functions, $f, g$, and $\lambda \in \mathbb{C} \setminus \{0\}$, we have*

$$\text{(a) } T_{f^n}(r) = nT_f(r) \tag{17.1.24}$$

$$\text{(b) } T_{\lambda f}(r) \le T_f(r) + O(1) \tag{17.1.25}$$

$$\text{(c) } T_{fg}(r) \le T_f(r) + T_g(r) \tag{17.1.26}$$

$$\text{(d) } T_{f+g}(r) \le T_f(r) + T_g(r) + O(1) \tag{17.1.27}$$

$$\text{(e) } T_{1/f}(r) = T_f(r) + O(1) \tag{17.1.28}$$

**Remark.** $f(z) = e^z$, $g(z) = e^{-z}$, and our analysis in the last two examples shows that one can have equality in (17.1.27).

**Proof.** (a) Immediate since poles of $f^n$ are the same as poles of $f$ with multiplicities multiplied by $n$ and $\log_+(|f^n(z)|) = n \log_+(|f(z)|)$.

(b) Poles and multiplicities are unchanged. By (17.1.8),

$$\log_+(|\lambda f(z)|) \le \log_+(|f(z)|) + \log_+(|\lambda|) \tag{17.1.29}$$

implying (17.1.25).

(c) Poles of $f, g$ are a subset of the union of poles of $f$ and $g$ with multiplicities less than or equal to the sum (it can be smaller than the sum since zeros of one function can cancel poles of the other). Thus,

$$N_{fg}(r, \infty) \le N_f(r, \infty) + N_g(r, \infty) \tag{17.1.30}$$

By (17.1.8),

$$m_{fg}(r, \infty) \le m_f(r, \infty) + m_g(r, \infty) \tag{17.1.31}$$

which implies (c).

(d) One has

$$N_{f+g}(r, \infty) \le N_f(r, \infty) + N_g(r, \infty) \tag{17.1.32}$$

If the positions of the poles of $f$ and $g$ are distinct, one has equality in (17.1.32). If they have common poles, there will be duplications on the right side or even cancellations on the left. By (17.1.6),

$$m_{f+g}(r, \infty) \le m_f(r, \infty) + m_g(r, \infty) + \log 2 \tag{17.1.33}$$

This implies (17.1.27).

(e) This is (17.1.13). □

**Example 17.1.7.** Let $f$ be a rational function of degree $d$. Thus, $f \equiv Q/P$, where $P, Q$ are polynomials with no common zero and $\max(\deg P, \deg Q) = d$. As discussed in Section 10.1 of Part 2A, these are degree $d$ maps of $\widehat{\mathbb{C}}$ to $\widehat{\mathbb{C}}$. Suppose first $f(\infty) = \infty$, that is, $\deg Q = d$, $\ell \equiv \deg P < d$. Thus, $f(z) = Cz^{d-\ell} + O(z^{d-\ell-1})$. It follows that $m_f(r, \infty) = (d - \ell)\log r + O(1)$. If $P$ has zeros at $z_1, \dots, z_\ell$, counting multiplicity if $r_0 = \max(|z_j|)$, then $n(r, \infty) = \ell$ for $r \geq r_0$, and so, $N_f(r, \infty) = \ell \log r + O(1)$. Therefore, $T_f(r) = d \log r + O(1)$.

If $f(\infty) = a \neq \infty$, we look at $g(z) = (f(z) - a)^{-1}$, which is a rational function of degree $d$ with $g(\infty) = \infty$, so $T_g(r) = d\log r + O(1)$. By (d), (e) of Proposition 17.1.6, $T_f(r) = T_g(r) + O(1)$. We summarize in the result below. □

**Theorem 17.1.8.** *If $f$ is a rational function of degree $d$,*

$$T_f(r) = d \log r + O(1) \tag{17.1.34}$$

We are heading towards proving that for any entire meromorphic, non-rational function, $f$, $T_f(r)/\log r \to \infty$ as $r \to \infty$. We begin with

**Theorem 17.1.9.** *Let $f$ be an entire function on $\mathbb{C}$. Let*

$$M_f(r) \equiv \max_{\theta \in [0, 2\pi)} |f(re^{i\theta})| \tag{17.1.35}$$

*Then for a universal constant, $C$, we have*

$$T_f(r) \leq \log_+ M_f(r) \leq C T_f(2r) \tag{17.1.36}$$

**Remark.** If $P_r(\theta, \varphi)$ is the standard Poisson kernel, (5.3.14) of Part 2A,

$$C = \sup_{\theta, \varphi} P_{1/2}(\theta, \varphi) = 3 \tag{17.1.37}$$

**Proof.** Fix $r$. Any entire $f$ can be written $f(z) = g(z)B(z)$, where $g(z)$ is nonvanishing on $\mathbb{D}_{2r}(0)$ and $B(z)$ is a Blaschke-type product, so $|B(zre^{i\theta})| = 1$ and $|B(z)| \leq 1$ on $\mathbb{D}_{2r}(0)$. Since $\log|g(z)|$ is harmonic on $\mathbb{D}_{2r}(0)$, continuous in an extended sense on $\overline{\mathbb{D}_{2r}(0)}$ (i.e., $\log|g(z)| \to -\infty$ is allowed on $\partial\mathbb{D}_{2r}(0)$ but in an integrable way), we have (since $|B(z)| \leq 1$)

$$\begin{aligned} \log|f(re^{i\theta})| \leq \log|g(re^{i\theta})| &= \int P_{1/2}(\theta, \varphi) \log|g(2re^{i\varphi})| \, \frac{d\varphi}{2\pi} \\ &\leq \int P_{1/2}(\theta, \varphi) \log_+|f(2re^{i\varphi})| \, \frac{d\varphi}{2\pi} \qquad (17.1.38) \\ &\leq C T_f(2r) \qquad (17.1.39) \end{aligned}$$

In (17.1.39), $C$ is given by (17.1.37) and (17.1.38) uses the positivity of the Poisson kernel. This yields the second inequality in (17.1.36). □

**Example 17.1.10.** Recall that an entire function, $f$, is said to have finite order $\rho(f) < \infty$ if

$$\rho(f) = \limsup_{r\to\infty} \frac{\log(\log(M_f(r)))}{\log r} \tag{17.1.40}$$

By (17.1.36), this is equivalent to

$$\rho(f) = \limsup \frac{\log(T_f(r))}{\log r} \tag{17.1.41}$$

that is, if $h$ has order $\rho$, for all $\varepsilon > 0$ and all large $r$,

$$T_f(r) \le r^{\rho(f)+\varepsilon} \tag{17.1.42}$$

and for all $\varepsilon > 0$, there exists $r_n \to \infty$ so

$$T_f(r_n) \ge r_n^{\rho(f)-\varepsilon} \tag{17.1.43}$$

□

**Theorem 17.1.11.** *If $f$ is an entire meromorphic function which is not rational, then*

$$\lim_{r\to\infty} \frac{T_f(r)}{\log r} = \infty \tag{17.1.44}$$

*In particular, for any nonconstant entire meromorphic function,*

$$\lim_{r\to\infty} T_f(r) = \infty \tag{17.1.45}$$

**Remark.** In the next section, we'll prove that $T_f(r)$ is a convex function of $\log r$, so by Problem 5 of Section 5.3 of Part 1, we know $\lim T_f(r)/\log r$ exists.

**Proof.** Suppose first that $f$ has no poles. If (17.1.44) is false, there exists $r_n \to \infty$ and $A < \infty$ so that

$$T_f(r_n) \le A\log(r_n) \tag{17.1.46}$$

By (17.1.36),

$$M_f\left(\frac{r_n}{2}\right) \le \exp(CA\log(r_n)) = 2^{CA}\left(\frac{r_n}{2}\right)^{CA} \tag{17.1.47}$$

By a Cauchy estimate, if $f(z) = \sum_{k=0}^\infty b_k z^k$, then for all $n$,

$$|b_k| \le 2^{CA}\left(\frac{r_n}{2}\right)^{CA}\left(\frac{2}{r^n}\right)^{-k} \tag{17.1.48}$$

In particular, if $k > CA$, $b_k = 0$, that is, $f$ is a polynomial and so rational.

If $f$ has finitely many poles, $f = g/P$, where $P$ is a polynomial and $g = fP$ has no poles. Thus, by (17.1.26) and $T_P(r) = d\log r + O(1)$, we see (17.1.44) holds for $f$ if and only if it holds for $g$. If it fails for $g$, $g$ is a polynomial as above, so $f$ is rational.

Finally, if $f$ has infinitely many poles, for any $\rho$,

$$N_f(r,\infty) \geq n_f\left(\frac{r}{2},\infty\right)\log\left(\frac{r}{2}\right) \geq n_f(\rho,\infty)\log\left(\frac{r}{2}\right)$$

if $r > 2\rho$. Thus, for any $\rho$,

$$\liminf \frac{N_f(r,\infty)}{\log r} \geq n_f(\rho,\infty)$$

Since $n_f(\rho,\infty) \to \infty$, we have (17.1.44). Thus, in all three cases if $f$ is not rational, (17.1.44) holds. $\square$

**Corollary 17.1.12.** *For any nonconstant meromorphic entire function, $f$, we have for any $a \in \widehat{\mathbb{C}}$ that*

$$\limsup_{r\to\infty} \frac{N_f(r,a)}{T_f(r)} \leq 1 \tag{17.1.49}$$

**Proof.** Immediate from (17.1.45) and the FMT. $\square$

### Problems

1. For $f(z) = e^z$, prove that $m_f(2\pi n, 1)$, $n = 1,2,\dots$, goes to zero as $n \to \infty$. Prove more generally that $m_f(r,a) \to 0$ as $r \to \infty$ for any $a \neq 0,\infty$.

## 17.2. Cartan's Identity

This section is an interlude—it gives some insight, but we'll prove much stronger results in later sections. But the arguments here follow immediately from Jensen's formula. The main result we are heading towards is (17.0.12). We begin with

**Proposition 17.2.1** (Newton's Lemma). *For all $a \in \mathbb{C}$,*

$$\frac{1}{2\pi}\int_0^{2\pi} \log|a - e^{i\theta}|\,d\theta = \log_+(|a|) \tag{17.2.1}$$

**Remarks.** 1. I don't believe this was ever written down by Newton! I've given it this name because Newton proved if $d\omega$ is a normalized rotation-invariant measure on $S^2 \subset \mathbb{R}^3$, then for $x \in \mathbb{R}^3$,

$$\int |x-\omega|^{-1}\,d\omega = \max(|x|,1)^{-1} \tag{17.2.2}$$

He used this to show the gravitational attraction of a ball with uniform density (rotational-invariant density suffices) on particles outside the ball was the same as if the mass was all concentrated at the center of the ball. $|x|^{-1}$ is the Coulomb potential in three dimensions and $\log(|x|)$ in two, so (17.2.1) is the analog of (17.2.2).

2. This has many proofs (see Problems 1 and 2), one of which uses Jensen's formula.

**Proof.** Call the left side of (17.2.1) $f(a)$. Notice that $f$ is rotation invariant, that is, for all $a \in \mathbb{C}$, $\psi \in [0, 2\pi)$,

$$f(ae^{i\psi}) = f(a) \tag{17.2.3}$$

This comes from $|ae^{i\psi} - e^{i\theta}| = |a - e^{i(\theta-\psi)}|$.

For each $\theta$, $\log|a - e^{i\theta}|$ is harmonic in $a$ on $\mathbb{C} \setminus \{e^{i\theta}\}$ and these are uniformly bounded in $\theta$ for $a$ in each $\mathbb{D}_{1-\delta}(0)$. Thus, $f$ is harmonic on $\mathbb{D}$. It follows for any $a \in \mathbb{D}$,

$$0 = f(0) = \int_0^{2\pi} f(ae^{i\psi})\, \frac{d\psi}{2\pi} = f(a) \tag{17.2.4}$$

by (17.2.3), that is, $f \equiv 0$ on $\overline{\mathbb{D}}$ (using continuity of $f$), proving (17.2.1) if $a \in \overline{\mathbb{D}}$.

If $a \in \mathbb{C} \setminus \overline{\mathbb{D}}$, we use

$$\begin{aligned} \log|a - e^{i\theta}| &= \log|a| + \log|1 - a^{-1}e^{i\theta}| \\ &= \log|a| + \log|a^{-1} - e^{-i\theta}| \end{aligned} \tag{17.2.5}$$

which says

$$f(a) = \log|a| + f(a^{-1}) \tag{17.2.6}$$

This proves (17.2.1) if $|a| > 1$. □

**Theorem 17.2.2** (Cartan's Identity). *Suppose* $f(0) \neq 0$. *Then*

$$T_f(r) = \frac{1}{2\pi} \int_0^{2\pi} N_f(r, e^{i\theta})\, \frac{d\theta}{2\pi} + \log_+|f(0)| \tag{17.2.7}$$

**Proof.** By (17.1.5) for $f$ replaced by $f - e^{i\psi}$ ($\psi$ fixed), we have

$$\int_0^{2\pi} \log|f(re^{i\theta}) - e^{i\psi}|\, \frac{d\theta}{2\pi} = \log|f(0) - e^{i\psi}| + N_f(r, e^{i\psi}) - N_f(r, \infty) \tag{17.2.8}$$

for all $\psi$, with perhaps one exception if $|f(0)| = 1$.

Integrate $\frac{d\psi}{2\pi}$ and use (17.2.1) to get

$$T_f(r) = \log_+|f(0)| + \int_0^{2\pi} N_f(r, e^{i\psi})\, \frac{d\psi}{2\pi} \tag{17.2.9}$$

□

**Corollary 17.2.3.** $y \to T_f(e^y)$ *is convex and monotone in* $y$ *(i.e.,* $T_f(r)$ *is a convex function of* $\log r$*).*

**Remark.** If $f$ has no poles, this is just the Hardy convexity theorem (Theorem 5.1.2 of Part 3), but if $f$ has poles, the $m_f(r,\infty)$ term is no longer a convex function of $\log r$.

**Proof.** $\log_+(r/a)$ is convex and monotone in $\log r$, so $N_f(r,e^{i\theta})$ is convex and monotone in $\log r$ for each $e^{i\theta}$. Thus, (17.2.7) implies $T_f(r)$ is a convex monotone function of $\log r$. □

**Corollary 17.2.4.** *For any entire meromorphic function,*

$$\int_0^{2\pi} m_f(r,e^{i\theta})\,\frac{d\theta}{2\pi} \le \log 2 \tag{17.2.10}$$

**Proof.** By (17.1.16) with $a=e^{i\theta}$ and (17.1.13), if $f(0)\neq e^{i\theta}$,

$$|N_f(r,e^{i\theta}) + m_f(r,e^{i\theta}) + \log|f(0)-e^{i\theta}| - T_f(r)| \le \log 2 \tag{17.2.11}$$

Integrating $\frac{d\theta}{2\pi}$ and using (17.2.1), we get

$$\left|\int_0^{2\pi} N_f(r,e^{i\theta})\,\frac{d\theta}{2\pi} + \log_+|f(0)| - T_f(r) + \int_0^{2\pi} m_f(r,e^{i\theta})\,\frac{d\theta}{2\pi}\right| \le \log 2 \tag{17.2.12}$$

By Cartan's identity, (17.2.7), the first three terms on the left combine to zero and proves (17.2.10), given that $m_f(r,e^{i\theta})\ge 0$. □

This allows us to prove $\limsup N_f(r,z)/T_f(r)=1$ for $d^2z$-a.e. $z$. We'll prove, in the next section, that it holds for all but a countable set of $z$'s.

**Theorem 17.2.5.** *Let $r_n$ be a sequence $r_{n+1}>r_n$, so*

$$\sum_{n=1}^\infty \frac{1}{T_f(r_n)} < \infty \tag{17.2.13}$$

*Then for each $\rho\neq 0$, for a.e. $\theta$,*

$$\lim_{n\to\infty} \frac{m_f(r_n,\rho e^{i\theta})}{T_f(r_n)} = 0 \tag{17.2.14}$$

*In particular, if (17.2.14) holds,*

$$\limsup_{r\to\infty} \frac{N_f(r,\rho e^{i\theta})}{T_f(r)} = 1 \tag{17.2.15}$$

**Proof.** Suppose first $\rho=1$. By Corollary 17.2.4,

$$\int_0^{2\pi} \left[\sum_{n=1}^\infty \frac{m_f(r_n,e^{i\theta})}{T_f(r_n)}\right] \frac{d\theta}{2\pi} < \infty \tag{17.2.16}$$

Thus, the sum is finite for a.e. $\theta$, so (17.2.14) holds for a.e. $\theta$.

(17.2.14) and $m_f \geq 0$ says (17.2.14) implies $\liminf_{r\to\infty} m_f/T_f = 0$. Since $(m_f + N_f)/T_f \to 1$, this yields (17.2.15).

By (17.1.25) and $T_f(r_n) \to \infty$, (17.2.13) holds for $f$ if and only if it holds for $f/\rho$. Using this and a simple calculation relating $m_{f/\rho}(r, e^{i\theta})$ and $m_f(r, \rho e^{i\theta})$ (Problem 3), we get the results for general $\rho$. $\square$

**Notes and Historical Remarks.** Cartan's formula is from Cartan [**74**] from his thesis (Section 11.5 of Part 2A has a capsule biography).

**Problems**

1. Prove (17.2.1) by checking Jensen's formula for $f(z) = z - a$.

2. Let $g(a)$ be the left side of (17.2.1).
   (a) Prove $g$ is rotation invariant and harmonic in $\mathbb{C} \setminus \partial\mathbb{D}$.
   (b) Prove $g(a) = 0$ if $|a| < 1$.
   (c) If $|a| \geq 1$, prove $g(a) = \alpha \log|a| + \beta$ for some $\alpha, \beta$.
   (d) Prove $\alpha = 1$ by looking at $g$ for large $a$.
   (e) Prove $g(1) = 0$ and conclude that $\beta = 0$.

3. (a) For any $\rho \in (0, \infty)$ and any $f$, prove that

$$m_f(r, \rho e^{i\theta}) \leq m_{f/\rho}(r, e^{i\theta}) + \log_+(\rho^{-1}) \qquad (17.2.17)$$

   (b) If (17.2.13) holds, prove that (17.2.11) holds if $e^{i\theta}$ is replaced by $\rho e^{i\theta}$.

## 17.3. The Second Main Theorem and Its Consequences

In this section, we'll state the second main theorem of Nevanlinna theory and discuss some of its consequences. We'll prove the SMT in the next section. We begin by discussing critical points. If $f$ is an entire meromorphic function, $z_0$ is called a *critical point* of $f$ if either

(a) $f(z_0) < \infty$ and $f(z) - f(z_0)$ has a higher-order zero at $z_0$, say

$$f(z) - f(z_0) = C(z - z_0)^{\ell+1} + O(|z - z_0|^{\ell+2}) \qquad (17.3.1)$$

(b) $f(z_0) = \infty$ and $f$ has a pole of order $\ell + 1 \geq 2$.

In either case, $\ell$ is called the *order* of the critical point. As usual, $n_{1,f}(r)$ is the number of critical points, counting multiplicity, in $\overline{\mathbb{D}_r(0)}$ and $N_{1,f}(r)$ is defined via the analog of (17.0.2). We also say $f$ is *ramified* at $z_0$.

**Proposition 17.3.1.** *We have that*

$$N_{1,f}(r) = N_{f'}(r, 0) + 2N_f(r, \infty) - N_{f'}(r, \infty) \qquad (17.3.2)$$

**Proof.** If (17.3.1) holds, then $f'(z_0) = 0$ and $\ell$ is the order of this zero, so $N_{f'}(r, 0)$ counts the critical points with $f(z_0) \neq \infty$.

If $f(z_0) = \infty$ and $f(z_0) = D(z - z_0)^{-m}$, then $2N_f(r, \infty)$ contributes $2m \log_+(r/|z_0|)$, while $N_{f'}(r, \infty)$ contributes $(m + 1)\log_+(r/|z_0|)$. Thus, if $m = \ell + 1$, $2N_f - N_{f'}$ contributes $[2m - (m + 1)] \log_+(r/|z_0|) = (m - 1)\log_+(r/|z_0|) = \ell \log_+(r/|z_0|)$. $\square$

**Theorem 17.3.2** (SMT). *For any $\{a_j\}_{j=1}^q$, distinct points in $\widehat{\mathbb{C}}$, there is a subset $E \subset (0, \infty)$ of finite measure so that as $r \to \infty$, $r \notin E$,*

$$\sum_{j=1}^{q} m_f(r, a_j) + N_{1,f}(r) \leq 2T_f(r) + O(\log(rT_f(r))) \tag{17.3.3}$$

We'll prove this in the next section. Note that, by Theorem 17.1.11, if $f$ is not rational, then $\log r = o(T_f(r))$, so

$$f \text{ not rational} \Rightarrow O(\log(rT_f(r))) = o(T_f(r)) \tag{17.3.4}$$

In the rational case, one can prove (Problem 1) that if $\deg f = d$, then

$$(\text{LHS of } (17.3.3)) \leq \left(2 - \frac{1}{d}\right) T_f(r) + O(1) \tag{17.3.5}$$

so even in the rational case, the error term is $o(T_f(r))$.

Writing $m_f = T_f - N_f + O(1)$, (17.3.3) is often rewritten

$$\sum_{j=1}^{q} N_f(r, a_j) \geq (q - 2)T_f(r) + N_{1,f}(r) + O(\log(rT_f(r))) \tag{17.3.6}$$

One defines the *deficiency* of the value, $a$, by

$$\delta_f(a) = \liminf_{r\to\infty} \frac{m_f(r, a)}{T_f(r)} = 1 - \limsup_{r\to\infty} \frac{N_f(r, a)}{T(r)} \tag{17.3.7}$$

and the *critical deficiency*: Define $\overline{N}_f(r, a)$ as the counting function associated to solutions of $f(z) = a$ *not counting multiplicity* and

$$\Theta_f(a) = 1 - \limsup_{r\to\infty} \frac{\overline{N}_f(r, a)}{T(r)} \tag{17.3.8}$$

Here is a consequence of the SMT that we'll use to prove the signature results mentioned in the introduction to the chapter.

**Theorem 17.3.3.** *Let $f$ be an entire meromorphic function. We have for any $\{a_j\}_{j=1}^q$ that*

$$\sum_{j=1}^{q} \delta_f(a_j) \leq \sum_{j=1}^{q} \Theta_f(a_j) \leq 2 \tag{17.3.9}$$

**Remark.** This implies $\{a \mid \Theta_f(a) > \frac{1}{n}\}$ is finite, so $\{a \mid \Theta_f(a) \neq 0\}$ is countable and some authors write (17.3.9) as

$$\sum_{a\in\hat{\mathbb{C}}} \Theta_f(a) \leq 2 \tag{17.3.10}$$

**Proof.** Since $\overline{N}_f(r,a) \leq N_f(r,a)$, we have $\delta_f(a) \leq \Theta_f(a)$, so we only need the second inequality. Define

$$\overline{m}_f(r,a) = m_f(r,a) + N_f(r,a) - \overline{N}_f(r,a) \tag{17.3.11}$$

$N_f(r,a) - \overline{N}_f(r,a)$ counts critical points (including multiplicities) for points with $f(z) = a$. Thus,

$$\sum_{j=1}^{q} [N_f(r,a_j) - \overline{N}_f(r,a_j)] \leq N_{1,f}(r) \tag{17.3.12}$$

so that the SMT implies

$$\sum_{j=1}^{q} \overline{m}_f(r,a_j) \leq (2+o(1))T_f(r) \tag{17.3.13}$$

Dividing by $T_f(r)$ and using

$$\Theta(a) = \liminf_{r\to\infty} \frac{\overline{m}_f(r,a)}{T_f(r)} \tag{17.3.14}$$

we get (17.3.9). $\square$

By (17.3.12), (17.3.6) implies that

$$\sum_{j=1}^{q} \overline{N}_f(r,a_j) \geq (q-2)T_f(r) + o(T_f(r)) \tag{17.3.15}$$

**Theorem 17.3.4** (Picard's Theorem). *An entire nonrational meromorphic function takes every value with at most two exceptions infinitely often.*

**Proof.** If $a$ is taken only finitely many times, since $T_f(r)/\log r \to \infty$, we see $\delta_f(a) = 1$. Thus, if there were three values, $a_1, a_2, a_3$, each taken only finitely many times, then $\sum_{j=1}^{3} \delta(a_j) = 3$, violating (17.3.9). $\square$

**Theorem 17.3.5.** *Let $f$ be an entire meromorphic function that fails to take two distinct values, $a_1$ and $a_2$. Then for $a_3 \neq a_1, a_2$ off a set, $E$, of finite measure,*

$$\lim_{\substack{r\to\infty \\ r\notin E}} \frac{N_f(r,a_3)}{T_f(r)} = 1 \tag{17.3.16}$$

**Proof.** Since $N_f(r,a_j)=0$ for $j=1,2$, by the FMT, $m_f(r,a_j)=T_f(r)+O(1)$ for $j=1,2$. Thus, for $r\notin E$,

$$m_f(r,a_3)=o(T_f(r)) \tag{17.3.17}$$

(since rational functions can't miss more than one value). By the FMT again, we have (17.3.16). $\square$

The argument in Theorem 17.3.5 actually proves more: With at most two exceptions, $1-\delta_f(a)=\limsup[N_f(r,a)/T_f(r)]>0$. From (17.1.41) and Proposition 9.10.1 of Part 2A, we find (Problem 2):

**Theorem 17.3.6** (Borel's Theorem). *Let $f$ be an entire analytic function of order $\rho<\infty$. Then with at most one exception, the exponent of convergence (given by* (9.10.7) *of Part 2A), $\sigma(f-a)$ obeys*

$$\sigma(f-a)=\rho(f) \tag{17.3.18}$$

**Theorem 17.3.7.** *For each integer $M$, with at most $2M$ exceptions, $\delta(a)<1/M$. In particular, for all but a countable set of values of $a$, we have*

$$\limsup_{r\to\infty}\frac{N_f(r,a)}{T_f(r)}=1 \tag{17.3.19}$$

**Proof.** If $\delta(a_j)\ge 1/M$ for $2M+1$ values, $\sum_{j=1}^{2M+1}\delta(a_j)\ge 2+1/M$, violating (17.3.9). Thus, $\delta(a)=0$ for all but countably many $a$'s, that is, (17.3.19) holds. $\square$

**Theorem 17.3.8.** *Let $f$ be an entire meromorphic function. Call a value, $a$, totally ramified if every $z_0$ with $f(z_0)=a$ is a critical point. Then $f$ has at most four totally ramified values.*

**Proof.** The case $f$ rational is left to the Problems (Problem 3). If $a$ is a totally ramified value, $N_f(r,a)\ge 2\overline{N}_f(r,a)$, so $\delta(a)\ge 0$ implies $\Theta(a)\ge\frac12$. If $\{a_j\}_{j=1}^5$ are totally ramified, then $\sum_{j=1}^5\Theta(a_j)\ge\frac52$, violating (17.3.9). $\square$

**Example 17.3.9.** Let $\wp$ be a Weierstrass $\wp$-function. Let $e_1,e_2,e_3$ be the values at $\frac12\tau_1$, $\frac12(\tau_1+\tau_2)$ and $\frac12\tau_2$. By Theorem 10.4.5 of Part 2A, $\{e_j\}_{j=1}^3$ and $\infty$ are totally ramified points, so there can be four totally ramified values, showing Theorem 17.3.8 is optimal. $\square$

**Theorem 17.3.10.** *Let $f$ be an entire analytic function. Then $f$ has at most two totally ramified (finite) values.*

**Proof.** The polynomial case is left to the Problems (Problem 4). Since $f$ is entire, $N_f(r,\infty)=0$ so $\delta(\infty)=\Theta(\infty)=1$. For a totally ramified value, $a$, $\Theta(a)\ge\frac12$. If $\{a_j\}_{j=1}^3$ were three finite totally ramified values, taking $a_4=\infty$, we would have $\sum_{j=1}^4\Theta(a_j)\ge\frac52$, violating (17.3.9). $\square$

**Example 17.3.11.** Let $f(z) = \sin z$. $a = \pm 1$ are totally ramified values (and the only ramified values since $f'(z)^2 + f(z)^2 = 1$). Thus, two totally ramified values can occur, showing two in Theorem 17.3.10 is optimal. $\square$

Define the *value set* by

$$V_f(a) = \{z \mid f(z) = a\} = f^{-1}\bigl[\{a\}\bigr] \tag{17.3.20}$$

**Theorem 17.3.12.** *If $V_f(a_j) = V_g(a_j)$ for five distinct values of $a_j$, then either $f = g$ or both are constant.*

**Remark.** If $f \equiv c_1$, $g \equiv c_2$, then $V_f(a) = V_g(a) = \emptyset$ for all $a \in \widehat{\mathbb{C}} \setminus \{c_1, c_2\}$, so distinct constants obey the hypothesis.

**Proof.** Suppose $f \neq g$ and either $f$ or $g$ is not constant. $V_f(a_j) = V_g(a_j)$ implies $\overline{N}_f(r, a_j) = \overline{N}_g(r, a_j) \equiv \overline{N}_j(r)$. By (17.3.15) with $q = 5$,

$$T_f(r) \leq [\tfrac{1}{3} + o(1)] \sum_{j=1}^{5} \overline{N}_j(r) \tag{17.3.21}$$

and similarly for $T_g(r)$.

By (17.1.27),

$$T_{f-g} \leq T_f + T_g + O(1) \tag{17.3.22}$$

Since $f$ or $g$ nonconstant implies that

$$\sum_{j=1}^{5} \overline{N}_j(r) \to \infty \tag{17.3.23}$$

we have

$$T_{f-g}(r) \leq (\tfrac{2}{3} + o(1)) \sum_{j=1}^{5} \overline{N}_j(r) \tag{17.3.24}$$

Since $f = g$ on $\bigcup V_f(a_j)$, $f - g = 0$ there, so

$$\sum_{j=1}^{5} \overline{N}_j(r) \leq N_{f-g}(r, 0) \leq T_{f-g}(r) + O(1) \tag{17.3.25}$$

(17.3.23), (17.3.24), and (17.3.25) cannot all hold. This is a contradiction. $\square$

**Example 17.3.13.** Let $f(z) = e^z$, $g(z) = e^{-z}$. Then $V_f(a) = V_g(a)$ for $a = 0, \infty, 1, -1$, showing 5 is optimal in Theorem 17.3.12. $\square$

**Notes and Historical Remarks.** There are three limitations on deficiency indices that the reader should be aware of. First, they involve $N_f(r, a)$, not $n_f(r, a)$ directly.

Second, they involve $\limsup N_f/T_f$, not $\liminf$ which would be more natural—for example, the $\liminf$ is 1 if and only if the limit is 1. There has been literature on $\liminf$ going back to Valiron [**385, 387**], Littlewood [**264**], and Ahlfors [**6**]. As we noted earlier, Ahlfors proved $\liminf N_f/T_f$ for all $a$ outside a zero capacity set. In this regard, we note that Hayman [**184**] constructed meromorphic functions with $\limsup N_f/T_f < 1$ exactly on any given capacity zero set which is an $F_\sigma$.

Third is the surprising discovery that deficiency indices are not invariant under shift of origin. The original example is the 1947 one of Dugué [**109**], who studied

$$f(z) = \frac{e^{e^z} - 1}{e^{e^{-z}} - 1} \tag{17.3.26}$$

If $f_h(z) = f(z+h)$, Dugué proved that

$$\delta_{f_h}(0) = \frac{e^{-h}}{e^h + e^{-h}}, \qquad \delta_{f_h}(\infty) = \frac{e^h}{e^h + e^{-h}} \tag{17.3.27}$$

Hayman [**181**] considered the entire function (of infinite order)

$$f(z) = \prod_{n=1}^\infty \left\{1 + \left(\frac{z}{n}\right)^{3n}\right\}^{2^n} \tag{17.3.28}$$

With $f_h(z) = f(z+h)$, he proved

$$\delta_f(0) = 0, \qquad \delta_{f_1}(0) > 0 \tag{17.3.29}$$

With regard to the best possible nature of the SMT, in 1977, Drasin [**108**] settled a long open conjecture by showing if $\{a_j\}_{j=1}^\infty$ is an infinite sequence of distinct points in $\widehat{\mathbb{C}}$ and $\{\delta_j\}_{j=1}^\infty$ is a sequence of nonnegative numbers so that $\sum_{j=1}^\infty \delta_j \le 2$, then there is a meromorphic function $f$ with

$$\delta_f(a_j) = \delta_j, \qquad \delta_f(a) = 0 \text{ if } a \neq a_j \text{ for any } j \tag{17.3.30}$$

Valiron [**386**] and Bloch [**60**], using Nevanlinna theory, proved a result related to Theorem 17.3.8 as Montel's three-value theorem is related to Picard's theorem: Given any four numbers $\{a_j\}_{j=1}^4$ and region, $\Omega \subset \mathbb{C}$, the set of meromorphic functions $f\colon \Omega \to \widehat{\mathbb{C}}$ so that $f(z_0) = a_j \Rightarrow f'(z_0) = 0$ (if $a_j = \infty$, $f(z_0)^{-1} = 0 \Rightarrow [f(z_0)^{-1}]' = 0$) is a normal family. Bergweiler [**52**] proved this result and Theorem 17.3.8 (and much more!) using Zalcman's lemma (see Section 11.4 of Part 2A) in place of Nevanlinna theory.

**Problems**

1. Let $P$ be a polynomial of degree $d$.
   (a) Prove that $m_f(r, \infty) = T_f(r) = d \log r + O(1)$.
   (b) Prove that $N_{1,f}(r) = (d-1)\log r + O(1)$.

(c) Prove that $m_f(r,a) = 0$ for $a \neq \infty$, $r \geq R_a$.

(d) If $\{a_j\}_{j=1}^{\ell}$ are distinct points, prove that

$$\sum_{j=1}^{q} m_f(r,a_j) + N_{1,f} = O(1) + \begin{cases} (2-\frac{1}{d})T_f(r) & \text{if } \infty \text{ is an } a_j \\ (1-\frac{1}{d})T_f(r) & \text{if not } a_j \text{ is } \infty \end{cases}$$

(e) Extend this analysis to general rational $f$.

2. Prove Theorem 17.3.6. (*Hint*: See Proposition 9.10.2 of Part 2A.)

3. (a) Let $f$ be a rational function of degree $d$. Prove the number of critical points, counting multiplicity, is at most $2d-2$.

   (b) If $f$ has a totally ramified point, prove $d$ is even and this value accounts for at least $\frac{d}{2}$ critical points.

   (c) Prove that a rational function has at most three totally ramified points.

4. (a) Let $P$ be a polynomial of degree $d$. Prove that it has at most $d-1$ critical points.

   (b) Prove $P$ has at most one totally ramified value.

5. This problem will consider entire functions, $f$ and $g$, solving

$$f^d + g^d = 1 \tag{17.3.31}$$

   (a) Find a solution when $d = 2$.

   (b) If $g$ has a zero of order $k$ at $z_0$, prove that one of the multiplicities of the zeros of $\{f - e^{2\pi iq/d}\}_{q=0}^{n-1}$ is $kd$. Conclude that

$$\sum_{q=0}^{d-1} \bar{N}_f(r, e^{2\pi iq/d}) \leq \frac{1}{d}\sum_{q=1}^{d-1} N(r, e^{2\pi iq/d}) \leq T(r,f) + o(T(r,f)) \tag{17.3.32}$$

   (c) Prove that if (17.3.31) has a solution, then $d = 2$. (*Hint*: Use (17.3.15) with the $d+1$ points $\{\infty\} \cup \{e^{2\pi iq/d}\}_{q=0}^{d-1}$, so see $d+1-2 \leq 1$.)

   (d) If $f$ and $g$ are allowed to be meromorphic, prove there are no solutions if $d \geq 4$.

**Remark.** This result is explicitly in Gross [**169**] but it follows from general theorems of Picard. He also discusses solutions in meromorphic functions and finds one in terms of elliptic functions if $f, g$ are meromorphic and $d = 3$.

## 17.4. Ahlfors' Proof of the SMT

In Section 17.2, we saw averaging Jensen's formula for $f(z) - a$ over $a = e^{i\theta}$ was useful. In this section, we see averaging over more general $a$'s in $\widehat{\mathbb{C}}$ is even more useful. We'll begin by averaging over the area measure induced by the spherical metric (6.5.1) of Part 2A. This will lead to a reformulation of the FMT due to Ahlfors–Shimizu. Then we'll modify the metric to give extra weight to $a$'s near $\{a_j\}_{j=1}^{\ell}$ and this will get us Ahlfors' proof of the SMT. We begin with a preliminary proposition that explains where the set of finite measure, $E$, of the SMT comes from.

**Proposition 17.4.1.** *If $g$ is an increasing function on $[0, \infty)$ with $g(0) > 0$, $\lim_{x\to\infty} g(x) = \infty$, then for any $\varepsilon > 0$, we have that*

$$g'(x) \le g(x)^{1+\varepsilon} \tag{17.4.1}$$

*for all $x \notin E$, where $|E| < \infty$.*

**Proof.** Let $E$ be the set where (17.4.1) fails. On $E$, $g'/g^{1+\varepsilon} \ge 1$. Thus, letting $y = g(x)$,

$$|E| \le \int_E \frac{g'(x)}{g(x)^{1+\varepsilon}}\, dx \tag{17.4.2}$$

$$= \int_{g(E)} \frac{dy}{y^{1+\varepsilon}} \tag{17.4.3}$$

$$\le \int_{g(0)}^{\infty} \frac{dy}{y^{1+\varepsilon}} = \varepsilon^{-1} g(0)^{-\varepsilon} < \infty \tag{17.4.4}$$

□

Let $d\rho(w)$ be a probability measure on $\mathbb{C}$ ($w = u + iv$) of the form

$$d\rho(w) = G(w)\, du dv \tag{17.4.5}$$

where $G$ is strictly positive and continuous on $\mathbb{C} \setminus \{a_j\}_{j=1}^{\ell}$ for finitely many $a_j$ with $G \in L^1(\mathbb{C}, dudv)$. We assume the technical condition that there is $R_0 > 0$, $\delta > 0$, so that

$$\int_{|w|\ge R} d\rho(w) \le R^{-\delta} \tag{17.4.6}$$

for $R > R_0$.

The *antipotential* of $\rho$ is defined by

$$\Psi_\rho(w) = \int_{\mathbb{C}} \log|w - a|\, d\rho(a) \tag{17.4.7}$$

It is easy to see (Problem 1) that (17.4.6) implies $\int_{|a|\ge 1} \log|a|\, d\rho(a) < \infty$, which implies for any $R > 0$, $\int_{|w-a|\ge R} \log|w - a|\, d\rho(a) < \infty$, so that (17.4.7) either is a convergent integral or diverges to $-\infty$. Indeed, it is convergent and continuous except perhaps at the $\{a_j\}_{j=1}^{\ell}$.

By Jensen's formula, (17.1.5), for $f(z)$ replaced by $f(z)-a$, using that $d\rho(\{f(0)\})=0$ and $\eta(f)=f(0)-a$ for all but $a=f(0)$, we find that

$$\int \Psi_\rho(f(e^{i\theta}))\,\frac{d\theta}{2\pi} = \Psi_\rho(f(0)) + \int_{\mathbb{C}} N_f(r,a)\,d\rho(a) - N_f(r,\infty) \tag{17.4.8}$$

Here is a calculation relevant to the integral of $N$.

**Theorem 17.4.2.** *If $\rho$ has the form* (17.4.5), *then for any entire meromorphic function $f$,*

$$\int n_f(r,a)\,d\rho(a) = \int_{|z|\le r} |f'(z)|^2 G(f(z))\,dxdy \tag{17.4.9}$$

**Proof.** If $f$ is constant, it is easy to see (Problem 2) that both sides of (17.4.9) are 0, so suppose $f$ is not constant. Inside $\mathbb{D}_r(0)$, there are finitely many points $\{z_j\}_{j=1}^k$ so that on $\mathbb{D}_r(0)\setminus\{z_j\}_{j=1}^k$, $f$ is finite and $f'$ nonzero. We can cover this set with countably many disks, $\{D_j\}_{j=1}^\infty$ so that on each disk, $f$ is a bijection of $D_j$ and $f[D_j]$. Let

$$U_j = \Bigl[D_j \setminus \bigcup_{k=1}^{j-1} \overline{D}_k\Bigr] \cap \mathbb{D}_r(0) \tag{17.4.10}$$

so, since $d\rho$ is absolutely continuous,

$$\begin{aligned} \text{RHS of (17.4.9)} &= \sum_{j=1}^\infty \int_{U_j} |f'(z)|^2 G(f(z))\,dxdy \\ &= \sum_{j=1}^\infty \int_{f[U_j]} G(a)\,d^2a \end{aligned} \tag{17.4.11}$$

since $f$ is a bijection and $\det(\frac{\partial^2 f}{\partial x \partial y}) = |f'(z)|^2$ (see (2.2.22) of Part 2A).

The number of times $a_0$ appears in some $f[U_j]$ is, for almost all $a$'s, exactly $n(r,a)$. Thus, the sum in (17.4.11) is the left side of (17.4.9). $\square$

We first use

$$G_0(w) = \pi^{-1}(1+|w|^2)^{-2} \tag{17.4.12}$$

to recover a variant form of the FMT. By the calculation in Proposition 6.5.4 of Part 2A, with $f=1$, and (2.2.22) of Part 2A, this is the image of the area measure on $\widehat{\mathbb{C}}$ viewed as a sphere of radius $\frac{1}{2}$ under the stereographic projection $Q^{-1}$ with $Q$ given by (6.5.2) of Part 2A. The total area of such a sphere is $4\pi(\frac{1}{2})^2=\pi$, so (17.4.12) is normalized as can also be seen by direct integration.

We define the *Ahlfors–Shimizu characteristic* by

$$\mathring{T}_f(r) = \int_0^r \frac{A_f(t)}{t}\,dt = \int N_f(r,a)G_0(a)\,d^2a \tag{17.4.13}$$

where

$$A_f(t) = \frac{1}{\pi}\int_{|z|\le r} \frac{|f'(z)|^2}{(1+|f(z)|^2)^2}\,d^2z \tag{17.4.14}$$

which, we we've seen, is the average number of times a point in $\widehat{\mathbb{C}}$ is covered by $f\colon \mathbb{D}_r(0)\to\widehat{\mathbb{C}}$.

Next, we note, by Newton's lemma (Problem 3), that

$$\begin{aligned}\Psi(z) \equiv \Psi_{G,d^2a}(z) &= \frac{1}{\pi}\int \frac{\log|z-a|}{(1+|a|^2)^2}\,d^2a \\ &= \log\sqrt{1+|z|^2} = \log(\sigma(w,\infty)^{-1})\end{aligned} \tag{17.4.15}$$

where $\sigma$ is the chordal distance of (6.5.1) of Part 2A. Note $\sigma\le 1$, so

$$\log\sigma^{-1}\ge 0 \tag{17.4.16}$$

Thus, if we define the *Ahlfors–Shimizu proximity function* by

$$\mathring{m}_f(r,a) = \frac{1}{2\pi}\int_0^{2\pi} \log(\sigma[f(re^{i\theta}),a]^{-1})\,d\theta \tag{17.4.17}$$

we get:

**Theorem 17.4.3** (FMT in Ahlfors–Shimizu Form).

$$N_f(r,a) + \mathring{m}(r,a) = \mathring{T}_f(r) + \mathring{m}_f(0,a) \tag{17.4.18}$$

$$= \mathring{T}_f(r) + O(1) \tag{17.4.19}$$

*Moreover,*

$$|\mathring{m}_f(r,a) - m_f(r,a)| \le O(1) \tag{17.4.20}$$

$$\mathring{T}_f(r) = T_f(r) + O(1) \tag{17.4.21}$$

**Proof.** (17.4.8) becomes (17.4.18) for $a=\infty$. We claim that

$$|\mathring{m}_f(r,\infty) - m_f(r,\infty)| \le \tfrac{1}{2}\log 2 \tag{17.4.22}$$

This closely follows, by integrating $\frac{d\theta}{2\pi}$, from

$$\left|\log\sqrt{1+x^2} - \log_+(x)\right| \le \tfrac{1}{2}\log 2 \tag{17.4.23}$$

for $x>0$. If $0<x\le 1$, $\log_+(x)=0$ and $\sqrt{1+x^2}\le\sqrt{2}$, so (17.4.23) holds. If $x\ge 1$,

$$\log\sqrt{1+x^2} - \log_+(x) = \log\sqrt{1+x^{-2}} \le \log\sqrt{2} \tag{17.4.24}$$

proving (17.4.23), and so (17.4.22).

Thus,

$$\begin{aligned}\mathring{T}_f(r) &= N_f(r,\infty) + \mathring{m}(r,\infty) + O(1) \\ &= N_f(r,\infty) + m(r,\infty) + O(1) \\ &= T_f(r) + O(1)\end{aligned}$$

proving (17.4.21).

Fix $a \in \widehat{\mathbb{C}}$, $a \neq \infty$. Let $R$ be a fractional linear transformation that induces a rotation on $\widehat{\mathbb{C}}$ as a sphere so that $Ra = \infty$. Since $\mathring{T}$ is defined in terms of a rotationally invariant measure on $\widehat{\mathbb{C}}$, we have

$$\mathring{T}_{Rf}(r) = \mathring{T}_f(r) \tag{17.4.25}$$

Since $\sigma$ is rotation invariant,

$$\mathring{m}_f(r,a) = \mathring{m}_{Rf}(r,\infty) \tag{17.4.26}$$

Thus, (17.4.18) for $Rf$ and $a = \infty$ is (17.4.18) for $f$ and $a$, that is, we have proven (17.4.18) in general.

This in turn implies (17.4.19). The ordinary FMT subtracted from (17.4.19) using (17.4.21) implies (17.4.20). □

We now turn to Ahlfors' proof of the SMT. We use a measure $d\rho$ of the form (17.4.5) with

$$G(w) = p(w)^2 \pi^{-1} (1+|w|^2)^{-2} \tag{17.4.27}$$

where

$$\log(p(w)) = \sum_{j=1}^q \log(\sigma(w,a_j)^{-1}) - 2\log\biggl(\sum_{j=1}^q \log(\sigma(w,a_j)^{-1})\biggr) + C \tag{17.4.28}$$

The double log term makes the integral $\int G(w)\,d^2w$ finite at $w = a_j$ and $C$ is picked to make $d\rho$ a probability measure.

By (17.4.9), we have that

$$\int_{\widehat{\mathbb{C}}} N_f(r,a)\,d\rho(a) = \int_0^r \frac{dt}{t} \int_0^t \lambda(s)\,s\,ds \tag{17.4.29}$$

where

$$\lambda(r) = \int_0^{2\pi} p^2(w)\,\frac{|w'|^2}{(1+|w|^2)^2}\,\frac{d\theta}{2\pi} \qquad (w \equiv f(re^{i\theta})) \tag{17.4.30}$$

Since log is a concave function, Jensen's inequality (see Theorem 5.3.14 of Part 1) says for any $g \geq 0$,

$$\log\biggl(\int_0^{2\pi} g(re^{i\theta})\,\frac{d\theta}{2\pi}\biggr) \geq \int_0^{2\pi} \log(g(re^{i\theta}))\,\frac{d\theta}{2\pi} \tag{17.4.31}$$

Thus,

$$\log(\lambda(r)) \geq h_1(r) + h_2(r) + h_3(r) \tag{17.4.32}$$

$$h_1(r) = \frac{1}{\pi}\int_0^{2\pi} \log(p(w))\,d\theta; \quad h_2(r) = -\frac{1}{\pi}\int_0^{2\pi} \log(1+|w|^2)\,d\theta;$$

$$h_3(r) = \frac{1}{\pi}\int_0^{2\pi} \log|w'|\,d\theta \tag{17.4.33}$$

**Proposition 17.4.4.**

(a) $h_1(r) \geq 2\sum_{j=1}^{q} \mathring{m}_f(r,a_j) + O(\log(T_f(r)))$ (17.4.34)

(b) $h_2(r) = -4\mathring{m}_f(r,\infty)$ (17.4.35)

(c) $h_3(r) = 2N_{f'}(r,0) - 2N_{f'}(r,\infty) + O(1)$ (17.4.36)

**Proof.** (a) We put (17.4.28) into $h_1$. The first term in (17.4.28) gives $2\sum_{j=1}^{q} \mathring{m}_f(r,a_j)$ (since $h_1$ has $\frac{1}{\pi}$, not $\frac{1}{2\pi}$). The second term is controlled by

$$-4\int \log\biggl(\sum_{j=1}^{q} \log(\sigma(w,a_j)^{-1})\biggr)\frac{d\theta}{2\pi}$$

$$\geq -4\int \log_+\biggl(\sum_{j=1}^{q} \log(\sigma(w,a_j)^{-1})\biggr)\frac{d\theta}{2\pi} \tag{17.4.37}$$

$$\geq -4\log_+\int\biggl(\sum_{j=1}^{q} \log(\sigma(w,a_j)^{-1})\biggr)\frac{d\theta}{2\pi} \tag{17.4.38}$$

$$\geq O(\log(T_f(r))) \tag{17.4.39}$$

(17.4.38) uses Jensen's inequality again and (17.4.39) uses $\mathring{m}_f(r,a) \leq \mathring{T}_f(r) + O(1) = T_f(r) + O(1)$.

(b) This is immediate from $\sigma(w,\infty)^{-1} = \sqrt{1+w^2}$ (the 4 comes from $\frac{1}{2}$ from $\sqrt{\phantom{x}}$ and $\frac{1}{2}$ from $(\pi)^{-1}$ not $(2\pi)^{-1}$).

(c) This is just Jensen's formula for $f'(z_0)$. □

**Proof of Theorem 17.3.2 (SMT).** Putting together (17.4.32) and the last proposition,

$$2\sum_{j=1}^{q} \mathring{m}_f(r,a_j) + 2\{N_f(r,0) - N_{f'}(r,\infty) - 2\mathring{m}(r,\infty)\} \leq \log(\lambda(r)) + O(\log T_f(r))) \tag{17.4.40}$$

We note that, by the FMT, $\mathring{m}(r,\infty) = T_f(r) - N_f(r,\infty) + O(1)$ and then (17.3.3) implies that the quantity in $\{\dots\}$ is $N_{1,f}(r) - 2T_f(r) + O(1)$. Thus, (17.4.40) becomes

$$\sum_{j=1}^{q} m_f(r,a_j) + N_{1,f}(r) - 2T_f(r) \leq \tfrac{1}{2}\log(\lambda(r)) + O(\log(T_f(r))) \tag{17.4.41}$$

where we used (17.4.22).

We now return to (17.4.29) and use (17.4.8) which implies

$$\int_{\widehat{\mathbb{C}}} N_f(r,a)\,d\rho(a) = N_f(r,\infty) + \int_0^{2\pi} \Psi_\rho(f(re^{i\theta}))\,\frac{d\theta}{2\pi} + O(1) \tag{17.4.42}$$

By (17.1.8),

$$\log(|w-a|) \leq \log_+(|w|) + \log_+(|a|) \tag{17.4.43}$$

so since $\int \log_+(|a|)\,d\rho(a) < \infty$,

$$\Psi_\rho(f(re^{i\theta})) \leq \log_+(|f(re^{i\theta})|) + O(1) \tag{17.4.44}$$

Thus, (17.4.42) becomes

$$\int_0^r \frac{dt}{t} \int_0^t [\lambda(s)\,s]\,ds = T_f(r) + O(1) \tag{17.4.45}$$

By Proposition 17.4.1, off a set, $E_1$, of finite measure,

$$t^{-1} \int_0^t [\lambda(s)\,s]\,ds \leq (T_f(t) + O(1))^{1+\varepsilon} \tag{17.4.46}$$

Applying it again, we see that a set, $E_2$, of finite measure,

$$t\lambda(t) \leq \left( \int_0^t [\lambda(s)\,s]\,ds \right)^{1+\varepsilon} \tag{17.4.47}$$

so off $E = E_1 \cup E_2$,

$$t\lambda(t) \leq [t(T_f(t) + O(1))^{1+\varepsilon}]^{1+\varepsilon} \tag{17.4.48}$$

so

$$\log(\lambda(r)) \leq O(\log(rT_f(r))) \tag{17.4.49}$$

so (17.4.41) is (17.3.3). □

**Notes and Historical Remarks.** The idea of using complex geometry, that is, conformal metrics, to get the SMT is due to F. Nevanlinna [**290**], Rolf's brother. He used a Poincaré metric. The use of the spherical metric to restate the FMT is due independently to Ahlfors [**5**], who was R. Nevanlinna's student, and Shimizu [**348**] in 1929. The proof we give of the SMT is due to Ahlfors [**7**]. This proof can also be found in Hille [**191**].

R. Nevanlinna's proof of his SMT relies on showing $T_{f'/f}(r) = O(\log(rT_f(r)))$ off a set of $r$'s of finite Lebesgue measure (known as the "lemma on logarithmic derivative").

We use the nonstandard term antipotential because it is the negative of what is usually called the potential; see Section 3.2 of Part 3.

**Problems**

1. Prove that (17.4.6) implies $\int_{|a|\geq 1} \log|a|\, d\rho(a) < \infty$. (*Hint*: If $f$ is $C^1$ with $f(1) = 0$, show that $\int_{|a|\geq 1} f(a)\, d\rho(a) = \int_1^\infty f'(a)\rho(\{|x| \geq a\}\, da)$.)

2. If $f$ is constant, prove both sides of (17.4.9) are 0.

3. Prove that

$$\frac{1}{\pi}\int \frac{\log\left(|z-a|\right)}{\left(1+|a|^2\right)^2}\, d^2a = \log\left(\sqrt{1+|z|^2}\right)$$

   by first using Newton's lemma to write it as the sum of $\int_0^{|z|}$ and $\int_{|z|}^\infty$ and then doing the corresponding one-dimensional integrals.

4. Let $\Psi_0$ be the antipotential of (17.4.15) and $\Psi_1(z) = \log(\sqrt{1+|z|^2})$. This problem will prove that $\Psi_0 = \Psi_1$.
   (a) Prove that $\Delta(\Psi_0 - \Psi_1) = 0$.
   (b) Prove that $\int_0^\infty \frac{\log y}{(1+y)^2}\, dy = 0$. (*Hint*: $x = y^{-1}$.)
   (c) Prove that $\Psi_0(0) = \Psi_1(0) = 0$.
   (d) Prove $\Psi_0 - \Psi_1$ is invariant under $z \to ze^{i\theta}$ and conclude that $\Psi_0 = \Psi_1$.

# Bibliography


[1] N. H. Abel, *Untersuchungen über die Reihe:* $1+\frac{m}{1}x+\frac{m(m-1)}{1\cdot 2}x^2+\frac{m(m-1)(m-2)}{1\cdot 2\cdot 3}x^3+\ldots$ *u.s.w.*, J. Reine Angew. Math. **1** (1826), 311–339. (Cited on 43, 58.)

[2] N. H. Abel, *Sur les fonctions generatrices et leurs déterminantes*, published posthumously in 1839, In Ouvres Complétes **2**, pp. 67–81, Grondahl & Son, Christiania, Denmark, 1881. (Cited on 180.)

[3] M. J. Ablowitz and P. A. Clarkson, *Solitons, Nonlinear Evolution Equations and Inverse Scattering*, London Mathematical Society Lecture Notes, Cambridge University Press, Cambridge, 1991. (Cited on 151.)

[4] M. Abramowitz and I. A. Stegun (editors), *Handbook of Mathematical Functions with Formulas, Graphs, and Mathematical Tables*, p. 555, Dover, New York, 1965; available online at `https://archive.org/details/handbookofmathem1964abra`. (Cited on 133, 135.)

[5] L. V. Ahlfors, *Beiträge zur Theorie der meromorphen Funktionen*, In 7th Scand. Math. Congr. (Oslo, 1929), pp. 84–87. (Cited on 283.)

[6] L. V. Ahlfors, *Ein Satz von Henri Cartan und seine Anwendung auf die Theorie der meromorphen Funktionen*, Commentationes Helsingfors **5** (1931), 1–19. (Cited on 258, 276.)

[7] L. V. Ahlfors, *Über eine methode in der Theorie der meromorpher Funktionen*, Soc. sci. fennicae Comment. Phys.-Mat. **8** (1935), 1–14. (Cited on 283.)

[8] L. V. Ahlfors, *An extension of Schwarz's lemma*, Trans. Amer. Math. Soc. **43** (1938), 359–364. (Cited on 16.)

[9] L. V. Ahlfors, *The theory of meromorphic curves*, Acta Soc. Fennicae A (2) **3** (1941), 171–183. (Cited on 261.)

[10] G. Airy, *On the intensity of light in the neighbourhood of a caustic*, Trans. Cambridge Philos. Soc. **6** (1838), 379–402. (Cited on 149.)

[11] T. Alberts and S. Sheffield, *Hausdorff dimension of the SLE curve intersected with the real line*, Electron. J. Probab. **13** (2008), 1166–1188. (Cited on 254.)

[12] S. L. Altmann, *Hamilton, Rodrígues, and the quaternion scandal*, Math. Magazine **62** (1989), 291–308. (Cited on 135.)

[13] S. L. Altmann and E. L. Ortiz (editors), *Mathematics and Social Utopias in France: Olinde Rodrigues and His Times*, History of Mathematics, American Mathematical Society and London Mathematical Society, Providence, RI, 2005. (Cited on 135.)

[14] G. E. Andrews, *The Theory of Partitions*, Encyclopedia of Mathematics and Its Applications, Addison-Wesley, Reading, MA–London-Amsterdam, 1976. (Cited on 211, 212.)

[15] G. E. Andrews, R. Askey, and R. Roy, *Special Functions*, Encyclopedia of Mathematics and Its Applications, Cambridge University Press, Cambridge, 1999. (Cited on 98, 133, 134, 135.)

[16] G. E. Andrews, S. B. Ekhad, and D. Zeilberger, *A short proof of Jacobi's formula for the number of representations of an integer as a sum of four squares*, Amer. Math. Monthly **100** (1993), 274–276. (Cited on 53.)

[17] G. E. Andrews and K. Eriksson, *Integer Partitions*, Cambridge University Press, Cambridge, 2004. (Cited on 211.)

[18] T. M. Apostol, *Introduction to Analytic Number Theory*, Undergraduate Texts in Mathematics, Springer-Verlag, New York–Heidelberg, 1976. (Cited on 44, 67, 77, 84.)

[19] T. M. Apostol, *Modular Functions and Dirichlet Series in Number Theory*, 2nd edition, Graduate Texts in Mathematics, Springer-Verlag, New York, 1990. (Cited on 44, 211.)

[20] N. Aronszajn, *La théorie des noyaux reproduisants et ses applications I*, Proc. Cambridge Philos. Soc. **39** (1943), 133–153. (Cited on 25.)

[21] N. Aronszajn, *Reproducing and pseudo-reproducing kernels and their application to partial differential equations of physics*, Studies in Partial Differential Equations (Tech. Report **5**), Harvard University Graduate School of Engineering, Cambridge, 1948. (Cited on 25.)

[22] N. Aronszajn, *Theory of reproducing kernels*, Trans. Amer. Math. Soc. **68** (1950), 337–404. (Cited on 25.)

[23] R. Askey and G. Gasper, *Positive Jacobi polynomial sums. II*, Amer. J. Math. **98** (1976), 709–737. (Cited on 238.)

[24] W. N. Bailey, *Generalized Hypergeometric Series*, Cambridge Tracts in Mathematics and Mathematical Physics, Stechert–Hafner, New York, 1964. (Cited on 133.)

[25] G. A. Baker, Jr., *Essentials of Padé Approximants*, Academic Press, New York–London, 1975. (Cited on 167.)

[26] G. A. Baker, Jr. and P. Graves-Morris, *Padé Approximants*, 2nd edition, Encyclopedia of Mathematics and Its Applications, Cambridge University Press, Cambridge, 1996. (Cited on 167.)

[27] E. W. Barnes, *The asymptotic expansion of integral functions defined by Taylor's series*, Philos. Trans. Roy. Soc. London A **206** (1906), 249–297. (Cited on 181.)

[28] P. Barrucand, S. Cooper, and M. D. Hirschhorn, *Results of Hurwitz type for five or more squares*, Ramanujan J. **6** (2002), 347–367. (Cited on 55.)

[29] P. T. Bateman and H. G. Diamond, *A hundred years of prime numbers*, Amer. Math. Monthly **103** (1996), 729–741. (Cited on 92.)

[30] P. T. Bateman and H. G. Diamond, *Analytic Number Theory. An Introductory Course*. World Scientific, Hackensack, NJ, 2004. (Cited on 44.)

[31] R. Beals and R. Wong, *Special Functions: A Graduate Text*, Cambridge Studies in Advanced Mathematics, Cambridge University Press, Cambridge, 2010. (Cited on 98.)

[32] H. W. Becker and J. Riordan, *The arithmetic of Bell and Stirling numbers*, Amer. J. Math. **70** (1948) 385–394. (Cited on 210.)

[33] , V. Beffara, *Cardy's formula on the triangular lattice, the easy way* In Universality and renormalization, Fields Inst. Commun., pp. 39–45, American Mathematical Society, Providence, RI, 2007. (Cited on 255.)

[34] V. Beffara, *The dimension of the SLE curves*, Ann. Probab. **36** (2008), 1421–1452. (Cited on 254.)

[35] V. Beffara and H. Duminil-Copin, *Planar percolation with a glimpse of Schramm-Loewner evolution*, Probab. Surv. **10** (2013), 1–50. (Cited on 255.)

[36] E. T. Bell, *Exponential polynomials*, Ann. Math. **35** (1934), 258–277. (Cited on 210.)

[37] E. T. Bell, *Exponential numbers*, Amer. Math. Monthly **41** (1934), 411–419. (Cited on 210.)

[38] E. T. Bell, *Men of Mathematics: The Lives and Achievements of the Great Mathematicians from Zeno to Poincaré*, Simon and Schuster, New York, 1986 (originally published in 1937). (Cited on 210.)

[39] E. T. Bell, *The iterated exponential integers*, Ann. Math. **39** (1938), 539–557. (Cited on 210.)

[40] S. R. Bell, *Nonvanishing of the Bergman kernel function at boundary points of certain domains in* $\mathbf{C}^n$, Math. Ann. **244** (1979), 69–74. (Cited on 36.)

[41] S. R. Bell, *Biholomorphic mappings and the* $\bar{\partial}$*-problem*, Ann. Math. **114** (1981), 103–113. (Cited on 36.)

[42] S. R. Bell and H. P. Boas, *Regularity of the Bergman projection in weakly pseudoconvex domains*, Math. Ann. **257** (1981), 23–30. (Cited on 36.)

[43] S. R. Bell and S. G. Krantz, *Smoothness to the boundary of conformal maps*, Rocky Mountain J. Math. **17** (1987), 23–40. (Cited on 36.)

[44] S. R. Bell and E. Ligocka, *A simplification and extension of Fefferman's theorem on biholomorphic mappings*, Invent. Math. **57** (1980), 283–289. (Cited on 36.)

[45] E. A. Bender, *Asymptotic methods in enumeration*, SIAM Rev. **16** (1974), 485–515. (Cited on 210.)

[46] N. Berestycki and J. R. Norris, *Lectures on Schramm–Loewner Evolution*, available at `http://www.statslab.cam.ac.uk/~james/Lectures/sle.pdf`, 2014. (Cited on 255.)

[47] S. Bergman, *Über die Entwicklung der harmonischen Funktionen der Ebene und des Raumes nach Orthogonalfunktionen*, Math. Ann. **86** (1922), 238–271. (Cited on 25.)

[48] S. Bergman, *Über die Bestimmung der Verzweigungspunkte eines hyperelliptischen Integrals aus seinen Periodizitätsmoduln mit Anwendungen auf die Theorie des Transformators*, Math. Z. **19** (1924), 8–25. (Cited on 25.)

[49] S. Bergman, *Über unendliche Hermitesche Formen, die zu einem Bereiche gehören, nebst Anwendungen auf Fragen der Abbildung durch Funktionen von zwei komplexen Veränderlichen*, Math. Z. **29** (1929), 641–677. (Cited on 25.)

[50] S. Bergman, *Über die Kernfunktion eines Bereiches und ihr Verhalten am Rande, I*, J. Reine Angew. Math. **169** (1933), 1–42. (Cited on 25.)

[51] S. Bergman, *The Kernel Function and Conformal Mapping*, 2nd, revised edition, Mathematical Surveys, American Mathematical Society, Providence, RI, 1970; 1st edition, 1950. (Cited on 25.)

[52] W. D. Bergweiler, *A new proof of the Ahlfors five islands theorem*, J. Anal. Math. **76** (1998), 337–347. (Cited on 276.)

[53] W. D. Bergweiler and J. K. Langley, *Multiplicities in Hayman's alternative*, J. Aust. Math. Soc. **78** (2005), 37–57. (Cited on 261.)

[54] D. Bernard, *Stochastic Schramm-Loewner Evolution (SLE) from. Statistical Conformal Field Theory (CFT): An Introduction for (and by) Amateurs*, Available at `https://www.lpt.ens.fr/IMG/pdf/Bernard_SLENotes_MSRI2012.pdf`, 2012. (Cited on 255.)

[55] B. C. Berndt, *Number Theory in the Spirit of Ramanujan*, Student Mathematical Library, American Mathematical Society, Providence, RI, 2006. (Cited on 54, 55, 135, 211.)

[56] F. W. Bessel, *Analytische Auflosung der Kepler'schen Aufgabe*, Abh. Berliner Akad. (1816–17), published 1819. (Cited on 149.)

[57] L. Bieberbach, *Über die Koeffizienten derjenigen Potenzreihen, welche eine schlichte Abbildung des Einheitskreises vermitteln*, Sitzungsber. Preuss. Akad. Wiss. Phys-Math. Kl. (1916), 940–955. (Cited on 237.)

[58] J.-M. Bismut, *Large Deviations and the Malliavin Calculus*, Progress in Mathematics, Birkhäuser, Boston, 1984. (Cited on 180.)

[59] N. Bleistein and R. A. Handelsman, *Asymptotic Expansions of Integrals*, 2nd edition, Dover Publications, New York, 1986. (Cited on 161, 163.)

[60] A. Bloch, *Les fonctions holomorphes et méromorphes dans le cercle-unit*, Gauthiers-Villars, Paris, 1926. (Cited on 276.)

[61] J. Boos, *Classical and Modern Methods in Summability*, Oxford Mathematical Monographs, Oxford University Press, Oxford, 2000. (Cited on 167.)

[62] E. Borel, *Mémoire sur les séries divergentes*, Ann. Sci. Ecole Norm. Sup. (3) **16** (1899), 9–131. Available online at `http://www.numdam.org/item?id=ASENS_1899_3_16__9_0`. (Cited on 168.)

[63] E. Borel, *Lectures on Divergent Series*, Los Alamos Scientific Laboratory, University of California, Los Alamos, NM, 1975; French original: *Leçons sur les séries divergentes*, 2nd edition, Gauthier–Villars, Paris, 1928. (Cited on 167.)

[64] R. Bott, C. Earle, D. Hejhal, J. Jenkins, T. Jorgensen, A. Marden, and R. Osserman, *Lars Valerian Ahlfors (1907–1996)*, Notices Amer. Math. Soc. **45** (1998), 248–255. (Cited on 257.)

[65] L. Brillouin, *La mécanique ondulatoire de Schrödinger; une méthode générale de resolution par approximations successives*, C. R. Acad. Sci. Paris **183**, (1926), 24–26. (Cited on 226.)

[66] H. Buchholz, *The Confluent Hypergeometric Function with Special Emphasis on Its Applications*, Springer Tracts in Natural Philosophy, Springer-Verlag, New York, 1969. (Cited on 133.)

[67] F. Camia and C. M. Newman, *Two-dimensional critical percolation: the full scaling limit*, Comm. Math. Phys. **268** (2006), 1–38. (Cited on 254.)

[68] E. R. Canfield, C. Greenhill, and B. D. McKay, *Asymptotic enumeration of dense 0-1 matrices with specified line sums*, J. Combin. Theory Ser. A **115** (2008), 32–66. (Cited on 210.)

[69] C. Carathéodory, *Untersuchungen über die konformen Abbildungen von festen und veränderlichen Gebieten*, Math. Ann. **72** (1912) 107–144. (Cited on 250.)

[70] J. L. Cardy, *Critical percolation in finite geometries*, J. Phys. A **25** (1992), L201–L206. (Cited on 254.)

[71] F. Carlini, *Ricerehe sulla convergenza della serie che serve alla soluzione del problema di Keplero*, Milan, 1817. Translated into German by Jacobi in Astro. Nach. **30** (1850), 197–254. (Cited on 226.)

[72] B. C. Carlson, *Special Functions of Applied Mathematics*, Academic Press [Harcourt Brace Jovanovich, Publishers], New York–London, 1977. (Cited on 98.)

[73] H. Cartan, *Sur les systèmes de fonctions holomorphes à variétés linéaires lacunaires et leurs applications*, Ann. Sci. Ecole Norm. Sup. (3) **45** (1928), 255–346. (Cited on 261.)

[74] H. Cartan, *Sur la fonction de croissance attachée à une fonction méromorphe de deux variables, et ses applications aux fonctions méromorphes d'une variable*, C. R. Acad. Sci. Paris **189** (1929), 521–523. (Cited on 271.)

[75] H. Cartan, *Sur les zéros des combinaisons linéaires de p fonctions holomorphes données*, Mathematica (Cluj) **7** (1933), 80–103. (Cited on 261.)

[76] P. Cartier, *An introduction to zeta functions*, In "From Number Theory to Physics" (Les Houches, 1989), pp. 1–63, Springer, Berlin, 1992. (Cited on 54, 56.)

[77] A. L. Cauchy, *Mémoire sur divers points d'analyse*, Mémoires Acad. Sci. France **8** (1829), 130–138. (Cited on 180.)

[78] K. Chandrasekharan, *Introduction to Analytic Number Theory*, Die Grundlehren der mathematischen Wissenschaften, Springer-Verlag, New York, 1968. (Cited on 44.)

[79] J. Chazy, *Sur les équations différentielles dont l'intégrale générale est uniforme et admet des singularités essentielles mobiles*, C. R. Acad. Sci. Paris **149** (1910), 563–565. (Cited on 151.)

[80] J. Chazy, *Sur les équations différentielles dont l'intégrale générale possède une coupure essentielle mobile*, C. R. Acad. Sci. Paris **150** (1910), 456–458. (Cited on 151.)

[81] J. Chazy, *Sur les équations différentielles du troisième ordre et d'ordre supérieur dont l'intégrale générale a ses points critiques fixes*, Acta Math. **34** (1911), 317–385. (Cited on 151.)

[82] P. L. Chebyshev, *Sur la totalité des nombres premiers inférieurs à une limite donnée*, J. Math. Pures Appl. (1) **17** (1852), 341–365. (Cited on 92.)

[83] P. L. Chebyshev, *Mémoire sur les nombres premiers*, J. Math. Pures Appl. (1) **17** (1852), 366–390. (Cited on 92.)

[84] D. Chelkak and S. Smirnov, *Discrete complex analysis on isoradial graphs*, Adv. in Math. **228** (2011), 1590–1630. (Cited on 254.)

[85] D. Chelkak and S. Smirnov, *Universality in the 2D Ising model and conformal invariance of fermionic observables*, Invent. Math. **189** (2012), 515–580. (Cited on 254.)

[86] W. Cherry and Z. Ye, *Nevanlinna's Theory of Value Distribution. The Second Main Theorem and Its Error Terms*, Springer Monographs in Mathematics, Springer-Verlag, Berlin, 2001. (Cited on 261, 262.)

[87] C. Chester, B. Friedman, and F. Ursell, *An extension of the method of steepest descents*, Proc. Cambridge Philos. Soc. **53** (1957), 599–611. (Cited on 193.)

[88] J. M. Cook, *Convergence to the Møller wave-matrix*, J. Math. Phys. **36** (1957), 82–87. (Cited on 194.)

[89] J. B. Conway, *Functions of One Complex Variable. II*, Graduate Texts in Mathematics, Springer-Verlag, New York, 1995. (Cited on 238.)

[90] S. Cooper, *Sums of five, seven and nine squares*, Ramanujan J. **6** (2002), 469–490. (Cited on 55.)

[91] E. T. Copson, *Asymptotic Expansions*, reprint of the 1965 original, Cambridge Tracts in Mathematics, Cambridge University Press, Cambridge, 2004. (Cited on 163, 166, 193.)

[92] O. Costin, *Asymptotics and Borel Summability*, Chapman & Hall/CRC Monographs and Surveys in Pure and Applied Mathematics, CRC Press, Boca Raton, FL, 2009. (Cited on 167.)

[93] H. Cramèr, *Sur un nouveau théorème-limite de la théorie des probabilités*, Actualités Scientifiques et Industrialles, **736** (1938), 5–23; Colloque Consecré à la Théorie des Probabilités **3**, Hermann, Paris. (Cited on 180.)

[94] L. de Branges, *A proof of the Bieberbach conjecture*, Acta Math. **154** (1985), 137–152. (Cited on 238.)

[95] N. G. de Bruijn, *Asymptotic Methods in Analysis*, corrected reprint of the 3rd edition, Dover Publications, New York, 1981. (Cited on 163, 210.)

[96] P. Debye, *Näherungsformeln für die Zylinderfunktionen für große Werte des Arguments und unbeschränkt veränderliche Werte des Index*, Math. Ann. **67** (1909), 535–558. (Cited on 210.)

[97] C.-J. de la Vallée Poussin, *Recherches analytiques la théorie des nombres premiers*, Ann. Soc. Scient. Bruxelles **20** (1896), 183–256. (Cited on 78, 92, 93.)

[98] P. Deligne, *Équations Différentielles à Points Singuliers Réguliers*, Lecture Notes in Mathematics, Springer-Verlag, Berlin–New York, 1970. (Cited on 98.)

[99] A. Dembo and O. Zeitouni, *Large Deviations Techniques and Applications*, corrected reprint of the second (1998) edition, Stochastic Modelling and Applied Probability, Springer-Verlag, Berlin, 2010. (Cited on 180.)

[100] J. Derbyshire, *Prime Obsession. Bernhard Riemann and the Greatest Unsolved Problem in Mathematics*, Joseph Henry Press, Washington, DC, 2003. (Cited on 37, 76.)

[101] H. G. Diamond, *Elementary methods in the study of the distribution of prime numbers*, Bull. Amer. Math. Soc. **7** (1982), 553–589. (Cited on 92, 93.)

[102] L. E. Dickson, *History of the Theory of Numbers. Vol. II: Diophantine Analysis*, Chelsea, New York 1966. (Cited on 55.)

[103] R. B. Dingle, *Asymptotic Expansions: Their Derivation and Interpretation*, Academic Press, London–New York, 1973. (Cited on 166.)

[104] P. G. L. Dirichlet, *Beweis des Satzes, dass jede unbegrenzte arithmetische Progression, deren erstes Glied und Differenz ganze Zahlen ohne gemeinschaftlichen Factor sind, unendlich viele Primzahlen enthält*, [There are infinitely many prime numbers in all arithmetic progressions with first term and difference coprime], Abh. Königlich. Preuss. Akad. Wiss. **48** (1837), 45–81; translation available online at `http://arxiv.org/abs/0808.1408`. (Cited on 84.)

[105] P. G. L. Dirichlet, *Lectures on Number Theory*, (with supplements by R. Dedekind), translated from the 1863 German original, History of Mathematics, American Mathematical Society, Providence, RI; London Mathematical Society, London, 1999. German original of 1871 second edition available online at `http://gdz.sub.uni-goettingen.de/dms/load/img/?IDDOC=159253`. (Cited on 67.)

[106] G. Doetsch, *Handbuch der LaplaceTransformation. Band I: Theorie der Laplace-Transformation*, Verbesserter Nachdruck der ersten Auflage 1950, Lehrbücher und Monographien aus dem Gebiete der exakten Wissenschaften, Mathematische Reihe, Birkhäuser Verlag, Basel–Stuttgart, 1971. (Cited on 181.)

[107] M. D. Donsker and S. R. S. Varadhan, *Asymptotic evaluation of certain Markov process expectations for large time. I, II, III, IV*, Comm. Pure Appl. Math. **28** (1975), 1–47; 279–301; **29** (1976), 389–461, **36** (1983), 183–212. (Cited on 180.)

[108] D. Drasin, *The inverse problem of Nevanlinna theory*, Acta. Math. **138** (1977), 83–151. (Cited on 276.)

[109] D. Dugué, *Le défaut au sens de M. Nevanlinna dépend de l'origine choisie*, C. R. Acad. Sci. Paris **225** (1947), 555–556. (Cited on 276.)

[110] J. J. Duistermaat, *Fourier Integral Operators*, Progress in Mathematics, Birkhäuser, Boston, 1996. (Cited on 192.)

[111] P. L. Duren, *Univalent Functions*, Grundlehren der Mathematischen Wissenschaften, Springer-Verlag, New York, 1983. (Cited on 238, 250.)

[112] C. J. Earle and R. S. Hamilton, *A fixed point theorem for holomorphic mappings*, In Global Analysis (Berkeley, CA, 1970), Proc. Sympos. Pure Math. 1968, pp. 61–65, American Mathematical Society, Providence, RI. (Cited on 14.)

[113] H. M. Edwards, *Riemann's Zeta Function*, reprint of the 1974 original, Dover Publications, Mineola, NY, 2001. (Cited on 76, 77.)

[114] R. S. Ellis, *Entropy, Large Deviations, and Statistical Mechanics*, reprint of the 1985 original, Classics in Mathematics, Springer-Verlag, Berlin, 2006. (Cited on 180.)

[115] W. J. Ellison, *Waring's problem*, Amer. Math. Monthly **78** (1971), 10–36. (Cited on 55.)

[116] V. Enss, *Asymptotic completeness for quantum mechanical potential scattering. I. Short range potentials*, Comm. Math. Phys. **61** (1978), 285–291. (Cited on 193.)

[117] A. Erdélyi, *Asymptotic Expansions*, Dover Publications, New York, 1956. (Cited on 163, 166.)

[118] A. Erdélyi, W. Magnus, F. Oberhettinger, and F. G. Tricomi, *Higher Transcendental Functions, I*, McGraw–Hill, New York–Toronto–London, 1953. (Cited on 133.)

[119] P. Erdős, *On an elementary proof of some asymptotic formulas in the theory of partitions*, Ann. Math. **43** (1942), 437–450. (Cited on 211.)

[120] A. Eremenko, *Meromorphic solutions of algebraic differential equations*, Uspekhi Mat. Nauk **37** (1982), 53–82 (in Russian); English translation, Russian Math. Surveys **37** (1982), 61–95. (Cited on 261.)

[121] A. Eremenko, *Lectures on Nevanlinna Theory*, available at `http://www.math.purdue.edu/~eremenko/dvi/weizmann.pdf`. (Cited on 261.)

[122] A. E. Eremenko and M. L. Sodin, *On meromorphic functions of finite order with maximal deficiency sum*, Teor. Funktsiĭ Funktsional. Anal. i Prilozhen. **55** (1991), 84–95; translation in J. Soviet Math. **59** (1992), 643–651. (Cited on 261.)

[123] A. E. Eremenko and M. L. Sodin, *On the distribution of values of meromorphic functions of finite order*, Dokl. Akad. Nauk. SSSR (N.S.) **316** (1991) 538–541; translation in Soviet Math. Dokl. **43** (1991), 128–131. (Cited on 261.)

[124] L. Euler, *Variae observationes circa series infinitas*, Comm. Acad. Sci. Petropolitanae **9** (1737), 160–188, available online at `http://eulerarchive.maa.org/pages/E072.html`. (Cited on 76.)

[125] L. Euler, *Demonstratio theorematis Fermatiani omnem numerum primum formae $4n+1$ esse summam duorum quadratorum*, Novi Comm. Acad. Sci. Petropolitanae **5** (1760) 3–13. This is based on work presented to the Berlin Academy on October 15, 1750 and described in part in a letter to Goldbach dated May 6, 1747. (Cited on 53.)

[126] L. Euler, *De seriebus divergentibus*, Novi Comm. Acad. Sci. Petropolitanae **5** (1754/55) 1760, 205–237. (Cited on 166.)

[127] L. Euler, *Remarques sur un beau rapport entre les séries des puissances tant directes que réciproques*, [written in 1749, presented in 1761, published in 1768], Mémoires de l'Academie des Sciences de Berlin **17** (1768), 83–106. English trasnslation with notes available online at `http://eulerarchive.maa.org/docs/translations/E352.pdf`. (Cited on 77.)

[128] L. Euler, *Institutiones calculi integralis*, volumen secundum, Imperial Academy of Sciences, St. Petersburg, 1769. (Cited on 114, 133.)

[129] L. Euler, *Novae demonstrationes circa resolutionem numerorum in quadrata*, Nova Acta Eruditorum (1773), 193–211; English translation in arXiv:0806.0104. (Cited on 53.)

[130] J. Farkas, *Sur les fonctions itératives*, J. Math. Pures Appl. (3) **10** (1884), 101–108. (Cited on 14.)

[131] C. L. Fefferman, *The Bergman kernel and biholomorphic mappings of pseudoconvex domains*, Invent. Math. **26** (1974), 1–65. (Cited on 36.)

[132] C. L. Fefferman, *The Bergman kernel in quantum mechanics*, In Analysis and Geometry in Several Complex Variables (Katata, 1997), pp. 39–58, Trends Math., Birkhäuser, Boston, 1999. (Cited on 25.)

[133] M. Fekete and G. Szegő, *Eine Bemerkung über ungerade schlichte Funktionen*, J. London Math. Soc. **2** (1933), 85–89. (Cited on 238.)

[134] J. Feng and T. G. Kurtz, *Large Deviations for Stochastic Processes*, Mathematical Surveys and Monographs, American Mathematical Society, Providence, RI, 2006. (Cited on 180.)

[135] B. Fine and G. Rosenberger, *Number Theory. An Introduction Via the Distribution of Primes*, Birkhäuser, Boston, 2007. (Cited on 44.)

[136] N. J. Fine, *Basic Hypergeometric Series and Applications*, Mathematical Surveys and Monographs, American Mathematical Society, Providence, RI, 1988. (Cited on 133.)

[137] A. S. Fokas, A. R. Its, A. A. Kapaev, and V. Yu. Novokshenov, *Painlevé Transcendents. The Riemann–Hilbert Approach*, Mathematical Surveys and Monographs, American Mathematical Society, Providence, RI, 2006. (Cited on 151.)

[138] M. I. Freidlin and A. D. Wentzell, *Random Perturbations of Dynamical Systems*, 2nd edition, Grundlehren der Mathematischen Wissenschaften, Springer-Verlag, New York, 1984. (Cited on 180.)

[139] F. G. Frobenius, *Über die Integration der linearen Differentialgleichungen durch Reihen*, J. Reine Angew. Math. 76 (1873), 214–235. (Cited on 113.)

[140] F. G. Frobenius, *Über den Begriff der Irreducibilität in der Theorie der linearen Differentialgleichungen*, J. Reine Angew. Math. 76 (1873), 236–270. (Cited on 113.)

[141] N. Fröman and P. O. Fröman, *JWKB Approximation. Contributions to the Theory*, North-Holland, Amsterdam, 1965. (Cited on 227.)

[142] N. Fröman and P. O. Fröman, *Physical Problems Solved by the Phase-Integral Method*, Cambridge University Press, Cambridge, 2002. (Cited on 227.)

[143] L. I. Fuchs, *Zur Theorie der linearen Differentialgleichungen mit veränderlichen Coefficienten*, Jahrsber. Gewerbeschule, Berlin, Ostern, 1865. (Cited on 113.)

[144] L. I. Fuchs, *Zur Theorie der linearen Differentialgleichungen mit veränderlichen Coefficienten*, J. Reine Angew. Math. **66** (1866), 121–160. (Cited on 113.)

[145] L. I. Fuchs, *Zur Theorie der linearen Differentialgleichungen mit veränderlichen Coefficienten*, (Ergänzungen zu der im $66^{\text{sten}}$ Bande dieses Journals enthaltenen Abhandlung), J. Reine Angew. Math. **68** (1868), 354–385; available online at `http://www.degruyter.com/view/j/crll.1868.issue-68/crll.1868.68.354/crll.1868.68.354.xml`. (Cited on 113.)

[146] Z. Gao, B. D. McKay, and X. Wang, *Asymptotic enumeration of tournaments with a given score sequence containing a specified digraph*, Random Structures Algorithms **16** (2000), 47–57. (Cited on 210.)

[147] G. Gasper and M. Rahman, *Basic Hypergeometric Series*, 2nd edition, Encyclopedia of Mathematics and Its Applications, Cambridge University Press, Cambridge, 2004. (Cited on 133, 134.)

[148] C. F. Gauss, *Disquisitiones Arithmeticae*, 1801; English translation for sale at `http://yalepress.yale.edu/yupbooks/book.asp?isbn=9780300094732`. (Cited on 133.)

[149] C. F. Gauss, *Disquisitiones generales circa seriam infinitam* $1 + \frac{\alpha\beta}{1\cdot\gamma}\, x + \frac{\alpha(\alpha+1)\beta(\beta+1)}{1\cdot 2\cdot\gamma(\gamma+1)}\, x\, x +$ *etc.*, Comm. Soc. Göttingen, **2** (1813), 123–162. (Cited on 133.)

[150] J. Glimm and A. Jaffe, *Quantum Physics. A Functional Integral Point of View*, 2nd edition, Springer-Verlag, New York, 1987. (Cited on 169.)

[151] A. A. Goldberg and I. V. Ostrovskii, *Value Distribution of Meromorphic Functions*, translated from the 1970 Russian original, Translations of Mathematical Monographs, American Mathematical Society, Providence, RI, 2008. (Cited on 261.)

[152] M. L. Goldberger and K. M. Watson, *Collision Theory*, corrected reprint of the 1964 edition, Robert E. Krieger Publishing, Huntington, NY, 1975. (Cited on 149.)

[153] D. Goldfeld, *The elementary proof of the prime number theorem: An historical perspective*, In Number Theory (New York, 2003), pp. 179–192, Springer, New York, 2004. (Cited on 93.)

[154] S. Gong, *The Bieberbach Conjecture*, AMS/IP Studies in Advanced Mathematics, American Mathematical Society, Providence, RI; International Press, Cambridge, MA, 1999. (Cited on 238.)

[155] E. Goursat, *Sur l'équation différentielle linéaire, qui admet pour intégrale la série hypergéométrique*, Ann. Sci. Ecole Norm. Sup. (2) **10** (1881), 3–142. (Cited on 134.)

[156] I. S. Gradshteyn and I. M. Ryzhik, *Table of Integrals, Series, and Products*, 7th edition, Elsevier/Academic Press, Amsterdam, 2007. (Cited on 133.)

[157] S. Graffi, V. Grecchi, and B. Simon, *Borel summability: application to the anharmonic oscillator*, Phys. Lett. B **32** (1970), 631–634. (Cited on 169.)

[158] H. Grauert, *Holomorphe Funktionen mit Werten in komplexen Lieschen Gruppen*, Math. Ann. **133** (1957), 450–472. (Cited on 100.)

[159] J. Gray, *Worlds Out of Nothing. A Course in the History of Geometry in the 19th Century*, Springer Undergraduate Mathematics Series, Springer-Verlag, London, 2007. (Cited on 1.)

[160] J. Gray, *Linear Differential Equations and Group Theory from Riemann to Poincaré*, reprint of the 2000 second edition, Modern Birkhäuser Classics, Birkhäuser, Boston, 2008. (Cited on 114.)

[161] J. Gray, *Henri Poincaré. A Scientific Biography*, Princeton University Press, Princeton, NJ, 2013. (Cited on 12.)

[162] G. Green, *On the motion of waves in a variable canal of small depth and width*, Trans. Cambridge Philos. Soc. **6** (1837), 457–462. (Cited on 226.)

[163] P. A. Griffiths, *Entire Holomorphic Mappings in One and Several Complex Variables*, Annals of Mathematics Studies, Princeton University Press, Princeton, NJ; University of Tokyo Press, Tokyo, 1976. (Cited on 261.)

[164] P. Griffiths and J. King, *Nevanlinna theory and holomorphic mappings between algebraic varieties*, Acta Math. **130** (1975), 145–220. (Cited on 261.)

[165] A. Grigis and J. Sjöstrand, *Microlocal Analysis for Differential Operators. An Introduction*, London Mathematical Society Lecture Notes, Cambridge University Press, Cambridge, 1994. (Cited on 192.)

[166] G. Grimmett, *Probability on Graphs. Random Processes on Graphs and Lattices*, Institute of Mathematical Statistics Textbooks, Cambridge University Press, Cambridge, 2010. (Cited on 255.)

[167] V. I. Gromak, I. Laine, and S. Shimomura, *Painlevé Differential Equations in the Complex Plane*, de Gruyter Studies in Mathematics, Walter de Gruyter, Berlin, 2002. (Cited on 151.)

[168] T. H. Gronwall, *Some remarks on conformal representation*, Ann. Math. (2) **16** (1914), 72–76. (Cited on 237.)

[169] F. Gross, *On the equation $f^n + g^n = 1$*, Bull. Amer. Math. Soc. **72** (1966), 86–88. (Cited on 277.)

[170] V. Guillemin and S. Sternberg, *Geometric Asymptotics*, Mathematical Surveys, American Mathematical Society, Providence, RI, 1977. (Cited on 192.)

[171] J. Hadamard, *Sur la distribution des zéros de la fonction $\zeta(s)$ et ses conséquences arithmétiques*, Bull. Soc. Math. France **24** (1896), 199–220. (Cited on 78, 92, 93.)

[172] H. Hankel, *Die Euler'schen Integrale bei unbeschränkter Variabilität des Arguments*, Z. Math. Phys. **9** (1864), 1–21. (Cited on 159.)

[173] G. H. Hardy, *On the representation of a number as the sum of any number of squares, and in particular of five*, Trans. Amer. Math. Soc. **21** (1920), 255–284. (Cited on 53, 55.)

[174] G. H. Hardy, *Ramanujan: Twelve Lectures on Subjects Suggested by His Life and Work*, Cambridge University Press, Cambridge; Macmillan, New York, 1940. (Cited on 211.)

[175] G. H. Hardy, *Divergent Series*, Clarendon Press, Oxford, 1949; 2nd edition, AMS Chelsea, Providence, RI, 1991. (Cited on 167.)

[176] G. H. Hardy and J. E. Littlewood, *A new solution of Waring's problem*, Quart. J. Math. **48** (1919), 272–293. (Cited on 212.)

[177] G. H. Hardy and S. Ramanujan, *Asymptotic formulae in combinatory analysis*, Proc. London Math. Soc. **17** (1918), 75–115. (Cited on 211.)

[178] G. H. Hardy and M. Riesz, *The General Theory of Dirichlet's Series*, Cambridge Tracts in Mathematics and Mathematical Physics, Stechert–Hafner, New York, 1964; available online at `http://ebooks.library.cornell.edu/cgi/t/text/text-idx?c=math;idno=01480002` and via print on demand from Amazon.com. (Cited on 67.)

[179] E. M. Harrell, *Double wells*, Comm. Math. Phys. **75** (1980), 239–261. (Cited on 228.)

[180] E. M. Harrell and B. Simon, *The mathematical theory of resonances whose widths are exponentially small*, Duke Math. J. **47** (1980), 845–902. (Cited on 228.)

[181] W. K. Hayman, *An integral function with a defective value that is neither asymptotic nor invariant under change of origin*, J. London Math. Soc. **28** (1953), 369–376. (Cited on 276.)

[182] W. K. Hayman, *Picard values of meromorphic functions and their derivatives*, Ann. Math. (2) **70** (1959), 9–42. (Cited on 261.)

[183] W. K. Hayman, *Meromorphic Functions*, Oxford Mathematical Monographs, Clarendon Press, Oxford, 1964. (Cited on 261.)

[184] W. K. Hayman, *On the Valiron deficiencies of integral functions of infinite order*, Ark. Mat. **10** (1972), 163–172. (Cited on 276.)

[185] W. K. Hayman, *The local growth of power series: a survey of the Wiman–Valiron method*, Canad. Math. Bull. **17** (1974), 317–358. (Cited on .)

[186] W. K. Hayman, *Multivalent Functions*, 2nd edition, Cambridge Tracts in Mathematics, Cambridge University Press, Cambridge, 1994. (Cited on 238.)

[187] E. Heine, *Theorie der Kugelfunctionen und der verwandten Functionen*, Reimer, Berlin, 1878. (Cited on 135.)

[188] H. Helson, *Dirichlet Series*, Henry Helson, Berkeley, CA, 2005. (Cited on 67.)

[189] D. Hilbert, *Beweis für die Darstellbarkeit der ganzen Zahlen durch eine feste Anzahl $n^{\text{ter}}$ Potenzen (Waringsches Problem)*, Math. Ann. **67** (1909), 281–300. (Cited on 55.)

[190] E. Hille, *Ordinary Differential Equations in the Complex Domain*, reprint of the 1976 original, Dover Publications, Mineola, NY, 1997. (Cited on 95, 98, 103, 114.)

[191] E. Hille, *Analysis. Vol. II*, reprint of the 1964 original with corrections, R. E. Krieger Publishing, Huntington, NY, 1979. (Cited on 283.)

[192] M. D. Hirschhorn, *A simple proof of Jacobi's four-square theorem*, J. Aust. Math. Soc. Ser. A **32** (1982), 61–67. (Cited on 53.)

[193] M. D. Hirschhorn, *A simple proof of Jacobi's two-square theorem*, Amer. Math. Monthly **92** (1985), 579–580. (Cited on 53.)

[194] M. D. Hirschhorn, *A simple proof of Jacobi's four-square theorem*, Proc. Amer. Math. Soc. **101** (1987), 436–438. (Cited on 53.)

[195] E. Hlawka, *Über Integrale auf konvexen Körpern. I*, Monatsh. Math. **54** (1950), 1–36. (Cited on 193.)

[196] H. Hochstadt, *Special Functions of Mathematical Physics*, Athena Series: Selected Topics in Mathematics, Holt, Rinehart and Winston, New York 1961. (Cited on 98.)

[197] E. Hopf, *Elementare Bemerkungen über die Lösungen partieller Differentialgleichungen zweiter Ordnung vom elliptischen Typus*, Sitzungsber. Preuss. Akad. Wiss. Berlin (1927) 147–152. (Cited on 36.)

[198] L. Hörmander, *Fourier integral operators. I*, Acta Math. **127** (1971), 79–183. (Cited on 192.)

[199] L. Hörmander, *The existence of wave operators in scattering theory*, Math. Z. **146** (1976), 69–91. (Cited on 192.)

[200] L. Hörmander, *The Analysis of Linear Partial Differential Operators. IV. Fourier Integral Operators*, Grundlehren der Mathematischen Wissenschaften, Springer-Verlag, Berlin, 1985. (Cited on 192.)

[201] A. Hurwitz, *Einige Eigenschaften der Dirichlet'schen Funktionen $F(s) = \sum(\frac{D}{n}) \cdot \frac{1}{n^s}$, die bei der Bestimmung der Klassenzahlen binärer quadratischer Formen auftreten*, Z. Math. Phys. **27** (1882), 86–101. (Cited on 67, 77.)

[202] A. Hurwitz, *Sur la décomposition des nombres en cinq carrés*, C. R. Acad. Sci. Paris **98** (1884), 504–507. (Cited on 55.)

[203] A. Hurwitz, *Über die Zahlentheorie der Quaternionen*, Nachr. Kön. Ges. Wiss. Göttingen, Math.-phys. Klasse **1896** (1896), 313–340. (Cited on 53.)

[204] A. Hurwitz, *Über die Anwendung der elliptischen Modulfunktionen auf einen Satz der allgemeinen Funktionentheorie*, Zürich. naturf. Ges. **49** (1904), 242–253. (Cited on 237.)

[205] A. Hurwitz, *Vorlesungen über die Zahlentheorie Quaternionen*, Springer, Berlin, 1919. (Cited on 53.)

[206] S. Ikehara, *An extension of Landau's theorem in the analytic theory of numbers*, J. Math. Phys. **10** (1931), 1–12. (Cited on 92.)

[207] Y. Ilyashenko and S. Yakovenko, *Lectures on Analytic Differential Equations*, Graduate Studies in Mathematics, American Mathematical Society, Providence, RI, 2008. (Cited on 98.)

[208] E. L. Ince, *Ordinary Differential Equations*, Dover Publications, New York, 1944; reprint of the English edition which appeared in 1926, Longmans, Green and Co., London. (Cited on 98.)

[209] A. E. Ingham, *On Wiener's method in Tauberian theorems*, Proc. London Math. Soc. (2) **38** (1935), 458–480. (Cited on 92.)

[210] A. E. Ingham, *Some Tauberian theorems connected with the prime number theorem*, J. London Math. Soc. **20** (1945), 171–180. (Cited on 92.)

[211] A. Ivić, *The Riemann Zeta-Function. The Theory of the Riemann Zeta-Function with Applications*, John Wiley & Sons, New York, 1985. (Cited on 77.)

[212] J. Ivory, *On the attractions of an extensive class of spheriods*, Philosophical Transactions of the Royal Society **102**, (1812), 46–82. (Cited on 135.)

[213] J. Ivory, *On the figure requisite to maintain the equlibrium of a homogeneous fluid mass that revolved upon an axis*, Philos. Trans. Roy. Soc. London **114**, (1824), 85–150. (Cited on 135.)

[214] H. Iwaniec and E. Kowalski, *Analytic Number Theory*, American Mathematical Society Colloquium Publications, American Mathematical Society, Providence, RI, 2004. (Cited on 44.)

[215] K. Iwasaki, H. Kimura, S. Shimomura, and M. Yoshida, *From Gauss to Painlevé. A Modern Theory of Special Functions*, Aspects of Mathematics, Friedr. Vieweg & Sohn, Braunschweig, 1991. (Cited on 151.)

[216] C. G. Jacobi, *Über den Ausdruck der verschiedenen Wurzeln einer Gleichung durch bestimmte Integrale*, J. Reine Angew. Math. **2** (1827), 1–8. (Cited on 135.)

[217] C. G. Jacobi, *Note sur la décomposition d'un nombre donné en quatre carrés*, J. Reine Angew. Math. **3** (1828), 191. (Cited on 53.)

[218] C. G. Jacobi, *Fundamenta nova theoriae functionum ellipticarum*, 1829; Available online at `https://openlibrary.org/books/OL6506700M/Fundamenta_nova_theoriae_functionum_ellipticarum`. (Cited on 53.)

[219] H. Jeffreys, *On certain approximate solutions of linear differential equations of the second order*, Proc. London Math. Soc. (2) **23** (1925), 428–436. (Cited on 227.)

[220] H. Jeffreys, *The effect on love waves of heterogeneity in the lower layer*, Geophys. Suppl. MNRAS **2** (1928), 101–111. (Cited on 149.)

[221] J. L. Jensen, *Om Räkkers Konvergens*[On convergence of series], Tidsskrift for mathematik (5) **II** (1884), 63–72. (Cited on 67.)

[222] J. L. Jensen, *Sur une généralisation d'un théorème de Cauchy*, C. R. Acad. Sci. Paris **106** (1888), 833–836. (Cited on 67.)

[223] L. P. Kadanoff, *Loewner evolution: Maps and shapes in two dimensions*, available online at `http://jfi.uchicago.edu/~leop/TALKS/MyFractalShapes.pdf`, 2004. (Cited on 231.)

[224] W. Kager and B. Nienhuis, *A Guide to Stochastic Loewner Evolution and its Applications*, J. Statist. Phys. **115** (2004), 1149–1229. (Cited on 255.)

[225] A. A. Karatsuba and S. M. Voronin, *The Riemann Zeta-Function*, de Gruyter Expositions in Mathematics, Walter de Gruyter, Berlin, 1992. (Cited on 76.)

[226] Lord Kelvin (W. Thomson), *On the waves produced by a single impulse in water of any depth, or in a dispersive medium*, Proc. Royal Soc. London **42** (1887), 80–83. (Cited on 192.)

[227] H. Koch, *Introduction to Classical Mathematics. I. From the Quadratic Reciprocity Law to the Uniformization theorem*, translated and revised from the 1986 German original, Mathematics and Its Applications, Kluwer, Dordrecht, 1991. (Cited on 84.)

[228] P. Koebe, *Über die Uniformisierung beliebiger analytischer Kurven, Erste Mitteilung, Zweite Mitteilung*, Nachr. Kön. Ges. Wiss. Göttingen, Math.-phys. Klasse **1907**, 191–210; 633–669. (Cited on 237.)

[229] P. Koebe, *Über die Uniformisierung der algebraischen Kurven durch automorphe Funktionen mit imaginärer Substitutionsgruppe (Fortsetzung und Schluss)*, Nachr. Kön. Ges. Wiss. Göttingen, Math.-phys. Klasse **1910**, 180–189. (Cited on 237.)

[230] J. Korevaar, *On Newman's quick way to the prime number theorem*, Math. Intelligencer **4** (1982), 108–115. (Cited on 92.)

[231] H. A. Kramers, *Wellenmechanik und halbzahlige Quantisierung*, Z. Phys. **39** (1926), 828–840. (Cited on 226.)

[232] S. G. Krantz, *Mathematical anecdotes*, Math. Intelligencer **12** (1990), 32–38. (Cited on 25.)

[233] S. G. Krantz, *Complex Analysis: The Geometric Viewpoint*, 2nd edition, Carus Mathematical Monographs, Mathematical Association of America, Washington, DC, 2004. (Cited on 3.)

[234] S. G. Krantz, *Geometric Function Theory: Explorations in Complex Analysis*, Birkhäuser, Boston, 2006. (Cited on 3.)

[235] S. G. Krantz, *Geometric Analysis of the Bergman Kernel and Metric*, Graduate Texts in Mathematics, Springer, New York, 2013. (Cited on 25.)

[236] E. E. Kummer, *Über die hypergeometrische Reihe* $1 + \frac{\alpha\cdot\beta}{1\cdot\gamma}\, x + \frac{\alpha(\alpha+1)\beta(\beta+1)}{1\cdot 2\cdot\gamma(\gamma+1)} x^2 + \frac{\alpha(\alpha+1)(\alpha+2)\beta(\beta+1)(\beta+2)}{1\cdot 2\cdot 3\cdot\gamma(\gamma+1)(\gamma+2)} x^3 + \dots$, J. Reine Angew. Math. **15** (1836), 39–83; 127–172. (Cited on 133, 134.)

[237] J. Lagrange, *Démonstration d'un théorème d'arithmétique*, Nouv. Mémoires de l'Academie des Sciences de Berlin, année 1770, Berlin 1772, 123–133; Oeuvres de Lagrange, Tome III, pp. 189–201, Gauthier–Villars, Paris, 1869. (Cited on 53.)

[238] I. Laine, *Nevanlinna Theory and Complex Differential Equations*, de Gruyter Studies in Mathematics, Walter de Gruyter, Berlin, 1993. (Cited on 261.)

[239] G. Lambert, *An Overview of Schramm Loewner Evolution*, available online at `http://www2.math.uu.se/~jakob/conformal/lambert_SLE.pdf`, 2012. (Cited on 255.)

[240] J. H. Lambert, *Observationes variae in mathesin puram*, Acta Helveticae **3** (1758), 128–168; available online at `http://www.kuttaka.org/~JHL/L1758c.pdf`. (Cited on 210.)

[241] E. Landau, *Über einen Satz von Tschebyschef*, Math. Ann. **61** (1905), 527–550. (Cited on 67.)

[242] E. Landau, *Handbuch der Lehre von der Verteilung der Primzahlen*, 2nd edition, Chelsea Publishing, New York, 1953; reprint of the original 1909 edition. (Cited on 67.)

[243] E. Landau, *Zur analytischen Zahlentheorie der definiten quadratischen Formen (Über die Gitterpunkte in einem mehrdimensionalen Ellipsoid)*, Sitzungsber. Preuss. Akad. Wiss. **31** (1915), 458–476. (Cited on 193.)

[244] J. Landen, *A disquisition concerning certain fluents, which are assignable by the arcs of the conic sections; wherein are investigated some new and useful theorems for computing such fluents*, Philos. Trans. Roy. Soc. London **61** (1771), 298–309. (Cited on 134.)

[245] J. Landen, *An investigation of a general theorem for finding the length of any arc of any conic hyperbola, by means of two elliptic arcs, with some other new and useful theorems deduced therefrom*, Philos. Trans. Roy. Soc. London **65** (1775), 283–289. (Cited on 134.)

[246] O. E. Lanford, III and D. Ruelle, *Observables at infinity and states with short range correlations in statistical mechanics*, Comm. Math. Phys. **13** (1969), 194–215. (Cited on 180.)

[247] S. Lang, *Introduction to Complex Hyperbolic Spaces*, Springer-Verlag, New York, 1987. (Cited on 261.)

[248] S. Lang and W. Cherry, *Topics in Nevanlinna Theory*, Lecture Notes in Mathematics, Springer-Verlag, Berlin, 1990. (Cited on 262.)

[249] R. E. Langer, *On the asymptotic solutions of ordinary differential equations, with an application to the Bessel functions of large order*, Trans. Amer. Math. Soc. **33** (1931), 23–64. (Cited on 193.)

[250] P.-S. Laplace, *Memoir on the probability of causes of events*, Mémoires de Mathématique et de Physique, Tome Sixième (1774); English translation by S. M. Stigler, Statist. Sci. **1** (1986), 364–378. (Cited on 180.)

[251] P.-S. Laplace, *Mémoire sur les approximations des formules qui sont fonctions de trés grands nombres*, Mémoires de l'Académie royale des sciences de Paris, (1782, 1785), 209–291; available online at `http://gallica.bnf.fr/ark:/12148/bpt6k775981/f218`. (Cited on 180.)

[252] P.-S. Laplace, *Mémoire sur les integrales definies, et leur application aux probabilities*, Mémoires de l'Académie royale des sciences de Paris, Ser XI (1810–11). (Cited on 135.)

[253] P.-S. Laplace, *Théorie analytique des probabilities*, Courcier, Paris, 1st edition 1812, 2nd edition 1814, 3rd enlarged edition 1820; available online at `http://gallica.bnf.fr/ark:/12148/bpt6k775950/f4` or `http://www.archive.org/details/thorieanalytiqu01laplgoog`. (Cited on 159, 166, 180.)

[254] G. F. Lawler, *Conformally Invariant Processes in the Plane*, Mathematical Surveys and Monographs, American Mathematical Society, Providence, RI, 2005. (Cited on 255.)

[255] G. F. Lawler, O. Schramm, and W. Werner, *The dimension of the planar Brownian frontier is 4/3*, Math. Res. Lett. **8** (2001), 401–411. (Cited on 254.)

[256] G. F. Lawler, O. Schramm, and W. Werner, *Values of Brownian intersection exponents. I. Half-plane exponents*, Acta Math. **187** (2001), 237–273. (Cited on 254.)

[257] G. F. Lawler, O. Schramm, and W. Werner, *Values of Brownian intersection exponents. II. Plane exponents*, Acta Math. **187** (2001), 275–308. (Cited on 254.)

[258] G. F. Lawler, O. Schramm, and W. Werner, *Values of Brownian intersection exponents. III. Two-sided exponents*, Ann. Inst. H. Poincaré Probab. Statist. **38** (2002), 109–123. (Cited on 254.)

[259] G. F. Lawler, O. Schramm, and W. Werner, *Analyticity of intersection exponents for planar Brownian motion*, Acta Math. **189** (2002), 179–201. (Cited on 254.)

[260] G. F. Lawler, O. Schramm, and W. Werner, *Conformal invariance of planar loop-erased random walks and uniform spanning trees*, Ann. Probab. **32** (2004), 939–995. (Cited on 254.)

[261] A.-M. Legendre, *Recherches sur l'attraction des sphéroïdes homogènes*, Mémoires de Mathématiques et de Physique, présentés à l'Académie royale des sciences (Paris) par sçavants étrangers **10** (1785),411–435 [*Note*: Legendre submitted his findings to the Academy in 1782, but they were published in 1785.]; available online (in French) at `http://edocs.ub.uni-frankfurt.de/volltexte/2007/3757/pdf/A009566090.pdf`. (Cited on 135.)

[262] A.-M. Legendre, *Essai sur la théorie des nombres, 1, 2*, Chez Duprat, Paris, 1798; Courcier, Paris, 1808. (Cited on 92.)

[263] J. Liouville, *Mémoire sur le développement des fonctions ou parties de fonctions en séries dont les divers termes sont assujettis à satisfaire à une même équation differéntielle du second ordre, contenant un paramètre variable*, J. Math. Pures Appl. **1** (1836), 253–265. (Cited on 226.)

[264] J. E. Littlewood, *On exceptional values of power series*, J. London Math. Soc. **5** (1930), 82–87. (Cited on 258, 276.)

[265] J. E. Littlewood and E. A. C. Paley, *A proof that an odd Schlicht function has bounded coefficients*, J. London Math. Soc. **3** (1932), 167–169. (Cited on 238.)

[266] J. J. Loeffel, A. Martin, B. Simon, and A. S. Wightman, *Padé approximants and the anharmonic oscillator*, Phys. Lett. B **30** (1969), 656–658. (Cited on 169.)

[267] C. Loewner, *Untersuchungen über schlichte konforme Abbildungen des Einheitskreises. I*, Math. Ann. **89** (1923), 103–121. (Cited on 250.)

[268] A. Majda, *Introduction to PDEs and Waves for the Atmosphere and Ocean*, Courant Lecture Notes in Mathematics, New York University, Courant Institute of Mathematical Sciences, New York; American Mathematical Society, Providence, RI, 2003. (Cited on 192.)

[269] S. Mandelbrojt, *Dirichlet Series. Principles and Methods*, D. Reidel Publishing, Dordrecht, 1972. (Cited on 67.)

[270] B. B. Mandelbrot, *The Fractal Geometry of Nature*, W. H. Freeman and Co., San Francisco, Calif., 1982. (Cited on 254.)

[271] Ju. I. Manin, *Moduli fuchsiani*, Ann. Scuola Norm. Sup. Pisa (3) **19** (1965), 113–126. (Cited on 98.)

[272] F. Mertens, *Über eine zahlentheoretische Funktion*, Akad. Wiss. Wien Math.-Natur. Kl. Sitzungsber. IIa **106** (1897), 761–830. (Cited on 68.)

[273] F. Mertens, *Über eine Eigenschaft der Riemann'schen $\zeta$-Function*, Akad. Wiss. Wien Math.-Natur. Kl. Sitzungsber. IIa **107** (1898), 1429–1434. (Cited on 78.)

[274] I. M. Milin, *Univalent Functions and Orthonormal Systems*, Translations of Mathematical Monographs, American Mathematical Society, Providence, RI, 1977. (Cited on 238.)

[275] P. D. Miller, *Applied Asymptotic Analysis*, Graduate Studies in Mathematics, American Mathematical Society, Providence, RI, 2006. (Cited on 192.)

[276] S. J. Miller and R. Takloo-Bighash, *The circle method*; available online at `http://web.williams.edu/Mathematics/sjmiller/public_html/OSUClasses/487/circlemethod.pdf`. (Cited on 212.)

[277] H. Milloux, *Les fonctions méromorphes et leurs dérivées*, Paris, 1940. (Cited on 261.)

[278] J. Milnor, *Hyperbolic geometry: The first 150 years*, Bull. Amer. Math. Soc. **6** (1982), 9–24. (Cited on 12.)

[279] A. F. Möbius, *Über eine besondere Art von Umkehrung der Reihen*, J. Reine Angew. Math. **9** (1832), 105–123. (Cited on 68.)

[280] L. J. Mordell, *On the representations of numbers as a sum of* $2^r$ *squares*, Quart. J. Math. Oxford **48** (1918), 93–104. (Cited on 53.)

[281] M. Morse, *The Calculus of Variations in the Large*, reprint of the 1932 original, American Mathematical Society Colloquium Publications, American Mathematical Society, Providence, RI, 1996. (Cited on 180.)

[282] P. M. Morse and H. Feshbach, *Methods of Theoretical Physics*, McGraw–Hill, New York–Toronto–London, 1953. (Cited on 149.)

[283] H. J. W. Müller-Kirsten, *Introduction to Quantum Mechanics. Schrödinger Equation and Path Integral*, World Scientific, Hackensack, NJ, 2006. (Cited on 227.)

[284] J. D. Murray, *Asymptotic Analysis*, 2nd edition, Applied Mathematical Sciences, Springer-Verlag, New York, 1984. (Cited on 192.)

[285] R. Narasimhan, *Une remarque sur* $\zeta(1+it)$, Enseign. Math. (2) **14** (1969), 189–191. (Cited on 78.)

[286] R. Narasimhan and Y. Nievergelt, *Complex Analysis in One Variable*, Second edition, Birkhäuser Boston, Inc., Boston, MA, 2001; first edition, 1985. (Cited on 18.)

[287] M. B. Nathanson, *Elementary Methods in Number Theory*, Graduate Texts in Mathematics, Springer-Verlag, New York, 2000. (Cited on 211.)

[288] P. A. Nekrasov, *The Lagrange series and approximate expressions of functions of very large numbers*, Mat. Sb. **12** (1885), 481–579. (Cited on 210.)

[289] F. Nevanlinna, *Zur Theorie der asymptotischen Potenzreihen*, Ann. Acad. Sci. Fenn. A **12** (1918), 1–81. (Cited on 168.)

[290] F. Nevanlinna, *Über die Anwendung einer Klasse uniformisierender Transzendenten zur Untersuchung der Wertverteilung analytischer Funktionen*, Acta Math. **50** (1927), 159–188. (Cited on 283.)

[291] R. Nevanlinna, *Zur Theorie der meromorphen Funktionen*, Acta Math. **46** (1925), 1–99. (Cited on 260.)

[292] R. Nevanlinna, *Le théorème de Picard–Borel et la théorie des fonctions méromorphes*, reprinting of the 1929 original, Chelsea, New York, 1974. (Cited on 260.)

[293] R. Nevanlinna, *Analytic Functions*, translated from the 1953 second German edition, Grundlehren der mathematischen Wissenschaften, Springer-Verlag, New York-Berlin, 1970. (Cited on 260.)

[294] D. J. Newman, *The evaluation of the constant in the formula for the number of partitions of* $n$, Amer. J. Math. **73** (1951), 599–601. (Cited on 211.)

[295] D. J. Newman, *A simplified proof of the partition formula*, Michigan Math. J. **9** (1962), 283–287. (Cited on 211.)

[296] D. J. Newman, *Simple analytic proof of the prime number theorem*, Amer. Math. Monthly **87** (1980), 693–696. (Cited on 92.)

[297] D. J. Newman, *A simplified version of the fast algorithms of Brent and Salamin*, Math. Comp. **44** (1985), 207–210. (Cited on 134.)

[298] D. J. Newman, *Analytic Number Theory*, Graduate Texts in Mathematics, Springer-Verlag, New York, 1998. (Cited on 44, 92, 211.)

[299] R. G. Newton, *Scattering Theory of Waves and Particles*, reprint of the 1982 second edition, Dover Publications, Mineola, NY, 2002. (Cited on 149.)

[300] P. W. Nicholson, *On Bessel functions of equal argument and order*, Philos. Magazine **16** (1909), 271–275. (Cited on 193.)

[301] B. Nienhuis and W. Kager, *Stochastic Löwner evolution and the scaling limit of critical models. Polygons, polyominoes and polycubes*, In Lecture Notes in Physics, pp. 425–467, Springer, Dordrecht, 2009. (Cited on 255.)

[302] J. Noguchi and J. Winkelmann, *Nevanlinna Theory in Several Complex Variables and Diophantine Approximation*, Grundlehren der Mathematischen Wissenschaften, Springer, Tokyo, 2014. (Cited on 261.)

[303] A. M. Odlyzko *Asymptotic enumeration methods*, In Handbook of Combinatorics, Vol. 1, 2, pp. 1063–1229, Elsevier, Amsterdam, 1995. (Cited on 210.)

[304] A. M. Odlyzko and H. J. J. te Riele, *Disproof of the Mertens Conjecture*, J. Reine Angew. Math. **357** (1985), 138–160. (Cited on 68.)

[305] A. B. Olde Daalhuis, *Chapter 15. Hypergeometric Function*, In NIST Digital Library of Mathematical Functions; available online at `http://dlmf.nist.gov/15` (Cited on 133.)

[306] F. W. J. Olver, *The asymptotic expansion of Bessel functions of large order*, Phil. Trans. Royal Soc. London A **247** (1954), 328–368. (Cited on 193.)

[307] F. W. J. Olver, *Chapter 9. Airy and Related Functions*, In NIST Digital Library of Mathematical Functions; available online at `http://dlmf.nist.gov/9`. (Cited on 149.)

[308] F. W. J. Olver and L. C. Maximon *Chapter 10. Bessel Functions*, In NIST Digital Library of Mathematical Functions; available online at `http://dlmf.nist.gov/10`. (Cited on 149.)

[309] C. Osgood, *Sometimes effective Thue–Siegel–Roth–Schmidt–Nevanlinna bounds, or better*, J. Number Theory **21** (1985), 347–389. (Cited on 261.)

[310] P. Painlevé, *Thèse d'Analyse. Sur les lignes singulières des fonctions analytiques*, Thesis, Paris, 1887. (Cited on 36.)

[311] E. Papperitz, *Über die Darstellung der hypergeometrischen Transcendenten durch eindeutige Functionen*, Math. Ann. **34** (1889), 247–296. (Cited on 133.)

[312] S. J. Patterson, *An Introduction to the Theory of the Riemann Zeta-Function*, Cambridge Studies in Advanced Mathematics, Cambridge University Press, Cambridge, 1988. (Cited on 76.)

[313] P. A. Perry, *Scattering Theory by the Enss Method*, Mathematical Reports, Part 1, Harwood Academic Publishers, Chur, 1983. (Cited on 193.)

[314] S. S. Petrova and A. D. Solov'ev, *The origin of the method of steepest descent*, Hist. Math. **24** (1997), 361–375. (Cited on 210.)

[315] J. F. Pfaff, *Disquisitiones analyticae maxime ad calculum integralem et doctrinam serierum pertinentes*, (1797); available in scanned form at `http://books.google.com/books/about/Disquisitiones_Analyticae_Maxime_Ad_Calc.html?id=vo8_AAAAcAAJ`. (Cited on 133.)

[316] H. Poincaré, *Théorie des groupes fuchsiens*, Acta Math. **1** (1882), 1–62. (Cited on 12.)

[317] H. Poincaré, *Sur les intégrales irrégulières des équations linéaires*, Acta Math. **8** (1886), 295–344. (Cited on 166.)

[318] H. Poincaré, *Sur la diffraction des ondes électriques: A propos d'un article de M. Macdonald*, Proc. Royal Soc. London **72** (1903), 42–52. (Cited on 192.)

[319] H. Poincaré, *Science and Hypothesis*, W. Scott Publ., London, 1905; available online at `https://openlibrary.org/books/OL6964685M/`; original edition, *La Science et L'Hypothése*, Paris, Flammarion, 1902. (Cited on 6.)

[320] C. Pommerenke, *Univalent Functions*, Studia Mathematica/Mathematische Lehrbücher, Vandenhoeck & Ruprecht, Göttingen, 1975. (Cited on 238.)

[321] A. G. Postnikov, *Introduction to Analytic Number Theory*, Translations of Mathematical Monographs, American Mathematical Society, Providence, RI, 1988. (Cited on 44, 211.)

[322] H. Rademacher, *On the partition function* $p(n)$, Proc. London Math. Soc. **43** (1938), 241–254. (Cited on 212.)

[323] H. Rademacher, *On the expansion of the partition function in a series*, Ann. Math. **44** (1943), 416–422. (Cited on 212.)

[324] Lord Rayleigh, *The problem of the whispering gallery*, Philos. Magazine **20** (1911), 1001–1004. (Cited on 193.)

[325] M. Reed and B. Simon, *Methods of Modern Mathematical Physics, III: Scattering Theory*, Academic Press, New York, 1979. (Cited on 193.)

[326] V. Retakh, *Israel Moiseevich Gelfand, Part II*, Notices Amer. Math. Soc. **60** (2013), 162–171. (Cited on 116.)

[327] G. F. B. Riemann, *Über die Anzahl der Primzahlen unter einer gegebenen Grösse*, Monatsber. Berlin Akad. (1858/60), 671–680; available online in German and in translation at `http://www.maths.tcd.ie/pub/HistMath/People/Riemann/Zeta/`. (Cited on 76, 79.)

[328] G. F. B. Riemann, *Sullo svolgimento del quoziente di due serie ipergeometriche in frazione continua infinita*, written in 1863, unpublished but appeared in his complete works. (Cited on 182, 210, 212.)

[329] J. F. Ritt, *On the conformal mapping of a region into a part of itself*, Ann. Math. **22** (1921), 157–160. (Cited on 14.)

[330] M. S. Robertson, *A remark on the odd schlicht functions*, Bull. Amer. Math. Soc. 42 (1936), 366–370. (Cited on 238.)

[331] D. W. Robinson and D. Ruelle, *Mean entropy of states in classical statistical mechanics*, Comm. Math. Phys. **5** (1967), 288–300. (Cited on 180.)

[332] R. M. Robinson, *A generalization of Picard's and related theorems*, Duke Math. J. **5** (1939), 118–132. (Cited on 18.)

[333] O. Rodrigues, *De l'attraction des sphéroïdes*, Correspondence sur l'École Impériale Polytechnique **3** (1816), 361–385. (Cited on 135.)

[334] S. Rohde and O. Schramm, *Basic properties of SLE*, Ann. Math. (2) **161** (2005), 883–924. (Cited on 254.)

[335] G.-C. Rota, *The number of partitions of a set*, Amer. Math. Monthly **71** (1964), 498–504. (Cited on 210.)

[336] M. Ru, *Nevanlinna Theory and Its Relation to Diophantine Approximation*, World Scientific Publishing Co., Inc., River Edge, NJ, 2001. (Cited on 262.)

[337] K. Sabbagh, *The Riemann Hypothesis: The Greatest Unsolved Problem in Mathematics*, Farrar, Straus, and Giroux, New York, 2003. (Cited on 76.)

[338] L. Sario, K. Noshiro, K. Matsumoto, and M. Nakai, *Value Distribution Theory*, Van Nostrand, Princeton, NJ–Toronto–London, 1966. (Cited on 261.)

[339] L. Schläfli, *Eine Bemerkungen zu Herrn Neumann's Untersuchungen über die Bessel'schen Functionen*, Math. Ann. **3** (1871), 134–149. (Cited on 159.)

[340] A. Schlissel, *The development of asymptotic solutions of linear ordinary differential equations, 1817–1920*, Arch. Hist. Exact Sci. **16** (1976/77), 307–378. (Cited on 226.)

[341] O. Schlömilch, *Über die Bessel'schen Function*, Z. Math. Phys. **2** (1857), 137–165. (Cited on 149.)

[342] O. Schramm, *Scaling limits of loop-erased random walks and uniform spanning trees*, Israel J. Math. **118** (2000), 221–288. (Cited on 254.)

[343] J. B. Seaborn, *Hypergeometric Functions and Their Applications*, Texts in Applied Mathematics, Springer-Verlag, New York, 1991. (Cited on 133.)

[344] S. L. Segal, *Nine Introductions in Complex Analysis*, revised edition, North-Holland Mathematics Studies, Elsevier, Amsterdam, 2008. (Cited on 238.)

[345] J.-P. Serre, *Linear Representations of Finite Groups*, Graduate Texts in Mathematics, Springer-Verlag, New York–Heidelberg, 1977. (Cited on 84.)

[346] B. Shawyer and B. Watson, *Borel's Methods of Summability. Theory and Applications*, Oxford Mathematical Monographs, The Clarendon Press, Oxford University Press, New York, 1994. (Cited on 167.)

[347] B. Shiffman, *On holomorphic curves and meromorphic maps in projective space*, Indiana U. Math. J. **28** (1979), 627–641. (Cited on 261.)

[348] T. Shimizu, *On the theory of meromorphic functions*, Japanese J. Math. **6** (1929), 119–171. (Cited on 283.)

[349] Y. Sibuya, *Linear Differential Equations in the Complex Domain: Problems of Analytic Continuation*, Translations of Mathematical Monographs, American Mathematical Society, Providence, RI, 1990. (Cited on 98, 100.)

[350] C. L. Siegel, *Über Riemanns Nachlaß zur analytischen Zahlentheorie*, Quellen Studien zur Geschichte der Math. Astron. und Phys. Abt. B, Studien 2 (1932), 45–80; reprinted in Gesammelte Abhandlungen **1**, Springer-Verlag, Berlin, 1966. (Cited on 76.)

[351] B. Simon, *Coupling constant analyticity for the anharmonic oscillator*, Ann. Phys. **58** (1970), 76–136. (Cited on 169.)

[352] B. Simon, *Fifty years of eigenvalue perturbation theory*, Bull. Amer. Math. Soc. **24** (1991), 303–319. (Cited on 169.)

[353] B. Simon, *Bounded eigenfunctions and absolutely continuous spectra for one-dimensional Schrödinger operators*, Proc. Amer. Math. Soc. **124** (1996), 3361–3369. (Cited on 228.)

[354] B. Simon, *Representations of Finite and Compact Groups*, Graduate Studies in Mathematics, American Mathematical Society, Providence, RI, 1996. (Cited on 84, 86.)

[355] B. Simon, *Convexity: An Analytic Viewpoint*, Cambridge Tracts in Mathematics, Cambridge University Press, Cambridge, 2011. (Cited on 250.)

[356] L. J. Slater, *Confluent Hypergeometric Functions*, Cambridge University Press, New York, 1960. (Cited on 133.)

[357] L. J. Slater, *Generalized Hypergeometric Functions*, Cambridge University Press, Cambridge, 1966. (Cited on 133.)

[358] N. Sloane et al., *Bell or exponential numbers: ways of placing n labeled balls into n indistinguishable boxes*, In Online Encyclopedia of Integer Sequences, available online at `http://oeis.org/A000110`. (Cited on 210.)

[359] N. Sloane et al., *$a(n)$ = number of partitions of n (the partition numbers)*, In Online Encyclopedia of Integer Sequences, available online at `http://oeis.org/A000041`. (Cited on 212.)

[360] A. B. Slomson, *An Introduction to Combinatorics*, Chapman and Hall Mathematics Series, Chapman and Hall, London, 1991. (Cited on 211.)

[361] S. Smirnov, *Critical percolation in the plane: conformal invariance, Cardy's formula, scaling limits*, C. R. Acad. Sci. Paris **333** (2001), 239–244. (Cited on 254.)

[362] C. D. Sogge, *Fourier Integrals in Classical Analysis*, Cambridge Tracts in Mathematics, Cambridge University Press, Cambridge, 1993. (Cited on 192, 193.)

[363] A. Sommerfeld, *Mathematische Theorie der Diffraction*, Math. Ann. **47** (1896), 317–374. (Cited on 159.)

[364] D. Stark and N. C. Wormald, *Asymptotic enumeration of convex polygons*, J. Combin. Theory Ser. A **80** (1997), 196–217. (Cited on 210.)

[365] E. M. Stein and R. Shakarchi, *Fourier Analysis. An Introduction*, Princeton Lectures in Analysis, Princeton University Press, Princeton, NJ, 2003. (Cited on 84, 211.)

[366] E. M. Stein and R. Shakarchi, *Complex Analysis*, Princeton Lectures in Analysis, Princeton University Press, Princeton, NJ, 2003. (Cited on 53.)

[367] T. Stieltjes, *Recherches sur quelques séries semi-convergentes*, Ann. Sci. Ecole Norm. Sup. (3) **3** (1886), 201–258. (Cited on 166.)

[368] T. Stieltjes, *Recherches sur les fractions continues*, Ann. Fac. Sci. Toulouse **8** (1894), J1–J122; **9** (1895), A5–A47. (Cited on 167.)

[369] J. Stillwell, *Mathematics and Its History*, 2nd edition, Undergraduate Texts in Mathematics, Springer-Verlag, New York, 2002. (Cited on 12.)

[370] G. G. Stokes, *On the numerical calculation of a class of definite integrals and infinite series*, Trans. Cambridge Philos. Soc. **9** (1856), 166–187 (read in 1850); available at Google Books. (Cited on 192.)

[371] G. G. Stokes, *On the discontinuity of arbitrary constants which appear in divergent developments*, Trans. Cambridge Philos. Soc. **10** (1857), 105–128. (Cited on 181.)

[372] G. G. Stokes, *Memoir and Scientific Correspondence, Vol. 1*, p. 62, Cambridge University Press, Cambridge, 1907. (Cited on 149.)

[373] W. Stoll, *Value Distribution Theory for Meromorphic Maps*, Aspects of Mathematics, Friedr. Vieweg & Sohn, Braunschweig, 1985. (Cited on 261.)

[374] J. Stopple, *A Primer of Analytic Number Theory. From Pythagoras to Riemann*, Cambridge University Press, Cambridge, 2003. (Cited on 44.)

[375] D. W. Stroock, *An Introduction to the Theory of Large Deviations*, Universitext, Springer-Verlag, New York, 1984. (Cited on 180.)

[376] O. Szász, *Introduction to the Theory of Divergent Series*, Hafner Publishing, New York, 1948. (Cited on 167.)

[377] G. Szegő, *Orthogonal Polynomials*, American Mathematical Society Colloquium Publications, American Mathematical Society, Providence, RI, 1939; 3rd edition, 1967. (Cited on 135.)

[378] L. A. Takhtajan, *Quantum Mechanics for Mathematicians*, Graduate Studies in Mathematics, American Mathematical Society, Providence, RI, 2008. (Cited on 227.)

[379] J. J. Tattersall, *Elementary Number Theory in Nine Chapters*, 2nd edition, Cambridge University Press, Cambridge, 2005. (Cited on 211.)

[380] A. Tauber, *Ein Satz aus der Theorie der unendlichen Reihen*, Monatsh. Math. **8** (1897), 273–277. (Cited on 43, 45.)

[381] N. M. Temme, *Special Functions. An Introduction to the Classical Functions of Mathematical Physics*, John Wiley & Sons, New York, 1996. (Cited on 149, 181.)

[382] G. Tenenbaum, *Introduction to Analytic and Probabilistic Number Theory*, Cambridge Studies in Advanced Mathematics, Cambridge University Press, Cambridge, 1995. (Cited on 44.)

[383] G. Teschl, *Ordinary Differential Equations and Dynamical Systems*, lecture notes available online at `http://www.mat.univie.ac.at/~gerald/ftp/book-ode/ode.pdf`. (Cited on 114.)

[384] J. V. Uspensky, *Asymptotic formulae for numerical functions which occur in the theory of partitions*, Bull. Acad. Sci. URSS **14** (1920), 199–218. (Cited on 211.)

[385] G. Valiron, *Sur la distribution des valeurs des fonctions méromorphes*, Acta Math. **47** (1925), 117–142. (Cited on 258, 276.)

[386] G. Valiron, *Familles normales et quasi-normales de fonctions méromorphes*, Gauthiers–Villars, Paris, 1929. (Cited on 276.)

[387] G. Valiron, *Lectures on the General Theory of Integral Functions*, Chelsea, New York, 1949. (Cited on 276.)

[388] O. Vallée and M. Soares, *Airy Functions and Applications to Physics*, Imperial College Press, London; distributed by World Scientific Publishing, Hackensack, NJ, 2004. (Cited on 149.)

[389] J. G. van der Corput, *Zahlentheoretische Abschätzungen*, Math. Ann. **84** (1921), 53–79. (Cited on 193.)

[390] V. S. Varadarajan, *Linear meromorphic differential equations: A modern point of view*, Bull. Amer. Math. Soc. **33** (1996), 1–42. (Cited on 98.)

[391] S. R. S. Varadhan, *Asymptotic probabilities and differential equations*, Comm. Pure Appl. Math. **19** (1966), 261–286. (Cited on 180.)

[392] S. R. S. Varadhan, *Large Deviations and Applications*, CBMS-NSF Regional Conference Series in Applied Mathematics, Society for Industrial and Applied Mathematics (SIAM), Philadelphia, PA, 1984. (Cited on 180.)

[393] S. R. S. Varadhan, *Large deviations*, Ann. Probab. **36** (2008), 397–419. (Cited on 180.)

[394] R. C. Vaughan, *The Hardy–Littlewood Method*, Cambridge Tracts in Mathematics, Cambridge University Press, Cambridge, 1981. (Cited on 212.)

[395] R. C. Vaughan and T. D. Wooley, *Waring's problem: A survey*, In Number Theory for the Millennium, III (Urbana, IL, 2000), pp. 301–340, A K Peters, Natick, MA, 2002. (Cited on 55.)

[396] P. Vojta, *Diophantine Approximations and Value Distribution Theory*, Lecture Notes in Mathematics, Springer-Verlag, Berlin, 1987. (Cited on 261.)

[397] H. von Mangoldt, *Beweis der Gleichung* $\sum_{k=1}^{\infty} \frac{\mu(k)}{k} = 0$, Berl. Ber. (1897), 835–852. (Cited on 68.)

[398] H. von Mangoldt, *Zur Verteilung der Nullstellen der Riemannschen Funktion $\xi(t)$*, Math. Ann. **60** (1905), 1–19. (Cited on 77.)

[399] J. von Neumann and E. Wigner, *Über merkwürdige diskrete Eigenwerte*, Z. Phys. **30** (1929), 465–467. (Cited on 228.)

[400] J. Wallis, *Arithmetica Infinitorum*, Oxford, 1656; English translation by J. A. Stedall, *John Wallis: The Arithmetic of Infinitesimals*, Sources and Studies in the History of Mathematics and Physical Sciences, Springer-Verlag, New York, 2004. (Cited on 133.)

[401] W. Walter, *Ordinary Differential Equations*, Graduate Texts in Mathematics, Readings in Mathematics, Springer-Verlag, New York, 1998. (Cited on 98.)

[402] W. Wasow, *Asymptotic Expansions for Ordinary Differential Equations*, Pure and Applied Mathematics, Interscience–John Wiley & Sons, New York–London–Sydney, 1965. (Cited on 98, 166.)

[403] G. N. Watson, *A theory of asymptotic series*, Phil. Trans. Royal Soc. London A **211** (1912), 279–313. (Cited on 168.)

[404] G. N. Watson, *The harmonic functions associated with the parabolic cylinder*, Proc. London Math. Soc. **17** (1918), 116–148. (Cited on 181.)

[405] G. N. Watson, *A Treatise on the Theory of Bessel Functions*, reprint of the 2nd (1944) edition, Cambridge Mathematical Library, Cambridge University Press, Cambridge, 1995; the original edition was published in 1922. (Cited on 149, 188.)

[406] G. N. Watson, *The Marquis and the Land-Agent; A Tale of the Eighteenth Century*, Math. Gazette **17**, (1933), 5–17; This was the retiring presidential address of the British Mathematical Association. (Cited on 134.)

[407] G. Wentzel, *Eine Verallgemeinerung der Quantenbedingungen für die Zwecke der Wellenmechanik*, Z. Phys. **38** (1926), 518–529. (Cited on 226.)

[408] W. Werner, *Random Planar Curves and Schramm-Loewner Evolutions* In Lectures on Probability Theory and Statistics, Lecture Notes in Mathematics, pp. 107–195, Springer, Berlin, 2004. (Cited on 255.)

[409] W. Werner, *Lectures on Two-Dimensional Critical Percolation. Statistical Mechanics*, In IAS/Park City Math. Ser., 16, pp. 297–360, American Mathematical Society, Providence, RI, 2009. (Cited on 254.)

[410] H. Weyl, *Meromorphic Functions and Analytic Curves*, Annals of Mathematics Studies, Princeton University Press, Princeton, NJ, 1943. (Cited on 261.)

[411] E. T. Whittaker and G. N. Watson, *A Course of Modern Analysis. An Introduction to the General Theory of Infinite Processes and of Analytic Functions; with an Account of the Principal Transcendental Functions*, reprint of the fourth (1927) edition, Cambridge Mathematical Library, Cambridge University Press, Cambridge, 1996. (Cited on 133, 182.)

[412] N. Wiener *Tauberian theorems*, Ann. Math. **33** (1932), 1–100. (Cited on 92.)

[413] R. Wong, *Asymptotic Approximations of Integrals*, corrected reprint of the 1989 original, Classics in Applied Mathematics, Society for Industrial and Applied Mathematics (SIAM), Philadelphia, PA, 2001. (Cited on 163.)

[414] B. H. Yandell, *The Honors Class. Hilbert's Problems and Their Solvers*, A K Peters, Natick, MA, 2002. (Cited on 115.)

[415] R. M. Young, *Excursions in Calculus. An interplay of the continuous and the discrete*, The Dolciani Mathematical Expositions, Mathematical Association of America, Washington, DC, 1992. (Cited on 87.)

[416] A. Zaccagnini, *Introduction to the circle method of Hardy, Ramanujan and Littlewood*, available online at `http://www.math.unipr.it/~zaccagni/psfiles/didattica/HRI.pdf`. (Cited on 212.)

[417] D. Zagier, *A one-sentence proof that every prime* $p \equiv 1 \pmod 4$ *is a sum of two squares*, Amer. Math. Monthly **97** (1990), 144. (Cited on 53.)

[418] D. Zagier, *Newman's short proof of the prime number theorem*, Amer. Math. Monthly **104** (1997), 705–708. (Cited on 92.)

# Symbol Index

# Subject Index

# Author Index

# Index of Capsule Biographies